A Mathematics Course for Political and Social Research

A Mathematics Course for Political and Social Research

Will H. Moore & David A. Siegel

PRINCETON UNIVERSITY PRESS
PRINCETON AND OXFORD

Published by Princeton University Press
41 William Street, Princeton, New Jersey 08540

In the United Kingdom: Princeton University Press
6 Oxford Street, Woodstock, Oxfordshire, OX20 1TW

ISBN 978-0-691-15995-9
ISBN (pbk.) 978-0-691-15917-1
Library of Congress Control Number: 2013935356

British Library Cataloging-in-Publication Data is available

This book has been composed in LaTeX

The publisher would like to acknowledge the authors of this volume for providing the camera-ready copy from which this book was printed.

Printed on acid-free paper. ∞

press.princeton.edu

Printed in the United States of America

10 9 8 7 6 5 4

To William Howard Moore, Jr., a father and engineer who finds it curious that people would try to cram multiple semesters of mathematics into a single semester, and Gabriel, who was just learning to hold up one finger for the number one when the first draft of this manuscript was completed.

Contents

List of Figures

List of Tables

Preface

The goal of a science of politics is to identify general patterns among abstract concepts. That is, political scientists think about politics and invent abstract concepts that help us describe the political world. But a science of politics requires more than inventive description. It involves the specification of expected causal relationships among the concepts and, ultimately, the determination of whether those expected relationships comport with evidence.

Mathematics is an abstract discipline. All of its concepts have been invented. It is nothing more than a formal language, a set of definitions and rules, a syntax. However, since we are not studying mathematics for the purpose of contributing to its development, we need not concern ourselves with that. Instead, we are interested in mathematics to help us discipline our conceptualization, our theoretical development, our specification of hypotheses, and our testing of hypotheses. In other words, we are interested in mathematics to help us develop rigorous theories of politics and rigorous tests of the implications of those theories.

This book is designed to provide aid in this endeavor. This is *not* a formal text in mathematics. We rarely define terms formally, and we do not offer formal proofs of theorems. Indeed, at times we sacrifice mathematical rigor for intuition in a way that would make a mathematician cringe.[1] And we certainly fail to include many topics that others may find important. We do this for two reasons. One, this book is intended primarily for political scientists, not mathematicians. We believe that many students of political science have had little prior experience with mathematics outside of high school, and further, may have developed a fairly serious case of math phobia. Before one can overcome these obstacles one needs to develop a strong intuition about what math is, what it can tell us, and how it can be useful. We aim to provide a practical text that provides these intuitions. Two, there is an abundance of other, more traditional alternatives for learning math more rigorously, e.g., all the courses offered by math departments.

The book is broken up into five parts. The first part, "Building Blocks," covers the preliminaries and reviews what should have been learned in most high school curricula, albeit at a higher level. Those with a strong math background can safely skim this section, though we would advise most not to skip it entirely: topics such as relations, proofs, utility representations, and the use of comput-

[1]Though footnotes and other asides will sometimes provide relevant formalism.

ers to aid in performing calculations might be new to many readers. Those whose background is more shaky should read this section carefully, preferably before beginning a math class or "camp," as this material provides the language on which the rest of the book rests. Chapter 1 (re)introduces variables, sets, operators, relations, notation, and proofs. Chapter 2 provides a review of basic algebra, including solving equations, and briefly discusses computational aids. Chapter 3 deals with functions, talks more about relations, and introduces utility representations. Chapter 4 discusses limits and continuity, as well as sequences and series and some related properties of sets not covered in Chapter 1.

The second part covers calculus in one dimension, including optimization. Those with a prior background in calculus can safely skim this section, or skip it entirely if the background is recent. calculus is used in the discussion of continuous distributions, and forms the bedrock for the final section on multivariate calculus and optimization. Chapter 5 introduces calculus and the derivative for functions of one variable. Chapter 6 offers rules of differentiation and provides derivatives of both common and special functions. Chapter 7 introduces the indefinite and definite integral, provides techniques of integration, and discusses the fundamental theorem of calculus. Chapter 8 defines a function's extrema, discusses higher-order derivatives and concave and convex functions, and illustrates techniques for unconstrained optimization in one dimension, including first- and second-order conditions.

The third part tackles probability, from its basics to discrete and continuous distributions. A brief tie-in to statistical inference is provided here, though the focus remains on probability and not on statistics. As this section is perhaps the most essential to the work of political scientists, we would advise a careful reading by all. Chapter 9 presents the basics of probability theory, illustrates how to calculate what are known as "simple" probabilities, and discusses conceptually the utility of probability theory in both statistical inference and formal theory. Chapter 10 introduces the fundamental concepts of a distribution, a random variable and associated probability distribution, and an expectation, and offers examples of relevant discrete probability distributions. Neither of these two chapters requires the material in Part II. Chapter 11 brings calculus back in, thereby permitting us to discuss continuous distributions and expectations of variables over these distributions; it also offers examples of relevant continuous probability distributions and elaborates slightly on statistical inference.

The fourth part is a primer on linear algebra. Linear algebra aids in everything from solving systems of equations to statistical modeling to understanding stochastic processes, yet most students are unlikely to require understanding of all these topics at once. We recommend Chapter 12 to all readers, as it covers the definitions of vectors and matrices, as well as how to perform nuts-and-bolts calculations with them. This is the material that students in statistics courses are most likely to require. For those seeking a somewhat deeper introduction to linear algebra, Chapter 13 delves into vector spaces and norms, spanning vectors, linear independence, matrix rank, and quadratic forms. As a concrete payoff to learning these concepts, it also includes an application of them to the solution of systems of linear equations. Neither of these chapters requires any of the material in Parts II or III of the course, save for the proof that the method of ordinary least squares (OLS) is the best linear unbiased estimator (BLUE) in Chapter 13. Chapter 14 covers, albeit briefly, three more advanced topics: eigenvalues and eigenvectors, matrix decomposition, and Markov processes. These are all useful in statistical analysis, particularly Bayesian statistics, and the last is growing in use in formal theory as well. We strongly recommend the study of probability distributions, discussed in Part III, prior to engaging with the material in Chapter 14.

The fifth and final part is the most complex in the book. It introduces selected topics in multivariate calculus, including constrained and unconstrained optimization and implicit differentiation. These topics generally prove useful in more advanced classes in statistics and game theory, and the reader, particularly the reader who has not previously been exposed to calculus, may desire to put most of these chapters off until a later date. We expect that many math classes, and most math camps, will not find time to cover all of these topics, and that this section may serve as a later reference as much as a guide to present instruction. However, the first chapter in this part offers more basic and essential information about multivariate functions and the partial derivative, and should be read on a first pass through the book. This material, covered in Chapter 15, briefly provides the multivariate analogues to the material in Chapters 3–7. It covers functions of several variables, multivariate calculus, and concavity and convexity in more than one dimension. Chapter 16 is technically more complex, and completes the extension of our coverage of calculus to more than one dimension by introducing multidimensional optimization. It then goes further, providing techniques to perform constrained optimization: optimization under constraints on the values the variables might take. For this topic we deviate from our habit of providing intuition into the method and concentrate on elucidating the procedure to perform constrained optimization. Finally, Chapter 17 discusses the concept of comparative statics and describes a tool for accomplishing these, the implicit function theorem. Again, we focus in this chapter on the method, leaving the intuition to more advanced courses.

There is no one right order in which to read the material covered in this book. We chose the order of the parts and chapters to match what we felt was a natural progression: each part from II through V delivers material matched to a different undergraduate math course, and someone reading straight through covers one-dimensional topics fully before moving to the complexity of multiple dimensions in Parts IV and V. Yet Part IV, which covers linear algebra, requires neither calculus nor probability theory, and thus might have been put right after Part I. In fact, we could have moved Chapters 12 and 13, covering vectors, matrices, and vector spaces, to follow Chapter 3, as neither Chapter 12 nor 13 makes use of limits. That ordering could be part of a larger progression that accommodates a department that gets very quickly to quantitative analysis but doesn't require calculus until later. That is, an instructor could begin with Chapters 1–3, move to 12 and 13, pick up limits in Chapter 4, then introduce the basics of probability and distributions in Chapters 9 and 10. Skipped chapters could be included according to taste.

We could imagine other combinations as well, tailored more closely to the needs of a particular group of students. For example, in a department with a course in probability, Part III could be skipped; this could be done in the book's progression or the alternative progression offered in the previous paragraph. The only resulting change that would need to be made would be to skip a couple of the topics in Part IV: the proof that OLS is BLUE in Chapter 13, and Markov processes in Chapter 14. Third, a brief math camp might cover only Chapters 1–8, 12, 13, and 15. We have tried to make the book modular to accommodate selective use, and have noted above the prior requirements of each new chapter. This modularity should also improve the book's utility as a reference, and we would be most pleased if people found themselves returning to it well after completing their coursework.

Throughout each chapter we attempt to provide *reasons* why a student should master particular areas of mathematics. These reasons typically focus on two broad areas of quantitative analysis commonly used in political science—statistics and formal theory. Statistics, loosely speaking, deals with the analysis of data, and is largely, though not entirely, used in the quantitative description of politics and to *test* hypotheses about politics via statistical inference. In political science, statistics is often subsumed under the rubric of "quantitative methods." As we noted above, we do not cover statistical inference in this book. However, as the chapters progress, we work through some of the background behind OLS, which is a method of minimizing the (squared) deviations between one's data and a best-fit line. OLS appears early in most students' statistical training in political science. Formal theory, of which game theory is by far the most commonly used, is the use of mathematical analysis to derive *theoretical* propositions about how the world might work. It is used in political science primarily to derive internally consistent theories and to highlight important interactions between actors. We do not cover game theory in this book, but as the chapters progress we introduce **utility** and **expected utility**, and discuss its **maximization**. This maximization is at the heart of game theory, as it allows game theorists to say what rational actors *should* do in a given situation. These actions form part of **equilibrium** behavior, meaning that each player is satisfied with her optimal action given everyone else's optimal actions, and has no incentive to change it.

Each chapter concludes with a set of exercises designed to develop mastery via practice. We also include worked examples throughout to highlight important concepts or techniques, aimed at fostering a practical knowledge of mathematics that may be brought to coursework and research. In support of this, we have made a special effort to reduce the typical level of bravado present in math texts. We eschew words like "clearly" and "obviously" and instead show the supposedly clear and obvious steps often left out. Our hope is that this reduces some of the unnecessary intimidation often associated with math texts. Further, we frequently reference online resources, as sometimes another perspective on a topic may be all that the stuck student needs to advance.

Finally, we offer a brief note about how we came to write this text. We began with the opinion that math texts generally seek to communicate their content with an emphasis on concise, elegant presentation. Moore, who has had considerably less mathematical training than Siegel, has often found it difficult to follow the presentation in math texts, and wanted to write a text that erred on the side of walking the reader through the material at a level of detail that would leave few, if any, lost. He wrote the first draft for the first twelve chapters.[2] Siegel then drafted the remaining five and revised (and in several cases substantially rewrote) Moore's drafts. Beginning in 2006, various versions of the text have been used in the semester-long course for PhD students in the Political Science Department at Florida State, and those students have been very helpful with their feedback, as has Chris Reenock, who has frequently taught this course. G. Jason Smith contributed to some chapters' exercises, and Xiaoli Guo provided considerable assistance with the figures and other aspects of manuscript preparation. We thank two anonymous reviewers (and their anonymous classes) for additional helpful feedback. Chuck Myers at Princeton University Press offered the project considerable support, and we very much appreciate the assistance of Kathleen Cioffi, Eric Henney, Steven Peter, Rob Tiempo, and others at the press who helped us bring this book to publication.

[2]Joseph K. Young kindly provided the first draft of Chapter 2, Section 1, and Andreas Beger revised the first draft of Chapter 12.

Part I

Building Blocks

Chapter One

Preliminaries

Math is a formal language useful in clarifying and exploring connections between concepts. Like any language, it has a syntax that must be understood before its meaning can be parsed. We discuss the building blocks of this syntax in this chapter. The first is the variables that translate concepts into mathematics, and we begin here. Next we cover groupings of these variables into sets, and then operators on both variables and sets. Most data in political science are ordered, and relations, the topic of our fourth section, provide this ordering. In the fifth section we discuss the level of measurement of variables, which will aid us in conceptual precision. In the sixth section we offer an array of notation that will prove useful throughout the book; the reader may want to bookmark this section for easy return. Finally, the seventh section discusses methods of proof, through which we learn new things about our language of mathematics. This section is the most difficult, is useful primarily to those doing formal theory or devising new methods in statistics, and can be put aside for later reading or skipped entirely.

1.1 VARIABLES AND CONSTANTS

Political scientists are interested in concepts such as participation, voting, democracy, party discipline, alliance commitment, war, etc. If scholars are to communicate meaningfully, they must be able to understand what one another is arguing. In other words, they must be specific about their theories and their empirical evaluation of the hypotheses implied by their theories.

A theory is a set of statements that involve concepts. The statements comprise assumptions, propositions, corollaries, and hypotheses. Typically, assumptions are asserted, propositions and corollaries are deduced from these assumptions, and hypotheses are derived from these deductions and then empirically challenged.[1] Concepts are inventions that human beings create to help them understand the world. They can generally take different values: high or low, present or absent, none or few or many, etc.

Throughout the book we use the term "concept," not "variable," when discussing theory. Theories (and the hypotheses they imply) concern relationships among abstract concepts. Variables are the indicators we develop to measure

[1] Of course, assumptions and the solution concepts from which deductions are made may be empirically challenged as well, but this practice is rarer in the discipline.

our concepts. Current practice in political science does not always honor this distinction, but it can be helpful, particularly when first developing theory, to speak of concepts when referring to theories and hypotheses, and reserve the term variables for discussion of indicators or measures.

We assign *variables* and *constants* to concepts so that we may use them in formal mathematical expressions. Both variables and constants are frequently represented by an upper- or lowercase letter. Y or y is often used to represent a concept that one wishes to explain, and X or x is often used to represent a concept that causes Y to take different values (i.e., vary). The letter one chooses to represent a concept is arbitrary—one could choose m or z or h, etc. There are some conventions, such as the one about x and y, but there are no hard-and-fast rules here.

Variables and constants can be anything one believes to be important to one's theory. For example, y could represent voter turnout and x the level of education. They differ only in the degree to which they vary across some set of cases. For example, students of electoral politics are interested in the gender gap in participation and/or party identification. Gender is a variable in the US electorate because its value varies across individuals who are typically identified as male or female.[2] In a study of voting patterns among US Supreme Court justices between 1850 and 1950, however, gender is a constant (all the justices were male).

More formally, a **constant** is a concept or a measure[3] that has a single value for a given set. We define sets shortly, but the sets that interest political scientists tend to be *the characteristics of* individuals (e.g., eligible voters), collectives (e.g., legislatures), and countries. So if the values for a given concept (or its measure) do not vary across the individuals, collectives, or countries, etc., to which it applies, then the value is a constant. A **variable** is a concept or a measure that takes different values in a given set. Coefficients on variables (i.e., the parameters that multiply the variables) are usually constants.

1.1.1 Why Should I Care?

Concepts and their relationships are the stuff of science, and there is nothing more fundamental for a political scientist than an ability to be precise in concept formation and the statement of expected relationships. Thinking abstractly in terms of constants and variables is a first step in developing clear theories and testable hypotheses.

---

[2]Definitions of concepts are, quite properly, contested in all areas of academia, and gender is no exception. Though it is not a debate that generates a great deal of interest among students of participation or party identification, it will be rather easy for you to find literature in other fields debating the value of defining gender as a binary variable.

[3]By measure we mean an operational indicator of a concept. For example, the concept gender might be measured with a survey question. The survey data provide a measure of the concept.

Table 1.1: Common Sets

Notation	Meaning
$\mathbb{N}$	Natural numbers
$\mathbb{Z}$	Integers
$\mathbb{Q}$	Rational numbers
$\mathbb{R}$	Real (rational and irrational) numbers
$\mathbb{C}$	Complex numbers
Subscript: $\mathbb{N}_+$	Positive (negative) values of the set
Superscript: $\mathbb{N}^d$	Dimensionality (number of dimensions)

1.2 SETS

This leads us naturally into a discussion of sets. For our purposes,[4] a *set* is just a collection of elements. One can think of them as groups whose members have something in common that is important to the person who has grouped them together. The most common sets we utilize are those that contain all possible values of a variable. You undoubtedly have seen these types of sets before, as all numbers belong to them. For example, the counting numbers $(0, 1, 2, \ldots,$ where $\ldots$ signifies that this progression goes on indefinitely) belong to the set of natural numbers.[5] The set of all natural numbers is denoted $\mathbb{N}$, and any variable n that is a natural number must come from this set. If we add negative numbers to the set of natural numbers, i.e., $\ldots, -3, -2, -1$, then we get the set of all integers, denoted $\mathbb{Z}$. All numbers that can be expressed as a ratio of two integers are called rational numbers, and the set of these is denoted $\mathbb{Q}$. This set is larger than the set of integers (though both are infinite!) but is still missing some important irrational numbers such as π and e. The set of all rational and irrational numbers together is known as the real numbers and is denoted $\mathbb{R}$.[6]

Political scientists are interested in general relationships among concepts. Sets prove fundamental to this in two ways. We have already discussed the association between concepts and variables. As the values of each variable, and so of each concept, are drawn from a set, each such set demarcates the range of possible values a variable can take. Some variables in political science have ranges of values equal to all possible numbers of a particular type, typically either integers, for a variable such as net migration, or real numbers, for a

[4]These purposes, you will recall, are to build intuition rather than to be exact. We play somewhat loosely with ordered sets in what follows, and ignore things like Russell's paradox.

[5]Some define the natural numbers without the zero. We are not precise enough in this book to make this distinction important.

[6]You may have occasion to use complex numbers, denoted $\mathbb{C}$. These have two components, a real and an imaginary part, and can be written $a + bi$, where a and b are both real numbers and $i = \sqrt{-1}$. These are beyond the scope of this book, though amply covered by classes in complex analysis.

variable such as GDP. More typically, variables draw their values from some subset of possible numbers, and we say the variable x is an element of a subset of $\mathbb{R}$. For example, population is typically an element of $\mathbb{Z}_+$, the set of all positive integers, which is a subset of all integers. (A + subscript typically signifies positive numbers, and a − negative.) The size and qualities of the subset can be informative. We saw this earlier for the gender variable: depending on the empirical setting, the sets of all possible values were either $\{\text{Male}, \text{Female}\}$ or $\{\text{Male}\}$.[7] The type of set from which a variable's values are drawn can also guide our theorizing. Researchers who develop a formal model, game theoretic or otherwise, must explicitly note the range of their variables, and we can use set notation to describe whether they are discrete or continuous variables, for example. A variable is discrete if each one of its possible values can be associated with a single integer. We might assign a 1 for a female and 2 for male, for instance. Continuous variables are those whose values cannot each be assigned a single integer.[8] We typically assume that continuous variables are drawn from a subset of the real numbers, though this is not necessary.

A *solution set* is the set of all solutions to some equation, and may be discrete or continuous. For example, the set of solutions to the equation $x^2 - 5x + 6 = 0$ is $\{2, 3\}$, a discrete set. We term a *sample space* a set that contains all of the values that a variable can take in the context of statistical inference. When discussing individuals' actions in game theory, we instead use the term *strategy space* for the same concept. For example, if a player in a one-shot game[9] can either (C)ooperate with a partner for some joint goal or (D)efect to achieve personal goals, then the strategy space for that player is $\{C, D\}$. This will make sense in context, as you study game theory.

Note that each of these is termed a *space* rather than a set. This is not a typo; spaces are usually sets with some structure. For our purposes the most common structure we will encounter is a metric—a measure of distance between the elements of the set. Sets like $\mathbb{Z}$ and $\mathbb{R}$ have natural metrics. These examples of sets form one-dimensional spaces: the elements in them differ along a single axis. Sets may also contain multidimensional elements. For example, a set might contain a number of points in three-dimensional space. In this case, each element can be written (x, y, z), and the set from which these elements are drawn is written $\mathbb{R}^3$. More generally, the superscript indicates the dimensionality of the space. We will frequently use the d-dimensional space $\mathbb{R}^d$ in this book. When $d = 3$, this is called Euclidean space. Another common multidimensional element is an ordered pair, written (a, b). Unlike elements of $\mathbb{R}^3$, in which each

[7] As explained below, curly brackets indicate that the set is discrete. Continuous sets are demarcated by parentheses and square brackets.

[8] Formally, a discrete variable draws values from a countable set, while a continuous variable draws from an uncountable set. We define countability shortly.

[9] A one-shot game is one that is played only once, rather than repeatedly. You will encounter unfamiliar terms in the reading you do in graduate school. It is important to get in the habit of referencing a good dictionary (online or printed) and looking up terms. A search on a site like Google is often a useful way to find definitions of terms that are not found in dictionaries.

of x, y, and z is a real number, each member of an ordered pair may be quite different. For example, an ordered pair might be (orange, lunch), indicating that one often eats an orange at lunch. Ordered pairs, or more generally ordered n-tuples, which are ordered pairs with n elements, are often formed via Cartesian products. We describe these in the next section, but they function along the lines of "take one element from the set of all fruit and connect it to the set of all meals."

Political scientists also think about sets informally (i.e., nonmathematically) on a regular basis. We may take as an example the article by Sniderman, Hagendoorn, and Prior (2004). The authors were interested in the source of the majority public's opposition to immigrant minorities and studied survey data to evaluate several hypotheses. The objects they studied were individual people, and each variable over which they collected data can be represented as a set. For example, they developed measures of people's perceptions of threat with respect to "individual safety," "individual economic well-being," "collective safety," and "collective economic well-being." They surveyed 2,007 people, and thus had four sets, each of which contained 2,007 elements: each individual's value for each measure.[10] In this formulation sets contain not the possible values a variable might take, but rather the realized values that many variables do take, where each variable is one person's perception of one threat. Thus, sets here provide us with a formal way to think about membership in categories or groups.

Given the importance of both ways of thinking about sets, we will take some time now to discuss their properties. A set can be finite or infinite, countable or uncountable, bounded or unbounded. All these terms mean what we would expect them to mean. The number of elements in a **finite set** is finite; that is, there are only so many elements in the set, and no more. In contrast, there is no limit to the number of elements in an **infinite set**. For example, the set $\mathbb{Z}$ is infinite, but the subset containing all integers from one to ten is finite. A **countable set** is one whose elements can be counted, i.e., each one can be associated with a natural number (or an integer). An **uncountable set** does not have this property. Both $\mathbb{Z}$ and the set of numbers from one to ten are countable, whereas the set of all real numbers between zero and one is not. A **bounded set** has finite size (but may have infinite elements), while an **unbounded set** does not. Intuitively, a bounded set can be encased in some finite shape (usually a ball), whereas an unbounded set cannot. We say a set has a lower bound if there is a number, u, such that every element in the set is no smaller than it, and an upper bound if there is a number, v, such that every element in the set is no bigger than it. These bounds need not be in the set themselves, and there may be many of them. The greatest lower bound is the largest such lower bound, and the least upper bound is the smallest such upper bound.

Sets contain **elements**, so we need some way to indicate that a given element

[10] One could also view this as four sets of ordered pairs, with each pair containing a variable name and a person's perceptions, or one set of ordered 5-tuples, each with a person's name and her responses to each question, in order.

is a member of a particular set. A "funky E" serves this purpose: $x \in A$ states that "x is an element of the set A" or "x is in A." You will find this symbol used when the author restricts the values of a variable to a specific range: $x \in \{1, 2, 3\}$ or $x \in [0, 1]$. This means that x can take the value 1, 2, or 3 or x can be any real number from 0 to 1, inclusive. It is also convenient to use this notation to identify the range of, say, a dichotomous dependent variable in a statistical analysis: $y \in \{0, 1\}$. This means that y *either can take a value of 0 or a value of 1*. So the "funky E" is an important symbol with which to become familiar. Conversely, when something is not in a set, we use the symbol $\notin$, as in $x \notin A$. This means that, for the examples in the previous paragraph, x does not take the values 1, 2, or 3 or is not between 0 and 1. As you may have guessed from our usage, curly brackets like $\{\}$ are used to denote discrete sets, e.g., $\{A, B, C\}$. Continuous sets use square brackets or parentheses depending on whether they are closed or open (terms we define in Chapter 4), e.g., $[0, 1]$ or $(0, 1)$, which are the sets of all real numbers between 0 and 1, inclusive and exclusive, respectively.

Much as sets contain elements, they also can contain, and be contained by, other sets. The expression $A \subset B$ (read "A is a **proper subset** of B") implies that set B contains all the elements in A, plus at least one more. More formally, $A \subset B$ if all x that are elements in A are also elements in B (i.e., if $x \in A$, then $x \in B$). $A \subseteq B$ (read "A is a **subset** of B"), in contrast, allows A and B to be the same. We say that A is a proper subset of B in the first case but not in the second. So the set of voters is a subset of the set of eligible voters, and is most likely a proper subset, since we rarely experience full turnout. We also occasionally say that a set that contains another set is a superset of the smaller one, but this terminology is less common. The **cardinality of a set** is the number of elements in that set. Note that proper subsets have smaller cardinalities than their supersets, finite sets have finite cardinalities, and infinite sets have infinite cardinalities.

A **singleton** is a set with only one element and so a cardinality of one. The power set of A is the set of all subsets of A, and has a cardinality of $2^{|A|}$, where $|A|$ is the cardinality of A. Power sets come up reasonably often in political science by virtue of our attention to bargaining and coalition formation. When one considers all possible coalitions or alliances, one is really considering all possible subsets of the overall set of individuals or nations. Power sets of infinite sets are always uncountable, but are not usually seen in political science applications. The **empty set** (or **null set**) is the set with nothing in it and is written $\emptyset$. The **universal set** is the set that contains all elements. This latter concept is particularly common in probability.

Finally, sets can be ordered or unordered. The **ordered set** $\{a, b, c\}$ differs from $\{c, a, b\}$, but the **unordered set** $\{a, b, c\}$ is the same as $\{c, a, b\}$. That is, when sets are ordered, the order of the elements is important. Political scientists primarily work with ordered sets. For example, all *datasets* are ordered sets. Consider again the study by Sniderman et al. (2004). We sketched four of the sets they used in their study; the order in which the elements of those sets is maintained is critically important. That is, the first element in each set must refer to the first person who was surveyed, the second element must refer to the second person, and the 1,232nd element must refer to the 1,232nd person surveyed, etc. All data analyses use ordered sets. Similarly, all equilibrium strategy sets in game theory are ordered according to player. However, this does not mean all sets used in political science are ordered. For example, the set of all strategies one might play may or may not be ordered.

1.2.1 Why Should I Care?

Sets are useful to political scientists for two reasons: (1) one needs to understand sets before one can understand relations and functions (covered in this chapter and Chapter 3), and (2) sets are used widely in formal theory and in the presentation of some areas of statistics (e.g., probability theory is often developed using set theory). They provide us with a more specific method for doing the type of categorization that political scientists are always doing. They also provide us with a conceptual tool that is useful for developing other important ideas. So a basic familiarity with sets is important for further study.

For example, game theory is concerned with determining what two or more actors should choose to do, given their goals (expressed via their utility) and their beliefs about the likelihood of different outcomes given the choices they might make and their beliefs about the expected behavior of the other actor(s). Sets play a central role in game theory. The choices available to each actor form a set. The best responses of an actor to another actor's behavior form a set. All possible states of the world form a set. And so on.

Those of you who are unfamiliar with game theory will find this brief discussion less than illuminating, but do not be concerned. Our point is not to explain sets of actions, best response sets, or information sets—each is covered in game theory courses and texts—but rather to underscore why it is important to have a functional grasp of elementary set theory if one wants to study formal models. Finally, we note that Riker's (1962) celebrated game theoretic model of political coalition formation makes extensive use of set theory to develop what he calls the size principle (see Appendix I, pp. 247–78, of his book). That is, of course, but one of scores of examples we might have selected.[11]

1.3 OPERATORS

We now have formalizations of concepts (variables) and ways to order and group these variables (sets), but as yet nothing to do with them. Operators, the topic of this section, are active mathematical constructs that, as their name implies, operate on sets and elements of sets. Some operators on variables have been familiar since early childhood: addition ($+$), subtraction ($-$), multiplication ($*$ or $\times$ or $\cdot$ or just placing two variables adjacent to each other as in xy),

[11]Readers interested in surveys of formal models in political science that are targeted at students might find Shepsle and Bonchek (1997) and Gelbach (2013) useful.

and division ($\div$ or /). We assume you know how to perform these operations. Exponentiation, or raising x to the power a (x^a), is likely also familiar, as is taking an nth root ($\sqrt[n]{x}$), and perhaps finding a factorial (!) as well.

Other useful basic operators include summation ($\sum_i x_i$), which dictates that all the x_i indexed by i should be added, and product ($\prod_i x_i$), which dictates that all the x_i be multiplied. These operators are common in empirical work, where each i corresponds to a data point (or observation). Here are a couple of examples:

$$\sum_{i=1}^{3} x_i = x_1 + x_2 + x_3,$$

and

$$\prod_{i=1}^{3} x_i = x_1 \times x_2 \times x_3.$$

Because they are just shorthand ways of writing multiple sums or products, each of these operators obeys all the rules of addition and multiplication that we lay out in the next chapter. So, for example, $\sum_{i=1}^{n} x_i^2$ does not generally equal $\left(\sum_{i=1}^{n} x_i\right)^2$ for the same reason that $(2^2+3^2) = 13$ does not equal $(2+3)^2 = 25$.[12] Other operators and their properties will be introduced as needed throughout the book. We present a collection of notation below in section 1.6 of this chapter.

You may be less familiar with operators on sets, though they are no less fundamental. We consider six here: *differences*, *complements*, *intersections*, *unions*, *partitions*, and *Cartesian products*. The **difference** between two sets A and B, denoted $A \backslash B$ (read "A difference B"), is the set containing all the elements of A that are not also in B: $x \in A \backslash B$ if $x \in A$ but $x \notin B$. This set comes up a great deal in game theory when one is trying to exclude individual players or strategies from consideration. The **complement** of a set, denoted A' or A^c, is the set that contains the elements that are not contained in A: $x \in A^c$ if x is not an element of A.[13] Continuing the example from above, the complement of the set of registered voters is the set of all people who are not registered voters.

Venn diagrams can be used to depict set relationships. Figure 1.1 illustrates the concepts of set difference and set complement. The shaded part of the left diagram is the set Registered Voters \ Registered Democrats, which is read "Registered Voters difference Registered Democrats." Or, in other words, all registered voters who are *not* registered Democrats. The shaded part of the right diagram illustrates the set Registered Voters^c, which is "the complement

[12] Summations and products can also be repeated; this is known as a double (or triple, etc.) summation or product. If x_{ij} is indexed by i and j, then we could write $\sum_i \sum_j x_{ij}$ or $\prod_i \prod_j x_{ij}$. Multiple summations may be useful, for example, when employing discrete distributions in more than one dimension, or when considering more than one random variable in game theory.

[13] One can also think of the complement of a set A as the difference between the universal set and A.

of Registered Voters." Or, in other words, people who are not registered voters, since the universal set in this case is the set of All People. Both diagrams illustrate the concept of a subset: the set Registered Voters is a (proper) subset of the set All People, and the set Registered Democrats is a (proper) subset of the set Registered Voters. And both diagrams illustrate another concept: the sets Registered Voters and Registered Voters^c are collectively exhaustive, in that together they constitute the set All People, which is the universal set in this case. In general, a group of sets is **collectively exhaustive** if together the sets constitute the universal set.[14]

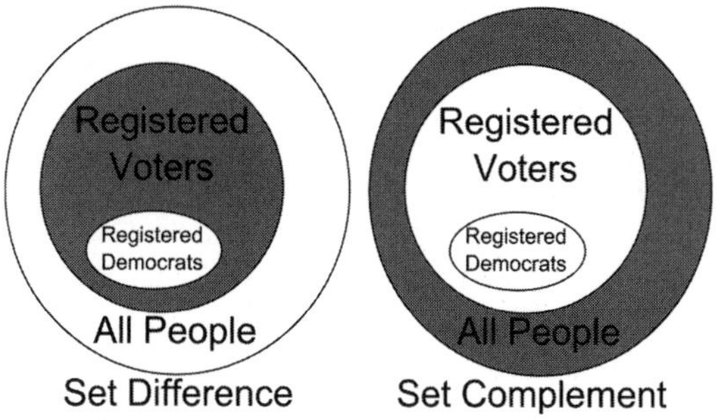

Figure 1.1: Set Difference and Complement

The **intersection** of two sets A and B, denoted $A \cap B$ (read "A intersection B"), is the set of elements common to both sets. In other words, $x \in A \cap B$ if $x \in A$ and $x \in B$. Thus, if set A consists of elected Democrats in the state of Florida and set B consists of legislators in the Florida House of Representatives, then the intersection of A and B is the set containing all Democratic House members in Florida.

The **union** of two sets is written $A \cup B$ (read "A union B") and is the set of all elements contained in either set. In other words, $x \in A \cup B$ if $x \in A$ or $x \in B$. Note that any x in both sets is also in their union. Continuing the example from above, the union of A and B is the set composed of all elected Democrats in Florida *and* all House members in Florida. Figure 1.2 shows the intersection of the sets House Members and Elected Democrats in the shaded part on the left, and their union in the shaded part on the right. The diagram on the left also illustrates the concept of mutually exclusive sets. **Mutually exclusive** sets are sets with an intersection equal to the empty set, i.e., sets with no elements in their intersection. In the diagram on the left, the two unshaded portions of the sets House Members and Elected Democrats are mutually exclusive sets. In fact, any two sets are mutually exclusive once their intersection has been removed, since they then must have an intersection that is empty.

A **partition** is a bit more complex: it is the collection of subsets whose union forms the set. The more elements a set has, the greater the number of partitions

[14]Strictly speaking, their union must equal the universal set. We discuss unions next.

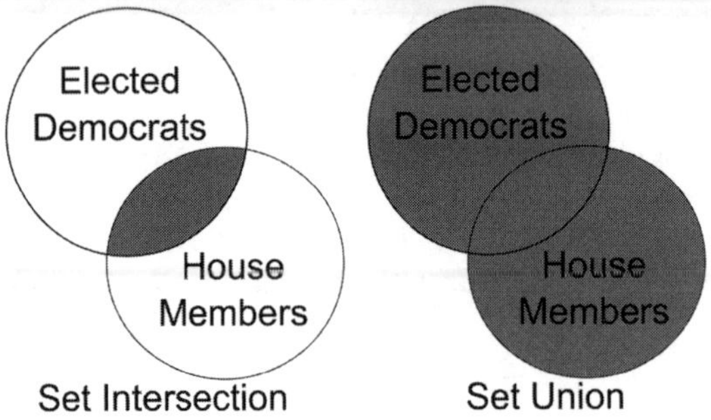

Figure 1.2: Set Intersection and Union

one can create. Let's consider the following example, the set of candidates for the 2004 US presidential election who received national press coverage:[15] $A = \{\text{Bush}, \text{Kerry}, \text{Nader}\}$. We can partition A into three subsets: {Bush}, {Kerry}, {Nader}; or we can partition it into two subsets: {Bush, Nader}, {Kerry}; or {Kerry, Nader}, {Bush}; or {Bush, Kerry}, {Nader}. Finally, the set itself is a partition: {Bush, Kerry, Nader}.

A **Cartesian product** is more complex still. Consider two sets A and B, and let $a \in A$ and $b \in B$. Then the Cartesian product $A \times B$ is the set consisting of all possible ordered pairs (a, b), where $a \in A$ and $b \in B$. For example, if $A =$ {Female, Male} and $B =$ {Income over \$50k, Income under \$50k}, then the Cartesian product is the set of cardinality four consisting of all possible ordered pairs: $A \times B =$ {(Female, Income over \$50k), (Female, Income under \$50k), (Male, Income over \$50k), (Male, Income under \$50k)}. Note that the type of element (ordered pairs) in the product is different from the elements of the constituent sets. Cartesian products are commonly used to form larger spaces from smaller constituents, and appear commonly in both statistics and game theory. We can extend the concept of ordered pairs to ordered n-tuples in this manner, and each element in the n-tuple represents a *dimension*. So x is one-dimensional, (x, y) is two-dimensional, (x, y, z) is three-dimensional, and so on. Common examples of such usage would be $\mathbb{R}^3 = \mathbb{R} \times \mathbb{R} \times \mathbb{R}$, which is three-dimensional space, and $S = S_1 \times S_2 \times \ldots \times S_n$, which is a strategy space formed from the individual strategy spaces of each of the n players in a game.

1.3.1 Why Should I Care?

Operators on variables are essential; without them we could not even add two numbers. Operators on sets are equally essential, as they allow us to manipulate sets and form spaces that better capture our theories, including complex inter-

[15]In August 2004 Project Vote Smart listed over ninety candidates for president of the United States, but working with the full set would be unwieldy, so we restrict attention to the subset of candidates who received national press coverage (`http://www.vote-smart.org/election_president_party.php?party_name=All`).

actions. They are also necessary for properly specifying functions of all sorts, as we shall see in Chapter 3.

1.4 RELATIONS

Now we have variables, conceptually informed groups of variables, and ways to manipulate them via operators, but we still lack ways to compare concepts and discern relationships between them. This is where relations enter. A mathematical **relation** allows one to compare constants, variables, or expressions of these (or, if you prefer, concepts). Binary relations (i.e., the relation between two constants/variables/expressions or concepts) are easiest to consider, so we will restrict the discussion to the two variable case, but the idea can be generalized to an n-ary relation. Similarly, we can define orders on sets, but these admit many possibilities and are less commonly observed in political science, so we will eschew this topic as well.

A binary **relation** can be represented as an ordered pair. So, if $a \in A$ is greater than $b \in A$, we can write the relation as (a, b). When constants or variables are drawn from the integers or real numbers, though, we have more familiar notation. Integers and real numbers have natural associated orders: three is greater than two is greater than one, and so on. When one is certain of the value of a concept, as one is with a constant, then we can write $3 > 2$, $1 < 4$, and $2.5 = 2.5$. The symbols $>$, $<$, and $=$ form the familiar relations of arithmetic. When one is less sure of the values of a concept, as one is with a variable, then we also have the relations $\geq$ and $\leq$, as in $x \geq z$. Algebra, reviewed in the next chapter, deals with the manipulation of these sorts of relations.

The concept of relations is more general than these orders, however. A relation exists between two sets (or concepts) when knowing one element provides information about the other element. So, for example, in networks the relation could be "linked," while in game theory it might be "like as well as." We will explore this latter idea more in Chapter 3. While relations can be specified quite generally,[16] typically we will only be concerned with a few types of relation. Inequalities are one, and preference relations, discussed in Chapter 3, are another. The most common relation we'll use, though, is a function, which is the topic of Chapter 3. In this context we want to know the mapping between sets A and B. In other words, we want to know how the function transforms an element of A into an element of B. In this case we call A the **domain** and B the **range**. Relations (and so functions) can have various properties, some of which we discuss in Chapter 3.

[16]A relation is a mathematical object that takes as input two sets A and B (called its domain in this context) and returns a subset of $A \times B$ (called its graph in this context).

1.4.1 Why Should I Care?

Relations are important because they help us describe the mapping of values across concepts. Relations such as "greater than" and "equal to" are critical to descriptive claims about the world as well as to making theoretical claims. Further, functions—a specific type of relation—are very common in both theoretical and empirical work in political science.

1.5 LEVEL OF MEASUREMENT

We now have most of the building blocks we need to describe relationships between concepts. These in turn allow us to distinguish among different levels of measurement: **nominal**, **ordinal**, **interval**, and **ratio**. Note that though levels of measurement tend to be associated with variables, they are equally applicable and important to conceptualization.[17] We briefly discuss each level of measurement in turn.

1.5.1 Differences of Kind

In some theories all we require of our concepts is that they distinguish one type from another. That is, some concepts are about differences of kind, but not differences of degree. Concepts that identify different types but do not order them on any scale are nominal, and they require only nominal level measurement of their indicators.[18]

Nominal level measurement does not establish mathematical relationships among the values. In other words, it does not make sense to assert that a case with a nominal value of 3 is greater than one with a nominal value of 1, or that two cases with a nominal value of 2 are equal. The symbols $<$, $\leq$, $=$, $\geq$, and $>$ have no meaning for variables measured at the nominal level. Gender is a good example of a nominal level variable. When entering data for a measure of gender into a computer a researcher might assign the values of 0 and 1 (or 1 and 2) to female and male, respectively. But she might also have assigned the values -64 and $3,241$. Or she might have assigned the values 1 and 0 (or 2 and 1) to female and male, respectively. The point is that higher values do not convey any meaning: the numerical values are placeholders that indicate a difference, but the numerical values do not tell us anything meaningful.

1.5.2 Differences of Degree

At other times we are interested in differences of degree. Whether one case has more, is stronger, etc., is important to us as we define concepts and then think

[17] Students interested in an extended discussion will find Cohen and Nagel (1934, pp. 223–44) useful.

[18] The four levels of measurement—nominal, ordinal, interval, and ratio—were proposed by Stevens (1946).

about ways to measure those concepts. In such cases, nominal level concepts and measures are inadequate for they do not imply mathematical relationships among the values.

Ordinal level measurement, on the other hand, does imply mathematical relationships among the values. More specifically, the symbols $<$, $\leq$, $=$, $\geq$, and $>$ have meaning for ordinal level concepts (variables). However, the distance between any two values does not measure a constant quantity across the values the variable might take. For example, a voting scholar might be interested in people's self-placement on an ideological scale. He might put together a survey that includes a question asking people to mark themselves as far left, moderate left, middle of the road, moderate right, far right. Such a concept makes "greater than," "less than," and "equal to" distinctions. For example, we can say that moderate left is further to the left on the scale than middle of the road. And when we assign numerical values we do not have the same freedom as with a nominal measure. That is, once we have assigned two values, we are constrained on others. For example, if we assign "middle of the road" the value 3 and "far left" the value 1, then we must assign "moderate left" a value greater than 1 and less than 3. If this were a nominal level variable, then we would not be so constrained and could assign any value we wish. But ordinal variables must use numerical values that retain the order of the concept's values because the order matters in the sense that it conveys meaning. So concepts with an ordinal level of measurement have ordered values that indicate "more than" and "less than."

The next level of measurement is **interval**. This requires that the distance between values be constant over the range of values. This property is important because it makes addition and subtraction meaningful. One cannot meaningfully add or subtract variables with nominal or ordinal values because the operation does not make sense. To see that this is so, consider that we can assign any values to a binary nominal variable: 0, 1; 1, 2; or -64 and $3,241$. We cannot meaningfully add or subtract the values of such a variable because the values do not have meaning as numerical values. Ordinal measures, on the other hand, have meaning up to "greater than" and "less than" operations, but they also cannot be added or subtracted. If one considers the example above, we might assign the numerical values 1, 2, 3, 4, and 5 to the ideology scale, or we might assign the numerical values -3, 2, 7, 44, and 1,324. Any set of numerical values that retains the order of the concept's values is valid. The distances in the first numerical value set are constant (they are each one unit apart), but the distances in the second set vary. As such, and because both sets of values are valid, the addition and subtraction of ordinal measures do not have meaning.

Interval level measures, on the other hand, have meaningful distances between values: the intervals between numbers are constant across the range of values. Put differently, a change of $\pm x$ on the scale is the same distance regardless of where one is on the scale.

Interval levels measures may be **discrete** or **continuous**. Discrete variables with interval level measurement are integers (or natural numbers). For example, a common survey item is the feeling thermometer, which asks respondents to identify the strength of their feelings toward a politician on a scale of 0 to 100, where 0 represents extremely cold and 100 represents extremely hot (e.g., Cain, 1978; Abramowitz, 1980). Most researchers submit that the respondent recognizes that a shift of +10 points is the same anywhere on the scale.[19] That is, the distance from 0 to 5 is equivalent to the distance from 26 to 31, from 47 to 52, from 83 to 88, etc. To the extent that this is so, the measure is interval. One can meaningfully add and subtract interval level measures.

Ratio level variables are interval level variables that have a meaningful zero value. The feeling thermometer variable has a zero value, but it does not represent the absence of feeling. Instead, it represents a very strong feeling: intense dislike. So zero is not a meaningful point on the scale. As such, while we can conduct meaningful addition and subtraction operations with such variables, we cannot conduct meaningful multiplication and division operations.

The label "ratio level" comes from the fact that the same ratio at two points on the scale conveys the same meaning. This is not terribly intuitive, so let us explain. On an interval level scale any distance x between two points has the same meaning, regardless of where we are on the scale. Ratio level measurement also has this property, but it has a constant ratio property that interval level measurement lacks: the ratio of two points on the scale conveys the same meaning regardless of where one is on the scale. A good example of a ratio level scale is a public budget. Imagine that a municipal government spends four times as much on public safety as it does on public health. This is a ratio of 4:1.[20] Thus, if the city spends $4.8 million on public safety, it must spend $1.2 million on public health. Similarly, if it spends $2 million on public safety, it must spend $0.5 million on public health. Ratios can only convey meaning (i.e., measure a constant ratio) when the scale over which they are measured has a 0 value that indicates the absence (i.e., none of) whatever is being measured.

To return to the feeling thermometer example, if the value 0 represents intense negative affect (i.e., dislike), 50 indicates an absence of affect (i.e., indifference), and 100 represents intense positive affect, then 0 is not an absence of affect. Thus, it is an interval level scale, not a ratio level scale, and we cannot conclude that the first member of two pairs of respondents with scores of 20 and 10, and 50 and 25, respectively, each have twice as much affect for a candidate as the second member of each pair. However, we could rescale the feeling thermometer to make it centered on zero, perhaps assigning the value of -50 to intense negative affect, 0 to the absence of affect (or indifference), and 50 to intense positive affect. Doing so would transform the level of measurement from interval to ratio.[21]

---

[19] Note that the respondents' (implicit) beliefs about the scale of the item are important in survey research.

[20] We discuss ratios in more detail in the first section of Chapter 2. You may want to skip ahead to there if you are unfamiliar with ratios.

[21] You may be thinking that this is a trivial transformation that is not consequential, but this is not the case. To see why, try the following. Arbitrarily select a ratio—perhaps 3:1—and select two pairs of points on the transformed feeling thermometer (the one with the proper

There are lots of examples of discrete ratio level variables in political science. Political scientists are often interested in the number of events that occur, and an event count has a meaningful constant distance between values and a meaningful zero point. Thus, they are ratio variables. Examples of event counts that have been used in political science include the number of seats a party holds in parliament, the number of vetoes issued by an executive, the number of unanimous decisions by a court, and the number of wars in which a country has participated.

Thus far we have restricted our attention to discrete variables. Continuous variables have an interval or ratio level of measurement, depending on whether the value 0 represents the absence of the concept. The vast majority of (empirical) concepts that political scientists have either created or borrowed from other disciplines are discrete, but some examples of continuous measures of interest to political scientists are income and GDP.[22]

You have likely noticed that each level of measurement subsumes the levels below it. That is, ordinal level measurement is also nominal, and an interval measure has ordinal and nominal properties. This suggests that whenever we have a concept at a high level of measurement we can reconceptualize and remeasure it at a lower level of measurement should we have cause to do so.

Some people mistakenly view the hierarchy of the levels of measurement as a means to judge the heuristic value of concepts. This is an error. Concepts can be evaluated on their clarity (vague concepts have little heuristic value), and one can make normative judgments about concepts (e.g., freedom, peace, order), but all sufficiently clear concepts are merely inputs to specific theories, and theories, not their concepts, should be judged. A proper discussion of this issue is beyond the scope of this book, but it is important to recognize that a nominal conceptualization may yield insights that a ratio conceptualization would miss and vice versa. Put differently, it would be an error to judge the levels of measurement as an ordinal scale with respect to their value to causal theory: it is nominal.

1.5.3 Why Should I Care?

Recognizing whether one is thinking about differences of kind (nominal) or degree (ordinal, interval, or ratio) is critical. If one is thinking about differences of degree, then how precise are those differences? Without a firm grasp on levels of measurement one cannot be precise about one's concepts, much less one's measures of one's concepts.

---

ratio scale where −50 is intense dislike, 0 is indifference, and 50 is strong positive affect) that have that ratio. Now transform the scale to the actual feeling thermometer (the one with the range from 0 to 100). Recalculate the ratios. They are different, right? The two scales do not produce the same ratio levels, and that means that one of them preserves ratios and the other does not. The one with the meaningful zero is the only scale that produces meaningful ratios. For a more detailed explanation, see Stevens (1946).

[22] If one rounds either to dollars, thousands of dollars, etc., then the values are integers (or natural numbers) and the measure is discrete.

1.6 NOTATION

Here we list, and in some cases briefly describe, common notation. This section is one you will likely refer to from time to time, but not everything might be clear now. Also, as a reference section it is heavier on the math and lighter on the intuition. It is important to read it once now, but if you find yourself unclear on some notation later, please refer back to this section. To make reference easier, we begin with the summary Table 1.2.

Operators take many forms, and are commonly used. We have already discussed some: $+$, $-$, $\times$, $/$, x^n, $\sqrt[n]{x}$, $\sum$, $\prod$, !. Some of these have multiple ways to represent them, others mean multiple things depending on context. For example, there are several ways to represent multiplication: $a \times b \times c = a * b * c = a \cdot b \cdot c = abc$. Of course, as we have seen, $\times$ can also mean a Cartesian product when applied to sets. Both $/$ and $\div$ mean divide; the mod operator, written 8 mod 3, means divide the first number by the second, and report the remainder: $8 \bmod 3 = 2$.

One can also use the product operator, $\prod$, to represent the product of a, b, and c: $\prod_a^c$.

> One reads that as *the product of a through c.*

More typically, the product operator is used by indexing a variable (this is accomplished by adding a subscript: x_i) and writing: $\prod_{i=k}^{l} x_i$.

> One reads that as *the product of x_i over the range from $i = k$ through $i = l$.*

When the product operator is used in an equation that is set apart from the text, it looks like this:

$$\prod_{i=k}^{l} x_i = x_k \times \ldots \times x_l.$$

The "..." here signals the reader to assume all interim values are included in the product. When used at the end of a list, e.g., $1, 2, 3, \ldots$, "..." signifies that the list (or product or sum) goes on indefinitely. In these cases you may also see ∞ as an end to the sequence instead, e.g., $1, 2, 3, \ldots, \infty$; ∞ is the symbol for infinity. In other words, ... means continue the progression until told to stop.

The summation operator, $\sum$, can be used to represent the addition of several numbers. For example, if we want to add together all members of a set indexed by i, then we can write: $\sum_i$. One reads that as *the sum over i*. You will also see summation represented over a range of values, say from value k through value l: $\sum_{i=k}^{l} x_i$.

> One reads that as *the sum of x_i over the range from $i = k$ through $i = l$.*

Table 1.2: Summary of Symbols and Notation

Symbol	Meaning
$+$	Addition
$-$	Subtraction
$*$ or $\times$ or $\cdot$	Multiplication
$/$ or $\div$	Division
$\pm$	Plus or minus
x^n	Exponentiation ("to the nth power")
$\sqrt[n]{x}$	Radical or nth root
$!$	Factorial
∞	Infinity
$\sum_{i=k}^{l} x_i$	Sum of x_i from index $i = k$ to $i = l$
$\prod_{i=k}^{l} x_i$	Product of x_i from index $i = k$ to $i = l$
$\ldots$	Continued progression
$\frac{d}{dx}$	Total derivative with respect to x
$\frac{\partial}{\partial x}$	Partial derivative with respect to x
$\int dx$	Integral over x
$\cup$	Set union
$\cap$	Set intersection
$\times$	Cartesian product of sets
$\setminus$	Set difference
A^c	Complement of set A
$\emptyset$	Empty (or null) set
$\in$	Set membership
$\notin$	Not member of set
$\mid$ or $:$ or $\ni$	Such that
$\subset$	Proper subset
$\subseteq$	Subset
$<$	Less than
$\leq$	Less than or equal to
$=$	Equal to
$>$	Greater than
$\geq$	Greater than or equal to
$\neq$	Not equal to
$\equiv$	Equivalent to or Defined as
$f()$ or $f(\cdot)$	Function
$\{\ \}$	Delimiter for discrete set
$(\ )$	Delimiter for open set
$[\ ]$	Delimiter for closed set
$\forall$	For all (or for every or for each)
$\exists$	There exists
$\Rightarrow$	Implies
$\Leftrightarrow$	If and only if
$\neg C$ or $\sim C$	Negation (not C)

Set apart from the text in an equation, the summation operator looks like this:

$$\sum_{i=k}^{l} x_i = x_k + \ldots + x_l.$$

The exponential operator, x^n (read "x to the nth power," or "x-squared" when $n = 2$ and "x-cubed" when $n = 3$), represents the power to which we raise the variable, x. The root operator, $\sqrt[n]{x}$ (read "the nth root of x," or "the square root of x" when $n = 2$ or "the cube root of x" when $n = 3$), represents the root of x.

Factorial notation is used to indicate the product of a specific sequence of numbers. Thus, $n! = n \times (n-1) \times (n-2) \ldots \times 2 \times 1$. So $5! = 5 \times 4 \times 3 \times 2 \times 1 = 120$, and $10! = 10 \times 9 \times \ldots 3 \times 2 \times 1 = 3,628,800$. This notation is especially useful for calculating probabilities.

You may not be familiar with some of the operators used in calculus. The derivative of x with respect to t is represented by the operator $\frac{dx}{dt}$. The operator ∂ indicates the partial derivative, and $\int$ indicates the integral. These will be the focus of Parts II and V of this book.

Though it's not an operator, one more symbol is useful to mention here: $\pm$. Read as "plus or minus," this symbol implies that one cannot be sure of the sign of what comes next. For example, $\sqrt{4} = \pm 2$, because squaring either 2 or -2 would produce 4.

Sets, as we have seen, have a good deal of associated notation. There are the set operators $\cap$, $\cup$, $\times$, and $\setminus$, plus the complement of A (A^c or A'). There are also the empty set $\emptyset$, set membership $\in$, set nonmembership $\notin$, proper subset $\subset$, and subset $\subseteq$. To these we add $|$, : , or $\ni$, which are each read as "such that." These are typically used in the definition of a set. For example, we define the set $A = \{x \in B | x \leq 3\}$, read as "the set of all x in B such that x is less than or equal to 3." In other words, the $|$ indicates the condition that defines the set. It serves the same purpose in conditional probabilities ($P(A|B)$), as we will see in Part III of the book. Sets also make use of delimiters, described below.

Relations include $<$, $\leq$, $=$, $\geq$, $>$. They also include $\neq$, which means "not equal to," and $\equiv$, which means "exactly equivalent to" or, often, "defined as." Relations between variables or constants typically have a left-hand side, to the left of the relation symbol, and a right-hand side, to the right of the relation symbol. These are often abbreviated as LHS and RHS, respectively. Functions are typically written as $f()$ or $f(\cdot)$, both of which imply that f is a function of one or more variables and constants. The "$\cdot$" here is a placeholder for a variable or constant; do not confuse it with its occasional use as a multiplication symbol, which occurs only when there are things to multiply.

Delimiters are used to indicate groups. Sometimes the groups are used to identify the order of the operations that are to be performed: $(x + x^2)(x - z)$. One performs the operations inside the innermost parentheses first and then moves outward. Square braces and parentheses are also used to identify closed and open sets, respectively. The open set (x_1, x_n) excludes the endpoint values

Table 1.3: Greek Letters

Upper-case	Lower-case	English	Upper-case	Lower-case	English
A	α	alpha	N	ν	nu
B	β	beta	Ξ	ξ	xi
Γ	γ	gamma	O	o	omicron
Δ	δ	delta	Π	π	pi
E	ϵ	epsilon	P	ρ	rho
Z	ζ	zeta	Σ	σ	sigma
H	η	eta	T	τ	tau
Θ	θ	theta	Υ	υ	upsilon
I	ι	iota	Φ	ϕ	phi
K	κ	kappa	X	χ	chi
Λ	λ	lambda	Ψ	ψ	psi
M	μ	mu	Ω	ω	omega

x_1 and x_n, whereas the closed set $[x_1, x_n]$ includes the endpoint values x_1 and x_n. Curly braces are used to denote set definitions, as above, or discrete sets: $\{x_1, x_2, \ldots, x_n\}$. Parentheses are also often used for ordered pairs or n-tuples, as we have seen; for example, $(2, 3, 1)$. They are also often used in vectors, which have a similar meaning. Both parentheses and square braces are used interchangeably to indicate the boundaries of matrices. We will discuss both vectors and matrices in Part IV of the book.

Proofs, the topic of the next section, have their own notation, which may pop up in other sections as well. The symbol $\forall$ means "for all," so $\forall x \in A$ means the associated statement applies for all x in the set A. The symbol $\exists$ means "there exists," typically used in the context of $\exists$ some $x \in A$ such that $x < 3$. The symbol $\Rightarrow$ is read as "implies" and is used as $C \Rightarrow D$, which means that whenever statement C is true, D is too. One can also use the reverse, $C \Leftarrow D$, which means that C is true if D is true. The symbol $\Leftrightarrow$ means that both implications are true and is read as "if and only if," so $C \Leftrightarrow D$ means that C is true if D is true, and only if D is true. In other words, C and D are equivalent statements. The symbol $\neg$ denotes negation, so $\neg C$ means statement C is not true. You will also sometimes see $\sim C$ used to mean C is not true.

People sometimes use **Greek letters** to represent variables, particularly in formal theory; they are often used to represent constants (aka parameters) in statistical analysis. Table 1.3 lists the Greek alphabet. If you have never encountered the Greek alphabet you may want to make a copy of this page, cut out the table, and tape it to the wall where you study for this and other courses that use math. Or just save it to your preferred portable electronic device.

1.6.1 Why Should I Care?

Notation that you cannot read is a serious stumbling block to understanding!

1.7 PROOFS, OR HOW DO WE KNOW THIS?

As we progress through this book, we will offer up a great many pieces of information as fact, often without explaining how we knew they were true. As noted in the preface to this book, we do this in order to focus on intuition rather than mathematical formalism. However, it is certainly fair to wonder—more than fair, really—how one comes to these conclusions. The answer, as we discuss briefly here, is that they have been proven to be true.

How does this work? Mathematics is not an empirical science; there are no experiments, and no data except insofar as experience shapes the thought of mathematicians. Rather, the progress of math begins with axioms and **assumptions**, which are stated up front with clarity and taken to be true.[23] One then conjectures a **proposition**, which is just a statement that is thought to be true given the assumptions made. From these assumptions, along with any previously proved theorems, one deductively proves, or disproves, the proposition. A proven proposition is often referred to as a **theorem**, unless it is of little interest in and of itself and is intended to be used only as a stepping stone, in which case it is called a **lemma**. A **corollary** is a type of proposition that follows directly from the proof of another proposition and does not require further proof. You will see assumptions and propositions commonly in pure and applied game theory, and lemmas, theorems, and corollaries somewhat less commonly. Propositions, though deductively derived, are often empirically testable given appropriate measures for the variables used in the proposition. In other words, a proposition might state that y is increasing in x_1 and decreasing in x_2. To test this empirically, one needs measures for y, x_1, and x_2. In some scientific fields it is common to distinguish propositions from hypotheses, with the former referring to statements of expected relationships among concepts and the latter referring to expected relationships among variables. In such contexts propositions are more general statements than hypotheses. At present, these distinctions are not widely used among political scientists.

It is not difficult to make assumptions, though learning to specify them clearly and to identify the implicit assumptions you may be making takes practice. Nor is it difficult to state propositions that may be true, though similar caveats apply. The tricky part is in proving the proposition. There is no one way to prove all propositions, though the nature of the proposition can suggest the appropriate alternative. We will consider a few commonly observed methods here, but this is far from a complete accounting.

We begin by considering four statements: A, B, C, D. A statement can be anything, e.g., A could be $x < 3$ or "all red marbles are in the left urn" or "democracies are characterized primarily by elections." Let's assume that A and B are assumptions. We take them to be true at the start of our proof and

[23]Political scientists rarely specify axioms, which tend to be more significant and wide-ranging assumptions than what are called simply assumptions. The following discussion uses terms as they are commonly observed in political science, which may elide mathematical nuance.

will not deduce them in any way from other statements. Of course, if they are not empirically true, then our conclusions may very well be incorrect empirically, but, as you can guess by the repeated use of the word "empirically," this is an empirical question and not a mathematical one. Let's further assume that C is an interim statement—that is, a deduced statement that is not our intended conclusion—and that D is that conclusion. Thus our goal is to derive D from A and B. This is the general goal of mathematical proofs.

More precisely, in this case we are seeking to show that A and $B \Rightarrow D$ (A and B imply D). This is a sufficiency statement: A and B are *sufficient* to produce D. We also can call this an *if* statement: D is true if A and B are true. This is not the only possible implication we could have written (implications are just a type of mathematical statement). We could instead have stated that A and $B \Leftarrow D$ (A and B are implied by D). This is a statement of necessity: A and B are *necessary* to produce D, since every time D is true, so are A and B. We can also call this an *only if* statement: D is true only if A and B are. Take a moment to think about the difference between these two ideas, as it is fairly central to understanding theory in political science, and it is not always obvious how different the statements are.

Ready? There is also a third common implication we could have written, a necessary *and* sufficient statement: A and $B \Leftrightarrow D$. This is also called an *if and only if* statement, as D is true if and only if A and B are true. In other words, A and B are entirely equivalent logically to D, and one can replace one statement with the other at will. This is one way one uses existing theorems to help in new proofs, by replacing statements with other statements proven to be equivalent. (One can also use if or only if propositions on their own in new proofs.)

In addition to using existing theorems, pretty much any mathematical procedure accepted as true can be used in a proof. We'll cover many in this book, but the most basic of these may be the tools of formal **logic**, which has much in common with set theory. Negation of a statement is much the same as the complement of a set. For example, you cannot be both true and not true, nor can you be both in and outside a set. You can also take the equivalent of a union and an intersection of sets for statements; these are called disjunction and conjunction, or, in symbols, *or* ($\vee$) and *and* ($\wedge$), respectively. Note that the *and* symbol looks like the intersection symbol. This is not accidental—*and* means that both statements are true, which is like being in both sets, which is like the intersection of the sets. Likewise, *or* means that at least one statement is true, which is like being in either set, which is like the union between the sets. Let's call a compound statement anything that takes any two simpler statements, such as A and B, and combines them with a logical operator, such as $\neg$, $\vee$, or $\wedge$. We can therefore write the implication we're trying to prove as $A \wedge B \Rightarrow D$.

De Morgan's laws prove handy for manipulating both sets and logical statements.[24] We'll present these in terms of logical statements, but they are true for sets as well after altering the notation. The best way to remember them is that the negation of a compound statement using *and* or *or* is the compound statement in which the *and* is switched for *or*, or vice versa, and each of the simpler statements is negated. So, for example, $\neg(A \wedge B)$ is $(\neg A) \vee (\neg B)$ and $\neg(A \vee B)$ is $(\neg A) \wedge (\neg B)$. In words, if both statements aren't true, then at least one of them must be false. Similarly, if it's not the case that at least one statement is true, then both statements are false.

We can use logic to obtain several important variants of our implications that might be useful. A negated implication just negates all the statements that are part of the implication. So the negation of our implication becomes $\neg(A \wedge B) \Rightarrow \neg D$, which by De Morgan's law is $(\neg A) \vee (\neg B) \Rightarrow \neg D$. Even when the statement is true, the negation might not be. Having two democracies may mean you're at peace (for the sake of this argument), but letting at least one of them not be a democracy does not automatically imply war.

The **converse** of an implication switches a necessary statement to a sufficient one, or vice versa. Thus the converse of A and $B \Rightarrow D$ is A and $B \Leftarrow D$ or $D \Rightarrow A$ and B. As noted above, just because an implication is true does not mean the converse is true—something may be necessary without being sufficient. However, negating the converse, called taking the contrapositive, *does* always yield a true statement. The **contrapositive** of our implication is $(\neg A) \vee (\neg B) \Leftarrow \neg D$, or, as it's more typically written, $\neg D \Rightarrow (\neg A) \vee (\neg B)$. If a pair (dyad) of democracies never experiences war, then having a war (the opposite of peace) means that at least one of the pair is not a democracy.

Okay, back to our proof. Proofs are sometimes classed into broad groups of direct and indirect proofs. **Direct proofs** use deduction to string together series of true statements, starting with the assumptions and ending with the conclusion. In addition to the construction of a string of arguments, direct proofs commonly observed in formal theory include proof by exhaustion, construction, and induction. Let us see briefly how these work, starting with a **general deductive proof**.

Let A be the statement that $x \in \mathbb{Z}$ is even, and B be the statement that $y \in \mathbb{Z}$ is even, and D, which we're trying to prove, be the statement that the product xy is even. Well, if x and y are even (our assumptions), then they can be written as $x = 2r$ and $y = 2s$ for some $r, s \in \mathbb{Z}$. (Here we've used the definition of even.) In this case, we can write $xy = (2r)(2s) = 4rs$, which is our new statement C. Since $4rs = 2(2rs)$, xy is even (again using the definition of even), thus proving D. Now we know that the product of any two even integers is also even, and we could use this knowledge in further, more complex proofs.

Proof by exhaustion is similar, save that you also break up the problem into exhaustive cases and prove that your statement is true for each case. This comes up often in game theory as there will be different regions of the parameter space that may behave differently and admit different solutions. (The parameter space

[24]See `http://en.wikipedia.org/wiki/De_Morgan_laws`.

is the space, in the sense of a set with a measure, spanned by the parameters. We will discuss this concept more in Part III of the book.)

Proof by construction is similarly straightforward, and can be useful when trying to show something like existence: if you can construct an example of something, then it exists.

Proof by induction is a bit different and merits its own example. It is generally useful when you would like to prove something about a sequence (we cover sequences in Chapter 4) or a sequence of statements. It consists of three parts. First, one proves the base case, which in this example is the first element in the sequence. Second, one assumes that the statement is true for some n (the inductive hypothesis). Third, one proves that the statement is true for $n + 1$ as well (the inductive step). Thus, since the base case is true and one can always go one further in the sequence and have the statements remain true, the entire sequence of statements is true.[25] Let's see how this works with an example: show that $\sum_{i=1}^{n} i = \frac{n(n+1)}{2}$. We basically need to show this is true for each n, but since they occur in sequence, we'll use induction rather than exhaustion (which wouldn't be appropriate, given that the sequence is infinite anyway). First we try the base case, which is for $n = 1$. We can check this: $\sum_{i=1}^{1} i = 1 = \frac{1(2)}{2} = \frac{1(1+1)}{2}$. So the base case is true. Now we assume, somewhat counterintuitively, the statement that we're trying to prove: $\sum_{i=1}^{n} i = \frac{n(n+1)}{2}$. Finally, we show it remains true for $n + 1$, so we need to prove that $\sum_{i=1}^{n+1} i = \frac{(n+1((n+1)+1)}{2}$, where we've replaced n in the right-hand side of the statement we're trying to prove with $n + 1$. The sum in the left-hand side of this is $\sum_{i=1}^{n} i + (n + 1)$, where we've just split the sum into two pieces. The first piece equals $\frac{n(n+1)}{2}$ by step two in our proof. So now we have $\frac{n(n+1)}{2} + (n + 1) = \frac{n(n+1)}{2} + \frac{2(n+1)}{2} = \frac{n(n+1)+2(n+1)}{2} = \frac{(n+1)(n+2)}{2}$. This is just what we needed to show, so the $n + 1$ inductive step is true, and we've proved the statement.

Indirect proofs, in contrast, tend to show that something must be true because all other possibilities are not. **Proof by counterexample** and **proof by contradiction** both fall into this category. Counterexamples are straightforward. If the statement is that $A \wedge B \Rightarrow D$ and A and B are both true, then a single counterexample of $\neg D$ is sufficient to disprove the proposition. Proof by contradiction has a similar intent, but instead of finding a counterexample one starts by assuming the statement one is trying to prove is actually false, and then showing that this implies a contradiction. This proves the proposition because if it cannot be false, then it must be true. Although it may seem counterintuitive, proof by contradiction is perhaps the most common type of proof, and is usually worth trying first. Proving the contrapositive, since it indirectly

[25]Though this method of proof is called mathematical induction, it's important to note that it is a deductive method of theory building, not an inductive one. That is, it involves making assumptions and deducing conclusions from these, not stating conclusions derived from a series of statements that may only be probabilistically linked to the conclusion, as in inductive reasoning.

also proves the statement, as they are equivalent, is sometimes also considered an indirect proof, though it seems pretty direct to us.

1.8 EXERCISES

1.8.1 Constants and Variables and Levels of Measurement

1. Identify whether each of the following is a constant or a variable:
   a) Party identification of delegates at a political convention.
   b) War participation of the Great Powers.
   c) Voting record of members of Congress relative to the stated position of the president.
   d) Revolutions in France, Russia, China, Iran, and Nicaragua.
   e) An individual voter's vote choice in the 1992 presidential election.
   f) An individual voter's vote choice in the 1960–1992 presidential elections.
   g) Vote choice in the 1992 presidential election.
2. Identify whether each of the following is a variable or a value of a variable:
   a) The Tonkin Gulf Crisis.
   b) Party identification.
   c) Middle income.
   d) Exports as a percentage of GDP.
   e) Republican.
   f) Female.
   g) Veto.
   h) Ethnic fractionalization.
   i) International crisis.
3. Identify whether each of the following indicators is measured at a nominal, ordinal, interval, or ratio level. Note also whether each is a discrete or a continuous measure:
   a) Highest level of education as (1) some high school, (2) high school graduate, (3) some college, (4) college graduate, (5) postgraduate.
   b) Annual income.
   c) State welfare expenditures, measured in millions of dollars.
   d) Vote choice among Bush, Clinton, and Perot.
   e) Absence or presence of a militarized interstate dispute.
   f) Military personnel, measured in 1,000s of persons.
   g) The number of wars in which countries have participated.

1.8.2 Sets, Operators, and Proofs

4. As a brief illustration of one use of set theory, consider the following question: given three parties in a legislature with a supermajority rule required to pass a bill, what is the likely outcome of a given session? We can use set theory and some rational choice assumptions to get a pretty good handle on that question. Assume that no party has enough seats to pass the bill by itself and that all three parties prefer some outcome other than the status quo. For concreteness, let's define two dimensions over which to define policy: guns (i.e., defense spending) and butter (i.e., health, education, and welfare spending). We can now create a two-dimensional space where spending on guns is plotted on the vertical axis and spending on butter is plotted on the horizontal axis. Take out a sheet of paper and draw this. Let the axes range from 0% of the budget, marked where the axes intersect, to 100% of the budget, marked as the maximum value on each axis. Connect the two maximum values with a straight line. You now have a triangle, and the legislature cannot go outside the triangle: the line you just drew represents spending the entire budget on some mix of guns and butter. Let's assume that the legislators want to spend some money on non-guns and non-butter, and thus both parties' most preferred combination of guns and butters is somewhere inside the budget constraint. Pick some point inside the budget constraint and mark it as the status quo. Now select a most preferred combination for each party and mark each as Party 1, Party 2, and Party 3. Finally, pick a fifth point and label it a bill. Make a conjecture on whether the bill will pass or whether the status quo will be sustained. (For now this is just a conjecture, but we'll return to this in the exercises to Chapter 3, so save your answer.)

5. Let $A = \{1, 5, 10\}$ and $B = \{1, 2, \ldots, 10\}$.

   a) Is $A \subset B$, $B \subset A$, both, or neither?
   b) What is $A \cup B$?
   c) What is $A \cap B$?
   d) Partition B into two sets, A and everything else. Call everything else C. What is C?
   e) What is $A \cup C$?
   f) What is $A \cap C$?

6. Write an element of the Cartesian product $[0, 1] \times (1, 2)$.

7. Prove that $\sqrt{2}$ is an irrational number. That is, show that it cannot be written as the ratio of two integers, p and q.

8. Prove that the sum of any two even numbers is even, the sum of any two odd numbers is even, and the sum of any odd number with any even number is odd.

Chapter Two

Algebra Review

Of all the chapters in this book, this is the one most safely skipped. Most of this chapter is taken up by a review of arithmetic and algebra, which should be familiar to most readers. If you feel comfortable with this material, skip it. If it is only vaguely familiar, don't. The third section briefly discusses the utility of computational aids for performing calculations and checking work.

2.1 BASIC PROPERTIES OF ARITHMETIC

There are several properties of arithmetic that one uses when simplifying equations. These arise from the real numbers or integers for which the variables stand. In other words, because the variables we use in political science generally take values in $\mathbb{R}$ or $\mathbb{Z}$, these five properties generally apply. This will be true nearly throughout the book; however, in Part IV we will see that matrix variables can fail to commute under multiplication, for example, and do not always possess multiplicative inverses. But for variables that stand for real numbers or integers, these properties will always hold. Most of these are expressed in terms of addition and multiplication, but the first three properties apply to subtraction and division, respectively, as well, except for division by zero.

The **associative properties** state that $(a+b)+c = a+(b+c)$ and $(a\times b)\times c = a \times (b \times c)$. In words, the properties indicate that the grouping of terms does not affect the outcome of the operation.

The **commutative properties** state that $a + b = b + a$ and $a \times b = b \times a$. In words, the properties claim that the order of addition and multiplication is irrelevant.

The **distributive property** states that $a(b + c) = ab + ac$. In words, the property says that multiplication distributes over addition (and subtraction).

The **identity properties** state that there exists a zero such that $x + 0 = x$ and that there exists a one such that $x \times 1 = x$. In other words, there exist values that leave x unchanged under addition and multiplication (and subtraction and division, respectively, as well).

The **inverse property** states that there exists a $-x$ such that $(-x) + x = 0$. In other words, there exist values that when added to any x produce the identity under addition. We might also consider an inverse under multiplication, x^{-1}, such that $(x^{-1}) \times x = 1$. The existence of this inverse is a property of the real numbers (and the rational numbers), but not the integers, so one must be careful. For example, if $x = 2$, then $x^{-1} = 0.5$ in the real numbers, but no integer multiplied by two equals one. Whether or not an inverse exists will depend, therefore, on the set of values the variable can take.

It is useful to recall at this stage that division by zero is undefined. The expression $x/0 = \infty$ is true for any $x \neq 0$, but is completely undefined for $x = 0$. Other useful facts include that $x = 1x = x^1 = 1x^1$, and that $x^0 = 1$. Recall also that multiplication by a variable with a negative value changes the sign of the product: $-1 \times x = -x$. The product of two terms with negative signs is positive: $(-x) \cdot (-y) = xy$.

2.1.1 Order of Operations

The order of operations is also important and can trip people up. In arithmetic and algebra the order of operations is parentheses, exponents, multiplication, division, addition, subtraction. A common mnemonic device people use to memorize order of operations is PEMDAS, or Please Excuse My Dear Aunt Sally.

2.1.2 Ratios, Proportions, and Percentages

Ratios, proportions, and percentages sometimes give people trouble, so let's briefly review those. The **ratio** of two quantities is one divided by the other $\frac{x}{y}$ is the ratio of x to y. Ratios are also written as $x : y$. Keep in mind that one can only take the ratio of two variables measured at a ratio level of measurement (i.e., there is a constant scale between values, and a meaningful zero). Though a ratio may be negative, we typically consider ratio variables that range from 0 to ∞. To get this, we take the absolute value of the ratio, denoted $\left|\frac{x}{y}\right|$. All this does is turn any negative number positive. As an example, international relations scholars are often interested in the ratio of military power between two countries (e.g., Organski and Kugler, 1981).

The **proportion** of two variables, on the other hand, is the amount one variable represents of the sum of itself and a second variable: $\left|\frac{x}{x+y}\right|$. A proportion ranges from a minimum of 0 to a maximum of 1. Students of budgetary politics are often interested in the proportion of expenditures that is spent in a given category (e.g., health and welfare, pork barrel politics, defense spending; see Ames, 1990).

The **percentage** one variable represents of a total is the proportion represented over the range from 0 to 100. In other words, the percentage is a linear transformation of the proportion $\left|\frac{x}{x+y}\right| \times 100\%$. Many people find a percentage representation more intuitive than a proportion representation, but they provide the same information.

You will also encounter the **percentage change** in a variable, which is calculated as $\frac{(x_{t+1}-x_t)}{x_t}$, where the subscript t indicates the first observation and the subscript $t+1$ indicates the second observation. For instance, according to the Center for Defense Information's *Almanac*, the United States spent \$75.4 billion for military personnel wages in 2001 and an estimated \$80.3 billion in

2002. The expenditures in 2002 represented a 6.5% increase over 2001 expenditures: $\frac{(80.3-75.4)}{75.4} \simeq 6.5\%$. Note that the percentage change can range from $-\infty$ to ∞.

2.1.3 Why Should I Care?

You care about these properties because you need to know them to follow along. People who use mathematics to communicate their ideas, whether in formal theory or statistics, assume that you can do the operations allowed by these properties. They often "skip steps" when writing down manipulations and expect you to do them in your head. If you cannot do them, you will get lost.

2.2 ALGEBRA REVIEW

This section reviews the most common algebraic manipulations you will encounter. Most of you will be familiar with these; the trick is trying to minimize errors, which are easy to make. We note some common errors to avoid.

2.2.1 Fractions

Many students find fractions the most frustrating part of algebra. People generally find whole numbers more intuitive than fractions, and that makes calculations with fractions more difficult to perform. As such, whenever possible it is best to convert fractions to whole numbers. Recall that the number on the top of a fraction is the **numerator** and the number on the bottom of a fraction is the **denominator**

$$\frac{Numerator}{Denominator}.$$

Thus, one can convert to a whole number whenever the denominator divides evenly into the numerator.

Many people find mixed numbers such as $2\frac{3}{4}$ even more frustrating. To convert these mixed numbers to fractions, follow these two steps. First, multiply the denominator of the fraction by the whole number (i.e., multiply 4×2, which equals 8). Second, take this product and add it to the numerator and place that sum over the original denominator (add 8 to 3, which equals 11, and place that over 4 for the final fraction $\frac{11}{4}$). These two quantities are equivalent.

Two common algebraic manipulations relating to fractions that often trouble students are cancellations and adding fractions.

2.2.1.1 Cancellations

The reason we want to reduce fractions is to make them easier to use (if the fraction can be converted to a whole number, this is ideal). For example, $\frac{10x}{2}$ can be reduced to $5x$. One that you might encounter in game theory could look like this: $\frac{7+3x}{2x}$.

One of the most common mistakes made is to cancel the xs and simplify $\frac{7+3x}{2x}$ to $\frac{10}{2}$, and then simplify this quantity to 5. However, $7+3x \neq 10x$, so $\frac{7+3x}{2x} \neq 5$.

In this example $\frac{7+3x}{2x}$ can be simplified to $\frac{7}{2x} + \frac{3x}{2x}$. The fraction $\frac{7}{2x}$ is in its simplest form. The fraction $\frac{3x}{2x}$ can be simplified to $\frac{3}{2}$, as long as $x \neq 0$.[1]

Therefore, $\frac{7+3x}{2x} = \frac{7}{2x} + \frac{3}{2}$.

2.2.1.2 Adding Fractions

Adding or subtracting fractions can be a bit frustrating as they do not follow the same rules as whole numbers. More specifically, you can only add the numerators of two or more fractions when the denominators of each fraction are the same (i.e., you cannot add fractions with different denominators). You can add $\frac{4}{\beta} + \frac{\alpha}{\beta}$, which equals $\frac{4+\alpha}{\beta}$. When two fractions have different denominators, such as $\frac{\beta}{4} + \frac{\alpha}{2}$, one must transform one or both of the denominators to make addition possible: the numerators of all fractions can be added once their denominators are made equal.

To pursue the above example, $\frac{\beta}{4} + \frac{\alpha}{2}$, if we multiply $\frac{\alpha}{2}$ by $\frac{2}{2}$ (which equals one; you can always multiply by things equal to one, or add things equal to zero because of the identity property), it becomes $\frac{2\alpha}{4}$. Since the two fractions now have the same denominator, we can add their numerators:[2] $\frac{\beta}{4} + \frac{2\alpha}{4} = \frac{2\alpha+\beta}{4}$.

Unlike addition, multiplication does *not* require a common base, and one does multiply both numerator and denominator: $\frac{2}{3} \times \frac{1}{4} = \frac{2}{12} = \frac{1}{6}$.

Another common mistake people make when adding fractions is to assume that all aspects of fractions follow the same rules of addition. For example, they assert that $\frac{1}{\Delta+\Theta}$ is equal to $\frac{1}{\Delta} + \frac{1}{\Theta}$. It is not. To see why this is so, let's add real numbers to the expression. If we substitute 2 for Δ and 1 for Θ and sum the denominator, we get $\frac{1}{2+1}$, which is equal to $\frac{1}{3}$. If we split the fraction improperly, including the numerator over both parts of the denominator as above, we will conclude that $\frac{1}{2+1} = \frac{1}{2} + \frac{1}{1}$, which equals $1\frac{1}{2}$, or 1.5,[3] not $\frac{1}{3}$.

2.2.2 Factoring

Factoring involves rearranging the terms in an equation to make further manipulation possible or to reveal something of interest. The goal is to make the expression simpler. One uses the properties described above rather extensively when factoring.

A standard algebraic manipulation involves combining like terms in an expression. For example, to simplify $\delta + \delta^2 + 4\delta - 6\delta^2 + 18\delta^3$, we combine all like terms. In this case we combine all the δ terms that have the same exponent, which gives us $18\delta^3 - 5\delta^2 + 5\delta$.

[1] Remember, anything divided by itself is one, and anything multiplied by one equals itself.

[2] Note that we do **not** take the sum of the denominators. One only adds the numerators.

[3] If you have forgotten how to convert fractions into decimals, the solution is to do the division implied by the fraction (you can use a calculator if you wish): $\frac{1}{2} = 0.5$.

Another standard factoring manipulation involves separating a common term from unlike ones. We first establish what we want to pull out of the equation, then apply the distributive property of multiplication in reverse. For example, we might want to pull x out of the following: $3x + 4x^2 = x(3 + 4x)$. Another example is $6x^2 - 12x + 2x^3 = 2x(3x - 6 + x^2)$.

A more complex example is $12y^3 - 12 + y^4 - y$.

We can factor 12 out of the first two terms in the expression and y out of the next two terms.

The expression is then $12(y^3 - 1) + y(y^3 - 1)$, which can be regrouped as $(12 + y)(y^3 - 1)$.

2.2.2.1 Factoring Quadratic Polynomials

Quadratic polynomials are composed of a constant and a variable that is both squared and raised to the power of one: $x^2 - 2x + 3$, or $7 - 12x + 6x^2$.[4] Quadratic polynomials can be factored into the product of two terms: $(x \pm ?) \times (x \pm ?)$, where you need to determine whether the sign is $+$ or $-$, and then replace the question marks with the proper values.

Hopefully, it is apparent that one can multiply many products of two sums or two differences to get a quadratic polynomial; this is the reverse of factoring.[5]

2.2.2.2 Factoring and Fractions

We can also reduce fractions by factoring. Consider the fraction $\frac{x^2-1}{x-1}$. We can factor the numerator $x^2 - 1 = (x + 1)(x - 1)$. We can thus rewrite the fraction as follows

$$\frac{x^2 - 1}{x - 1} = \frac{(x + 1)(x - 1)}{x - 1}.$$

The term $x - 1$ is in both the numerator and the denominator and thus (as long as $x \neq 1$) cancels out, leaving $x + 1$. Thus, $\frac{x^2-1}{x-1} = x + 1$ for $x \neq 1$.

This factoring need not be accomplished in one step. Consider the expression

$$\frac{3\lambda^4 + 3\lambda^3 - 6\lambda^2}{6\lambda^2 + 12\lambda}.$$

First, we can factor out the common factor from both the numerator and denominator. All of the terms in the numerator are multiples of $3\lambda^2$ and both of the terms in the denominator are multiples of 6λ. This yields

---

[4] Another way of putting this is that a quadratic polynomial is a second-order polynomial in a single variable x. We discuss polynomial functions in the next chapter. Finally, given that the Latin prefix *quadri* is associated with four, you may be wondering why *quadratic* is used to describe equations with a term raised to the power of two. The reason is that the Latin term *quadratum* means "square." So an equation with a variable that is squared is a quadratic equation (Weisstein, N.d.).

[5] Note that this is true of some, *but not all*, products of two sums or two differences.

$$\frac{3\lambda^2(\lambda^2+\lambda-2)}{6\lambda(\lambda+2)}.$$

Next, we factor the quadratic polynomial in the numerator to get

$$\frac{3\lambda^2(\lambda+2)(\lambda-1)}{6\lambda(\lambda+2)}.$$

Then we factor out like terms. Both the numerator and denominator have $\lambda+2$, so (as long as $\lambda \neq -2$) they cancel out, leaving

$$\frac{3\lambda^2(\lambda-1)}{6\lambda}.$$

Finally, 3λ can be canceled (as long as $\lambda \neq 0$) from both the numerator and the denominator, leaving the expression in its simplest form

$$\frac{\lambda(\lambda-1)}{2}.$$

2.2.3 Expansion: The FOIL Method

Sometimes we need to simplify a complex expression. At other times we need to expand a simple expression. Here is a pop quiz:

Does $(\delta+\gamma)^2 = \delta^2+\gamma^2$?

The answer: no.

Why? The expression $(\delta+\gamma)^2 = (\delta+\gamma)(\delta+\gamma)$. This can then be expanded using the FOIL method. The expanded expression is $\delta^2+2\delta\gamma+\gamma^2$.

The FOIL method can be used to expand the product of two sums or differences. FOIL stands for first, outer, inner, last, and represents the products one must calculate.

F: Multiply the first terms: $(\underline{2\pi}+7)(\underline{4}+3\pi) = 2\pi \times 4 = 8\pi$.

O: Multiply the outer terms: $(\underline{2\pi}+7)(4+\underline{3\pi}) = 2\pi \times 3\pi = 6\pi^2$.

I: Multiply the inner terms: $(2\pi+\underline{7})(\underline{4}+3\pi) = 4 \times 7 = 28$.

L: Multiply the last terms: $(2\pi+\underline{7})(4+\underline{3\pi}) = 7 \times 3\pi = 21\pi$.

Add terms to get $8\pi+6\pi^2+28+21\pi$.

Finally, group like terms to get $6\pi^2+29\pi+28$.

To test yourself, factor the final expression and show it yields the simplified expression with which we started. This is one way to check your work for any careless mistakes.

2.2.4 Solving Equations

Solving an equation involves isolating a variable on one side (by convention, the left side of the equals sign) and all other variables and constants on the other side. One does so by performing the same calculations on both sides of the equation such that one ends up isolating the variable of interest. This often takes multiple steps and there is almost always more than one way to arrive at the solution. As an example, the equation $y = 2x$ is already solved for y. If we want to solve that equation for x, we need to do some algebra. Start with

$$y = 2x.$$

Divide both sides of the equation by 2, yielding

$$\tfrac{y}{2} = x.$$

Rewrite the equation:

$$x = \tfrac{y}{2}.$$

Note that we can go about this in a more convoluted fashion:

$$y = 2x.$$

Divide both sides by x, yielding

$$\tfrac{y}{x} = 2.$$

Divide both sides by y:

$$\tfrac{1}{x} = \tfrac{2}{y}.$$

Multiply both sides by x:

$$1 = x(\tfrac{2}{y}).$$

Multiply both sides by $\frac{y}{2}$:

$$\tfrac{y}{2} = x.$$

Now rewrite:

$$x = \tfrac{y}{2}.$$

That is hardly efficient, but the good news is that we ended up at the same place, though we would have had to be careful that neither x nor y was equal to zero when dividing by them. We also got some practice in manipulating an equation. Here are a few useful techniques for those rusty in their algebra.

1. **Focus on the variable of interest**. Work on isolating the variable you care about, and don't worry so much about what this does to the rest of the equation.

2. **Combine all like terms**. Simplifying equations is easiest when you sort out all the noise and add together like terms.

3. **Check your answer**. Substitute the value that you obtain into the original equation to make sure that your answer is correct and that you didn't make a careless mistake.

4. **Make use of identities**. Remember $\frac{a}{b} \times \frac{b}{a} = 1$ and $a - a = 0$. That means you can multiply by the first and add the second at all times, whenever convenient, without changing the equation.

5. **Operate on both sides in the same manner**. Adding the same number to each side or multiplying each side by the same number won't change the equation.

2.2.4.1 Solving Quadratics

Solving quadratic polynomials requires learning how to complete the square and/or knowing the quadratic equation.

Completing the Square

Many quadratic equations that you will face can be solved relatively easily by **completing the square**. The basic intuition for solving these is to isolate the variable and its square and then add a value to each side of the equation to "complete the square."

To see what we are trying to accomplish, it helps to begin with a simple example. Sometimes we are presented with a quadratic equation that factors into a squared term, i.e., $(x - n)^2 \pm c$, where c is some constant. Consider the quadratic $x^2 - 6x + 5$. We can factor this into $(x - 3)^2 - 4$ (use the FOIL method to verify). We can then rewrite this equation as $(x - 3)^2 = 4$. Finally, by taking the square root of both sides we, can solve for x:

$$(x - 3)^2 = 4 \Rightarrow$$
$$x - 3 = \pm 2 \Rightarrow$$
$$x = 5 \text{ or } x = 1.$$

Note that this quadratic equation will have two solutions in the real numbers, or zero, but not one.[6] In other words, the cardinality of the solution set for a quadratic equation will be zero or two. An example of a quadratic equation with no real solutions (i.e., no solutions in the real numbers) is $x^2 + 1 = 0$.[7]

Solving a quadratic by factoring it into a squared term $\pm$ a constant and then taking the square root is quick, but most quadratics cannot be factored in integers so easily. However, we can transform any quadratic using the following steps to "complete the square" (i.e., transform it into a squared term $\pm$ a constant) and then solve for x by taking the square roots.

[6] Quadratic equations with real coefficients will always have two solutions in complex numbers; if these solutions are complex they will come in pairs, i.e., $a \pm bi$.

[7] The solutions of this equation are $\pm i$.

1. Start with a quadratic in your variable of interest (we'll say it's x) and move the constant to the right-hand side. Divide through by the coefficient on x^2. So if you have $2x^2 - 4x - 2 = 0$, you get $x^2 - 2x = 1$.

2. Divide the coefficient on x by 2 and then square it. Add that value to both sides of the equation. So now you have $x^2 - 2x + 1 = 1 + 1$.

3. Factor the left-hand side into a "($x \pm$ some term) squared" form and simplify the right-hand side. So now you have $(x - 1)^2 = 2$.

4. Take the square root of both sides (remember that when you take the square root of a number, the solution is always $\pm$, because the square of a negative number is a positive number). So now you have $x - 1 = \pm\sqrt{2}$.

5. Solve for x. So the solutions are $x = 1 + \sqrt{2}$ and $x = 1 - \sqrt{2}$.[8]

Let's work another example. Consider the quadratic

$$x^2 + 8x + 6 = 0.$$

The first thing you might try is factoring to see if it yields the "($x \pm$ some term) squared" form. This quadratic does not, so we turn to completing the square. The first step is to isolate the squared term and variable by subtracting 6 from each side (note the coefficient on x^2 is already 1):

$$x^2 + 8x = -6.$$

Next we need to add the square of half of the value in front of x to both sides. The value in front of x is 8, so we divide 8 by 2 and then square the result $4^2 = 16$. Thus, to complete the square we need to add 16 to each side:

$$x^2 + 8x + 16 = 10.$$

We then perform step 3 and factor the left-hand side of the equation:

$$(x + 4)^2 = 10.$$

We now need to take the square root of each side, which gives us

$$x + 4 = \pm\sqrt{10}.$$

Now we can solve for x by subtracting 4 from each side. Our final answer is

$$x = -4 + \sqrt{10} \text{ and } x = -4 - \sqrt{10}.$$

[8] Irrational solutions such as $1 \pm \sqrt{2}$ will also always come in pairs.

The Quadratic Formula and Equation

Completing the square is one method for solving a quadratic equation, but it is not the only one, and you will sometimes encounter equations that are rather complicated to solve by completing the square. For example, if you are faced with $x^2 + \sqrt{15}x - 1 = 0$, you will not want to calculate half of $\sqrt{15}$, and then square it, add it to both sides, and try to factor the result. Instead, you will want to turn to the quadratic equation and formula.[9]

During your high school algebra courses you were probably required to memorize the quadratic equation and formula. The general form of a quadratic equation is[10]

$$ax^2 + bx + c = 0.$$

The general solutions to this equation are

$$x = \frac{-b \pm \sqrt{b^2 - 4ac}}{2a}.$$

These solutions are called the quadratic formula. The formula can be derived from the equation by completing the square (we ask you to do this below, in an exercise). When we obtain values for x, we call these values the roots of the equation. For our purposes, this formula is used when completing the square is made difficult by fractions, decimals, or large numbers. We refer to a and b as the coefficients and c as the constant.

What are the roots of the quadratic equation

$$1.4x^2 + 3.7x + 1.1 = 0?$$

To find the solutions, we first list the values for a, b, and c:

$$a = 1.4, b = 3.7, c = 1.1.$$

Then we plug the values for a, b, and c into the quadratic formula:

$$x = \frac{-3.7 \pm \sqrt{3.7^2 - 4 \times 1.4 \times 1.1}}{2.8}.$$

Using a calculator to solve, we find that $x = -.341$ or $x = -2.301$.

These problems can be cumbersome because it is somewhat more difficult to check them. Two pieces of advice, though, can help you minimize mistakes. First, make sure to follow order of operations (PEMDAS, remember?). Second, you can go online and use a quadratic equation solver and plug in the values for a, b, and c to verify the accuracy of your computations. We discuss this further in Section 3.

---

[9] It is possible to use either method—completing the square or the quadratic equation and formula—to solve a quadratic equation. But one is more likely to make errors using the latter than the former, and so many people find completing the square preferable as long as they are not faced with an unusual (e.g., radical) value in front of the x term.

[10] The equation is true when $a \neq 0$. When $a = 0$ and $b \neq 0$, it is a simple linear equation with solution $x = -\frac{c}{b}$. If both a and b are zero, the equation is false unless $c = 0$ as well.

2.2.5 Inequalities

To solve inequalities, we have to discuss a few extra properties.

First, all pairs of real numbers have exactly one of the following relations: $x = y$, $x > y$, or $x < y$.

Adding any number to each side of these relations will not change them; this includes the inequalities. That is, inequalities have the same addition and subtraction properties as equalities such that if $x > y$, then $x + a > y + a$ and $x - a > y - a$.

The properties for multiplication and division for inequalities are a bit different than for equalities. For multiplication, if a is positive and $x > y$, then $ax > ay$. If a is negative and $x > y$, then $ax < ay$. For division, if a is positive and $x > y$, then $\frac{x}{a} > \frac{y}{a}$. If a is negative and $x > y$, then $\frac{x}{a} < \frac{y}{a}$. Multiplying or dividing an inequality by zero is not allowed.

To summarize, this means that you flip the $<$ or $>$ sign when multiplying or dividing by a negative.

Example: Solve for y: $-4y > 2x + 12$. First, we want to isolate y by itself on the left side of the equation. We divide both sides by -4, which gives us $y < -\frac{x}{2} - 3$. Dividing by a -4 flips the $>$ sign to $<$. If we do not know the value of x, then we can leave it in this form.

2.2.6 Review: Avoiding Common Errors

We have included a list of some common mistakes people make when solving equations as a sort of help file for when you are struggling to find the right answer. Below this list, we've included some websites you may go to for extra examples or help.

That said, please remember that the World Wide Web is dynamic, and the links below will become dated. We found them using search engines, and you will be able to do the same.

Sign errors: Sign errors are probably the most common mistakes. Most people think of this ($-$) sign as a negative sign. This is part of the problem. Tackling math as if it were a foreign language is the best way to approach learning the fundamentals of mathematics. This sign ($-$) is best thought of as "the opposite of," or, in other words, "the additive inverse." (Recall that every integer and real number has an additive inverse that, when added to the number, produces zero.) When reading an equation such as $-x + y = 7$, you should say in your head, "the opposite of x added to y is 7." The reason for thinking of this sign as "the opposite of" is twofold. First, it can help you find mistakes in your work. Second, it will help you deal with situations such as $-x = 7$. You will easily interpret this as the opposite of x equals 7, so x must be the additive inverse of 7, which is -7.[11]

[11] If this is a bit confusing, remember that the number line has 0 in the middle, positive

Only changing one side or term in an equation: Think of an equation as a scale or seesaw. Whatever you do to one side you must do to the other.[12] An equation must be in equilibrium. If you divide one side by 12, you must divide the other by 12. In addition, you must divide all terms on both sides by 12.

Not distributing: Always distribute across addition (and subtraction). If you have an expression such as $4x(2 + 6y + 3t)$, many people simply multiply $4x$ by 2. Each term inside a parentheses must be multiplied by what is outside the parentheses. The correct expression is $8x + 24xy + 12xt$.

Distributing with radicals and exponents: Radicals and exponents have different rules, which we discuss in depth in the next chapter. They do not follow the same rules as multiplication and addition. For example, $\sqrt{9+16}$ is not the same as $\sqrt{9} + \sqrt{16}$. Also, as discussed above, $(\alpha + \beta)^2$ is not the same as $\alpha^2 + \beta^2$.

You can find other lists of common errors at several websites. For example, Eric Schecter maintains a page of the most common math errors by undergraduates (`http://atlas.math.vanderbilt.edu/~schectex/commerrs/`). See Schecter's page for entries on "multiplying by a negative one and other sign errors," "loss of invisible parentheses," "everything is additive," and "everything is commutative."

Other common algebra mistakes include canceling terms instead of factors, misunderstanding fractions, and misunderstanding negative and fractional components. See `http://tutorial.math.lamar.edu/pdf/algebra_Cheat_Sheet.pdf`.

Beyond this, some of you may be interested in more practice, especially with algebra. One of the authors finds Huettenmueller (2010) a useful resource, but there are a number of other self-teaching guides. You can also find a number of useful resources available on the Web. We recommend `http://www.purplemath.com/`, `http://mathworld.wolfram.com/`, and `http://math.com/`. Wikipedia (`http://en.wikipedia.org/`) also has many good entries for mathematical concepts, though many of these can substantially be found elsewhere.

For more information on set theory, see Peter Suber's "A Crash Course in the Mathematics of Infinite Sets" (`http://www.earlham.edu/~peters/writing/infapp.htm`). Oregon State's "Field Guide to Functions" (`http://oregonstate.edu/instruct/mth251/cq/FieldGuide/`) is a good guide to functions (the topic of the next chapter). R.H.B. Exell's page on relations is also useful for

integers falling to the right of 0 and negative integers falling to the left. Any number has an opposite on the number line that is equidistant from zero. So the opposite of 8 is -8 and the opposite of -9 is 9. Thinking of negatives in these terms will also help you deal with absolute values.

[12] Of course, as noted above, adding zero to one side or multiplying one side by one is acceptable, as these are identities and leave the value of the expression unchanged. This may be may be useful when working with fractions.

a more detailed introduction (http://www.jgsee.kmutt.ac.th/exell/Logic/Logic42.htm).

2.2.7 Why Should I Care?

algebra is the set of rules one uses to manipulate equations that have variables in place of numerical values, whereas arithmetic is the set of rules we use to manipulate equations made of numerical values. arithmetic is thus essential for making specific calculations, but algebra is needed if we want to study general concepts. You care about algebra for the same reason you care about arithmetic: people use it to communicate their ideas precisely, and they often assume you can do algebraic operations in your head. To follow along, then, you need to do the algebra. This is true in both the study of statistics and the study of formal theory. If you do not master this basic algebra, you will get lost. Solving equations and simplifying inequalities in order to find the range of solutions also proves highly useful in both game theory and statistics. Indeed, as we explain in Chapter 12, which introduces vector algebra and matrix algebra, the algebra covered here (which is called scalar algebra) is a foundation for both vector and matrix algebra.

2.3 COMPUTATIONAL AIDS

Throughout this book we assume that you will be performing all mathematical manipulations by hand, or at most using a (simple) calculator. We believe this is pedagogically important: one needs to be able to do the relevant calculations oneself in order to understand them; if one doesn't understand them, then one doesn't really know what one is saying; and if one doesn't know what one is saying, there is very little point to formalizing one's concepts with mathematics at all. So this book is intended to lead you through doing the calculations yourself. That said, it is often helpful to have access to computational aids for arithmetic, algebra, and the later topics in this book. One reason simply is as a check for your work. We all make mistakes, and it is nice to have a second pair of eyes, so to speak, to check one's work. A second reason is to increase speed once one is sure of one's understanding. As the techniques of math become more familiar to you, the boundary of your skills will expand, and you will want to devote more of your time to the harder stuff rather than simple Algebraic manipulation. Computational aids can help with this. Finally, a third reason is to help with the writeup. Some aids allow output in formats that may be easily converted to word processors or typographical languages such as LaTeX (http://www.latex-project.org/).

There are many computational aids out there. Some are freeware, meaning you can download them from the Internet, or use them in your browser directly, with no further obligation. Some examples of these include Eigenmath (http://eigenmath.sourceforge.net/) and Maxima (http://maxima.sourceforge.net/), along with the website http://www.wolframalpha.com/, which allows you to input expressions directly into your browser. There are also some useful tools at http://www.math.com/students/tools.html, including a function plotter. LaTeXis a free typographical language that is very good at typesetting math; we wrote this book using it, and thus were able to deliver it typeset to Princeton University Press, retaining greater control over its "look and feel" and reducing production costs. Various options exist to make LaTeXmore user-friendly, e.g., LyX (http://www.lyx.org/).

Other tools are potentially more powerful, but they are also more expensive. However, they can have more functionality in some areas. If you are located at a university with access to them, they can be worthwhile to try. Mathematica (http://www.wolfram.com/mathematica/) and Maple (http://www.maplesoft.com/products/maple/) work well with symbolic math, and Matlab (http://www.mathworks.com/products/matlab/) is well suited to matrix algebra.

2.4 EXERCISES

2.4.1 Arithmetic Rules

Complete the following equations:

1. $x^1 =$ ______.
2. $-a \times (-b)^2 =$ ______.
3. $\sum_{i=1}^{4} x_i =$ ______.
4. $\prod_{m=6}^{9} x_m =$ ______.
5. $4! =$ ______.
6. $z^4 =$ ______.
7. $\sqrt[2]{9} =$ ______.
8. $\sqrt[3]{27} =$ ______.
9. $\left(\frac{3(2-4)}{2+3}\right)^3 =$ ______.

2.4.2 Ratios, Proportions, Percentages

10. Represent the following as a ratio, a proportion, and a percentage:
    a) Latinos relative to all others: African American 98,642; Asian 62,346; Caucasian 436,756; Latino 105,342; Other 32,654.
    b) Independent registered voters relative to Republicans: Democrat 432; Independent 221; Republican 312.

c) Republican relative to Democrat from no. 2.

11. If the Latino population shrunk to 100,322 in no. 1 above, what would be the percentage change in the Latino population?

12. If the other populations remained constant, what would be the percentage change in the proportion of Latinos to all others?

13. If voter turnout in the United States in 1996 was 56% and in 2000 it was 62%, what was the percentage change in turnout from 1996 to 2000?

14. Express these two quantities as a simplified ratio: 18 and 12.

2.4.3 Algebra Practice

15. Simplify into one term the following expressions:

    a) $xz + yz$.

    b) $mn + ln - pn$.

    c) $z \times y \times x - 2 \times y \times x$.

    d) $(z + x) \times y \times \frac{1}{z}$.

16. Simplify this expression as much as possible: $\frac{2x^2+20x+50}{2x^2-50}$.

17. Simplify this expression: $\frac{5+17x+4x+7}{42x}$.

18. Add these fractions: $\frac{2g+13}{3g} + \frac{4g-5}{4g}$.

19. Factor: $-7\theta^2 + 21\theta - 14$.

20. FOIL: $(2x - 3)(5x + 7)$.

21. Factor: $q^2 - 10q + 9$.

22. Factor and reduce: $\frac{\beta-\alpha}{\alpha^2-\beta^2}$.

23. Solve: $15\delta + 45 - 6\delta = 36$.

24. Solve: $.30\Omega + .05 = .25$.

25. Solve: $11 = (y + 1)2 + (6y - 12y)\frac{7}{2y}$.

26. Solve: $-4x^2 + 64 = 8x - 32$.

27. Complete the square and solve for x: $x^2 + 14x - 14 = 0$.

28. Complete the square and solve for y: $\frac{1}{3}y^2 + \frac{2}{3}y - 16 = 0$.

29. Solve using the quadratic formula: $2x^2 + 5x - 7$.

30. Derive the quadratic formula by completing the square for the equation $ax^2 + bx + c = 0$.

31. Solve: $-\delta > \frac{\delta+4}{7}$,

Chapter Three

Functions, Relations, and Utility

Hagle (1995, p. 7) opines that "functions are valuable to social scientists because most relationships can be modelled in the form of a function." We would add that functions are valuable for those political scientists who want to make *specific* theoretical claims and/or use statistics to test the implications of theories of politics. In other words, functions are valuable because they are explicit: they make very specific arguments about relationships. In addition, functions play a key role in developing statistical models.

What is a function? Functions may be defined in several ways, each developed more fully below. To get us started, functions provide a specific description of the association or relationship between two (or among several) concepts (in theoretical work) or variables (in empirical work). In other words, a function describes the relationship between ordered pairs (or n-tuples) arising from sets under special conditions (specified below).

That said, some students can come away from an introduction to relations and functions with a misguided notion that the key to developing sound theory is to master a wide array of functions and then see which one applies to a given theoretical or statistical problem. One might characterize this as a "toolbox" approach to political science, where different functions are hammers and wrenches to be tried here and there until one finds one that works. Perhaps such an approach would yield insight, but we are not sanguine: one's thinking about politics is unlikely to be usefully informed simply by mastery of different functional forms. Instead, a general working knowledge of functions can be used to sharpen one's thinking and bring greater specificity to one's theories of politics. In particular, learning to be able to translate verbal conjectures into graphs and/or equations that represent those conjectures is a valuable skill to develop. That skill is essential for anyone who wants to do formal modeling. Finally, such a working knowledge is critical to mastering the material in statistics courses and will help one select appropriate statistical models for hypothesis testing.

The first section discusses functions in general and elaborates on some of their properties. The second illustrates various functions of one variable; most of these can be readily generalized to multiple variables. The third section covers properties of relations in the milieu in which they are most typically seen in political science—individual preferences—and introduces the utility representation that underlies all of game theory. This serves as another example of the use of functions in political science (an empirical example appears at the end of the second section) and provides us with an opportunity to mention correspondences briefly as well. Readers with stronger math backgrounds should be able to skim the first two sections, but may not have seen the material in the third before.

3.1 FUNCTIONS

Recall from Chapter 1 that relations allow one to compare variables and expressions (or concepts). This is a general idea, but some relations are considerably more specific about the comparison. In particular, any relation that has a unique value in its range (we'll call these y values) for each value in its domain (we'll call these x values) is a function. Put differently, all functions are relations, but only some relations are functions. Another way to put this is that functions are subsets of relations. That said, political scientists do not often distinguish between relations and functions, and the term "function" is often used loosely to cover both relations and functions. Alternatively, you may encounter relations described as "set functions" and functions (as defined here) described as "point functions." More precisely, a relation that assigns one element of the range to each element of the domain is a **function**, while one that assigns a subset of the range to each element of the domain is a **correspondence**. We will focus largely on functions here, as they are the most commonly used by political scientists. However, correspondences are commonly used in game theory, and we discuss them briefly in Section 4.

More formally, a function maps the values measuring one characteristic of an object onto values measuring another characteristic of the object. Stated in set theoretic terms, a function is a relation such that (1) for all x in A, there exists a y such that (x, y) is an ordered pair in the function, and (2) if (x, y) and (x, z) are in the function, then $(y = z)$. In other words, if the value x is mapped to the value y by a function, and the value x is also mapped to the value z by the same function, then it follows that y and z are the same value. If $y \neq z$, then it is not a function but a correspondence.

Note that some equations with which you are familiar from middle school and high school math are either functions or correspondences. We review some examples below.

One can use both equations and graphs to describe functions. If you can develop an ability to translate your verbal conjectures into functions, you will have sharper, more explicit conjectures. Thus, developing the ability to work comfortably with both equations and their graphs will prove very valuable for developing your own theories about politics.

3.1.1 Equations

The linear equation $y = a + bx$ is the best-known and most frequently used function in political science.[1] We discuss it below. Here we want to remind you of the manner in which functions can be represented using equations. One often encounters equations of the form $x^2 + y^2 = 1$ or $\frac{y}{x} = 3$. We can use the rules covered previously to isolate y on the left-hand side (LHS),[2] yielding $y^2 = 1 - x^2$ and $y = 3x$. It turns out that the first of these equations is not a function while the second is, and we demonstrate that below where we introduce graphs.

You will hopefully recall the notation $y = f(x)$, which is read "y is a function of x." This is **implicit** notation that simply states that values of x are associated with singular values of y. Here we call x the **argument** of the function. But we do not know what the specific function is, so if we were given the values of x we could not produce the values of y. An **explicit** function describes the mapping of values in the domain to values in the range. For example, if we were given the explicit function $y = 3x$, then we could map the values of y for any given set of x values. In empirical work we typically refer to the x here as the independent or exogenous variable and the y as the dependent or endogenous variable, as it depends on and is affected by x.

3.1.2 Graphs

As noted above, we can graph relations and functions. If we plot the values of a set (or concept or variable) on the horizontal axis and the values of another set that shares ordered pairs with the first set on the vertical axis, then we can plot the intersection of each pair's values with a point in the space defined by the axes. Such a graph is known as a Cartesian, or xy, graph and is quite common. You will recall such graphs from arithmetic and algebra courses. The horizontal axis is also referred to as the x-axis (or domain) and the vertical axis is also known as the y-axis (or range).

The graph of the relation $x^2 + y^2 = 1$ forms a circle through the values 1, -1 on both axes, as depicted in Figure 3.1.[3] Note that this is not a function: all values in the domain (x) produce two different values in the range (y). If this were not true, it would not form a circle. If you do not find this apparent, select a value of x and plot the value for y in Figure 3.1.

Now consider the equation $y = 3x$, shown in Figure 3.2. The graph of this equation is a straight line moving through the origin and up to the right. No matter what x values we plug into the equation we get a unique value of y. As such, the equation is a function. Note that we can make use of a graph to determine whether an equation is or is not a function: if we can draw a vertical

---

[1] It may surprise you that though it is often referred to as a linear function, the linear equation is not a linear function, as strictly defined in mathematics. We discuss this below.

[2] Subtract x^2 from both sides in the first case and multiply both sides by x in the second.

[3] Because it makes a circle with a one-unit radius, it is known as the unit circle.

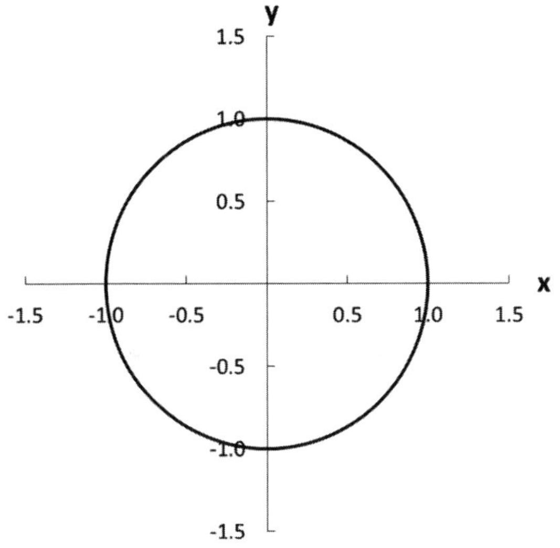

Figure 3.1: Graph of the Unit Circle

line at *any* point on the graph that intersects the curve at more than one point, then the equation is not a function.

3.1.3 Some Properties of Functions

As we go on, several properties of functions will be important. We cover inverse and identity functions, monotonic functions, and functions in more than one dimension, saving continuity for Chapter 4 and function maxima and minima, along with concave and convex functions, for Part II of the book. To begin, we expand our notation for a function slightly. We define the function f as $f(x) : A \to B$. This is often read as "f maps A into B." You've already seen the first part, which just means that the variable x is an input to the function $f(x)$, which spits out some value. Sometimes we assign this value to a variable y, as in $y = f(x)$, and sometimes we just leave it as $f(x)$, where it is understood that the function $f(x)$ may itself be a variable or a constant. For example, $f(x) = 3x$ is a variable, whereas $f(x) = 3$ is a constant.

The A and the B in the function's definition are new, but not conceptually. A here is the **domain** of the function, that is, the set of elements over which the function is defined. In other words, we draw our values of x from this set, and the function needs to produce a value for each element of x in this set. The most common domain political scientists use is the real numbers, $\mathbb{R}$, but there are numerous other domains you will see. B is known as the **codomain**, and it specifies the set from which values of $f(x)$ may be drawn. Depending on A

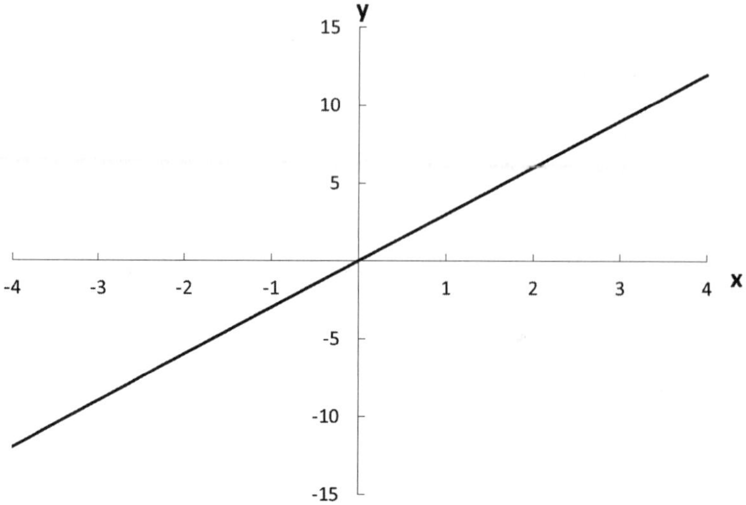

Figure 3.2: Graph of $y = 3x$

and f, though, not all the values in B may be reached. The set of all values actually reached by running each $x \in A$ through f is known as the **image**, or **range**, and it is necessarily a subset of B.

This may be confusing, so let's consider an example. Let $f(x) = x$. This function maps x to itself, and so does really nothing.[4] If $A = \mathbb{R}$, then $B = \mathbb{R}$, and the codomain and the range (or image) are exactly the same, since every real number is just mapped to itself. Now instead keep $B = \mathbb{R}$, indicating that the function f is real-valued, but let $A = (0, 1)$, or the set of all real numbers between zero and one, exclusive. In this case the image (or range) is just $(0, 1)$, which is the only part of B reached by the function, given the domain A.

One can chain multiple functions; this is called **function composition**. This is written either as $g \circ f(x)$ or $g(f(x))$ and is read as "g composed with f" or more commonly g of f of x. If we have $f(x)A \to B$ and $g(x) : B \to C$, then the full definition is $g \circ f(x) : A \to C$. Composition of functions is associative $(f \circ (g \circ h) = (f \circ g) \circ h)$, but not always commutative ($f \circ g$ does not always equal $g \circ f$). One takes a function composition in stages: first one computes $f(x)$ for each x to get a set of y, and then one takes $g(y)$ for each of these y. For more than two functions that are composed, first plug each x into the innermost function, then plug the output of this into the next innermost function, and so on until you've finished with all the functions. For example, if $f(x) = 2x$ and $g(x) = x^3$, then $g \circ f(x) = (2x)^3 = 8x^3$, whereas $f \circ g(x) = 2(x^3) = 2x^3$.

[4]This function is called the identity function, and we return to it below.

Table 3.1: Identity and Inverse Function Terms

Term	Meaning
Identity function	Elements in domain are mapped to identical elements in codomain
Inverse function	Function that when composed with original function returns identity function
Surjective (onto)	Every value in codomain produced by value in domain
Injective (one-to-one)	Each value in range comes from only one value in domain
Bijective (invertible)	Both surjective and injective; function has an inverse

3.1.3.1 Identity and Inverse Functions

Why does this all matter? To answer that, we need a couple more definitions as we need to introduce identity and inverse functions, as well as some other terms. Table 3.1 summarizes those terms.

A function is **surjective** or **onto** if every value in the codomain is produced by some value in the domain.[5] Our first example was surjective, because every point in $\mathbb{R}$ was reached by some point in the domain (the same point, in the example). The second was not surjective, as nothing outside $(0, 1)$ in the codomain was reached.

A function is **injective** or **one-to-one** if each value in the range comes from only one value in the domain.[6] We already knew that each $x \in A$ produced only one $f(x)$; otherwise it wouldn't be a function. This tells us that this property goes both ways: each $y \in f(x)$ comes from only one $x \in A$. Both of our examples for the identity function are injective; the function is just a straight line. In contrast, $f(x) = x^2$ would not be injective on the same domain as, for example, $y = 4$ is the result of plugging both $x = 2$ and $x = -2$ into the function (it would be injective if we confined ourselves to real numbers no less than zero, though).

If a function is both injective and surjective (one-to-one and onto), then it is **bijective**. A bijective function is **invertible**, and so has an inverse. This inverse is the payoff of our definitions, as it allows us to take a y and reverse our function to retrieve the original x. How do we do this? First we (re)define an **identity function**: $f(x) = x, f(x) : A \to A$, where we have made the domain and codomain identical, as we saw in our earlier example. This function merely returns what is put into it and is just like multiplying each element in our domain by one (or adding zero to each element), hence the use of the word identity.

[5] Formally, it is surjective if $\forall b \in B, \exists a \in A \ni f(a) = b$ (for all b in B there exists an a in A such that the function of a is b).

[6] Formally, $\forall a, c \in A, \forall b \in B$, if $f(a) = b$ and $f(c) = b$, then $a = c$.

The **inverse function** is the function that when composed with the original function returns the identity function. That is, it undoes whatever the function does, leaving you with the original variable again. The inverse is $f^{-1}(x) : B \to A$, and remember to be *very* careful not to confuse it with $(f(x))^{-1} = \frac{1}{f(x)}$. Thus, in symbols, the inverse is defined as the function $f^{-1}(x)$ such that $f^{-1} \circ f(x) = x$, or just $f^{-1}(f(x)) = x$. The inverse does commute with its opposite $f(f^{-1}(x)) = f^{-1}(f(x))$. For example, if $f(x) = 2x + 3$, a bijective mapping, then its inverse is $f^{-1}(x) = \frac{x-3}{2}$. We can check this both ways: $f^{-1}(f(x)) = \frac{(2x+3)-3}{2} = \frac{2x}{2} = x$ and $f(f^{-1}(x)) = 2\left(\frac{x-3}{2}\right) + 3 = x - 3 + 3 = x$.

3.1.3.2 Monotonic Functions

Some functions increase over some subset of their domains as x increases within this subset. Others decrease over the same subset, and the rest increase over some x and decrease over others, depending on the value of x. If a function never decreases and increases for at least one value of x on some set $C \subseteq A$, it is an **increasing function** of x on C, while if it never increases and decreases for at least one value of x on some set $C \subseteq A$, it is a **decreasing function** of x on C. If a function increases always as x increases within C it is a **strictly increasing function** on C; if it decreases always as x increases within C it is a **strictly decreasing function** on C. Strictly increasing and strictly decreasing functions are injective. We sometimes call a function that does not decrease (but may or may not increase ever) a **weakly increasing function**, and a function that does not increase (but may or may not decrease ever) a **weakly decreasing function**.

You will sometimes encounter the term **monotonic function** in statements such as "y increases monotonically as a function of x." Monotonicity is the characteristic of order preservation—it preserves the order of elements from the domain in the range. A monotonic function is one in which the explained variable either raises or retains its value as the explanatory variable(s) rises. Thus it is an increasing function across its entire domain. A strictly monotonic function is strictly increasing over its entire domain. Table 3.2 summarizes these concepts.

We provide several examples of monotonic functions in the next section. All affine and linear functions with positive coefficients on x are strictly monotonic, as are exponential functions, logarithms, cubic equations, etc. Ordered sets can also be monotonic or strictly monotonic. An example of two ordered sets with a monotonic, but not a strictly monotonic, relationship is $\{1, 2, 3, 4, 5\}$, $\{10, 23, 23, 46, 89\}$. Monotonic functions have many nice properties that will become apparent as you study both statistics and game theory.

3.1.3.3 Functions in More Than One Variable, and Interaction (Product) Terms

Thus far we have (primarily) simplified things by focusing on the idea that y was a function of one variable. Unfortunately, few (if any!) political relationships

Table 3.2: Monotonic Function Terms

Term	Meaning
Increasing	Function increases on subset of domain
Decreasing	Function decreases on subset of domain
Strictly increasing	Function always increases on subset of domain
Strictly decreasing	Function always decreases on subset of domain
Weakly increasing	Function does not decrease on subset of domain
Weakly decreasing	Function does not increase on subset of domain
(Strict) monotonicity	Order preservation; function (strictly) increasing over domain

are so simple that they can be described usefully as a function of one variable. As such, we need to be able to use functions of two or more variables, such as $y = f(x_1, x_2, x_3)$ or $z = f(x, y)$.

Graphs of the function of one variable are straightforward, and graphs of the function of two variables are feasible (though many of us begin to struggle once we have to start thinking in three dimensions). Consider two variables multiplied by one another, also known as a product term. Product terms are a commonly used nonlinear function. Consider the plot of $y = 3xz$, in Figure 3.3, and observe that it produces a plane with a changing slope rather than a plane with a constant slope.

Another way of saying the same thing is that the relationship of x on y is different (stronger or weaker) depending on the value of z. Further, the strength of the impact of z on y also depends on the value of x. That is what is meant by interaction: x and z interact with one another to produce y. You will learn in your statistics course how to properly specify statistical models to test interaction hypotheses.[7]

As another example, consider the three-dimensional plot of the linear function $y = 3x + z$, depicted in Figure 3.4.

Graphs of the function of three or more variables, however, become terribly complex and generally are not used, though there are some exceptions. Instead of using graphs, analysis of multiple variable functions focuses on equations.

Luckily, the specification of equations in more than one variable is not much more complicated than that in one variable. You've already seen some of the notation, e.g., $f(x, y)$. The rest just accounts for the more complex domain

[7]We note here that one would not want to estimate the model $y = \alpha + \beta(xz) + e$ because doing so would produce a biased estimate of β. Rather, one would want to include the variables x and z in the model as well (Blalock Jr., 1965; Friedrich, 1982; Braumoeller, 2004; Brambor, Clark, and Golder, 2006). This will be discussed in your statistics courses.

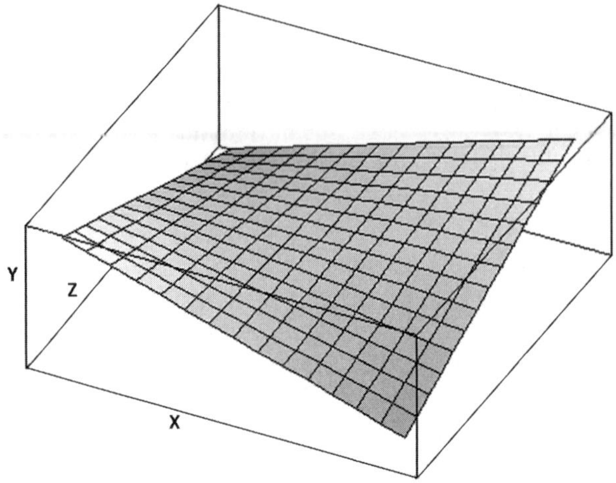

Figure 3.3: Graph of $y = 3xz$

that is present when there is more than one variable. If there are n variables, denoted x_1 through x_n, and the set from which each variable is drawn is called A_1 through A_n, respectively, then the domain of the function is the Cartesian product $A_1 \times A_2 \times \ldots \times A_n$. The formal definition of the function is $f(x_1, \ldots, x_n) : A_1 \times \ldots \times A_n \rightarrow B$. To get any value of f you just plug in the values of all the input variables. Most of the concepts discussed above are either directly applicable or have analogues in the multidimensional case, though there is more complexity involved. For example, properties such as continuity can be defined for each input variable independently. We save discussion of the properties of multi-dimensional functions most relevant to us until Part V of the book, however.

3.1.4 Why Should I Care?

A basic understanding of functions is critical to any political scientist who wants to be able to make specific causal conjectures. Making specific causal conjectures is useful because it increases one's ability to evaluate whether relevant evidence is at odds with one's theory (i.e., improves hypothesis testing; Popper, 1959, pp. 121–23) and it facilitates communication with other scholars (Cohen and Nagel, 1934, pp. 117–20). Vagueness is antithetical to science, and stating hypotheses as functions helps one eliminate vagueness. Further, statistical inference is a powerful tool for hypothesis testing, and functions are one of the building blocks on which statistics is constructed. Finally, game theory makes extensive use of functional forms to represent preferences and payoffs, as we'll see in Section

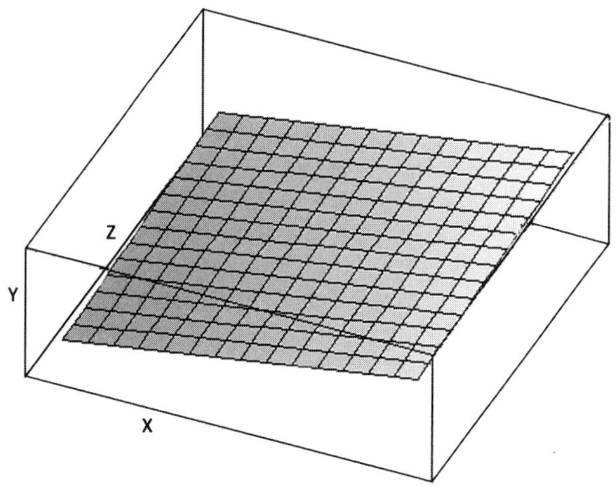

Figure 3.4: Graph of $y = 3x + z$

3 of this chapter. For these reasons, the properties of functions we discuss in this section are fundamental, as they have substantive meaning in the settings in which we are using the functions. Monotonocity in one's preferences, for example, means that someone always prefers more to less. This is very different from having what is known as an ideal point, in which case moving away from the ideal in *either* direction is not preferred.

3.2 EXAMPLES OF FUNCTIONS OF ONE VARIABLE

Political scientists are generally interested in the relationships among multiple variables. Nevertheless, in this section we begin with associations where y is a function of one x. These functions extend readily to more than one variable, as noted above.

3.2.1 The Linear Equation (Affine Function)

You encountered the additive linear equation back in algebra classes: $y = a + bx$. Technically, this is an **affine function**, though it is frequently referred to as a linear function. We discuss the technical distinction between the two below. For now, let's review some basics.

In the equation $y = a + bx$, a and b are constants.[8] The constant a is the

[8]Recall that in this equation, y is a function of only one variable, x. Therefore a and b cannot be variables and must be constants.

intercept, or in terms of the graph, where the function crosses the vertical (y) axis (i.e., the value of y when $x = 0$). The constant b is the slope of the line, or the amount that y changes given a one-unit increase in x. That is, a one-unit increase in x produces a $1b$-unit increase in y, a three-unit increase in x produces a $3b$-unit increase in y, etc.

One might conjecture that the probability that an eligible voter casts a ballot in a US presidential election is a linear function of education.[9] Let p_v represent the probability of voting and ed represent education level: $p_v = a + b(ed)$. In this function a represents the likelihood that someone without any formal education turns out to vote, and b indicates the impact of education on the probability of voting. Shaffer (1981, p. 82) estimates a model somewhat like this, and we can borrow his findings for illustrative purposes, yielding $p_v = 1.215 + 0.134 \times ed$. The intercept of 1.215 makes little sense,[10] but we ignore that for this example. Shaffer's education measure has four categories: 0–8 years of education, 9–11 years, 12 years, and more than 12 years. A slope (i.e., b) of 0.134 suggests that as we move from one category to another (e.g., from 0–8 years to 9–11 years, or from 12 years to more than 12 years), the probability that someone votes rises by 0.134. So if this linear model and its results are accurate, the typical adult with a college education has roughly a 0.4 greater probability of voting in a US presidential election than the typical adult without any high school education.[11]

The linear equation states that the size of the impact of x on y is constant across all values of x. For example, in the above example the impact of x on y is roughly 0.13. Since the relationship is linear, that means that a shift from 0–8 years of education to 9–11 years of education increases the probability of voting in a national election by $\sim$.13, and a shift from 9–11 years to 12 years also produces an increase of $\sim$.13, as do shifts from 12 years of education to more than 12 years of education. Nonlinear functions, which we discuss in the third subsection, specify that the size of the impact of x on y varies across values of x.

[9]We recognize that people are very unlikely to posit such a claim. We offer it not as a reasonable conjecture but simply as an illustration.

[10]A value of 1.2 is nonsense because the intercept represents the probability of voting when a person has had zero education. Since probabilities by definition have a range from zero to one, any probability above one is nonsense. This is but one reason that this may be an unrealistic example. In your statistics courses you will learn a number of reasons why this estimate is nonsense.

[11]You will learn how to do these sorts of calculations in your statistics courses. For those who want a brief description, you need to calculate two values and then determine the distance between them. More specifically, multiply the slope (0.134) by the first value in the comparison, someone with no high school education: $1 \times 0.134 = 0.134$. Now multiply the slope by the second value in the comparison, someone with a college education: $4 \times 0.134 = 0.536$. Finally, take the difference of these two probabilities of voting (i.e., subtract the probability that a citizen without a high school education votes from the probability that a college educated citizen votes) to get $0.536 - 0.134 = 0.402 \sim 0.4$.

3.2.2 Linear Functions

Mathematicians make distinctions that few political scientists employ. We review them for the purpose of helping you avoid confusion when you read "mathematically correct" presentations. In particular, we distinguish between affine functions (discussed above), linear equations, and linear functions (discussed here). As suggested above, a **linear equation** is an equation that contains only terms of order x^1 and $x^0 = 1$.[12] In other words, only x and 1, multiplied by constants, may appear on the right-hand side (RHS) of a linear equation. This means that the RHS of a linear equation is an affine function. Linear *functions* are not affine functions; e.g., they do not permit a translation (the x^0 term).

The formal definition of a **linear function** is any function with the following properties:

- Additivity (aka superposition): $f(x_1 + x_2) = f(x_1) + f(x_2)$,
- Scaling (aka homogeneity): $f(ax) = af(x)$ for all a.

Additivity states that the impact of a sum of variables is equivalent to the sum of the impacts of those variables. The scaling property, on the other hand, states that the size of the input is proportional to the size of the output.

Let's begin by comparing the linear function $y = \beta x$ with the affine function $y = \alpha + \beta x$ along these criteria. The additivity property states that $f(x_1 + x_2) = f(x_1) + f(x_2)$. So we substitute the RHS of each $y = \ldots$ equation for the parts in the parentheses (i.e., $f(\cdot)$) and see if that statement is true. If it is, the property is met. We begin with the linear function $y = f(x) = \beta x$. To determine whether it meets the additivity property, we need to replace x with $x_1 + x_2$, following the additivity property equation above, and determine whether the equality is true:

$$\begin{aligned} f(x_1 + x_2) = \beta(x_1 + x_2) &= \beta x_1 + \beta x_2, \\ \beta x_1 + \beta x_2 &= f(x_1) + f(x_2). \end{aligned}$$

As one can see, the equality is true. Now we'll try the linear equation (or affine function), under the assumption that $\alpha \neq 0$: $y = f(x) = \alpha + \beta x$. Again, we replace x with x_1 and x_2, in accord with the additive property equation, and see whether the equality is true:

$$\begin{aligned} f(x_1 + x_2) = \alpha + \beta(x_1 + x_2) &= \alpha + \beta x_1 + \beta x_2, \\ f(x_1) + f(x_2) &= (\beta x_1 + \alpha) + (\beta x_2 + \alpha), \\ \alpha + \beta x_1 + \beta x_2 &\neq 2\alpha + \beta x_1 + \beta x_2. \end{aligned}$$

It is not true; the RHS and LHS differ by α. So the linear equation (or affine function) does not have the additive property, but the linear function does.

[12]Order refers to the highest exponent in the polynomial.

Now let's consider the scaling property, which states that $f(ax) = af(x)$. Let's begin with the linear function $y = f(x) = \beta x$:

$$f(ax) = \beta(ax) = a\beta x,$$
$$a\beta x = af(x).$$

So, the linear function satisfies the scaling property. What about the linear equation (i.e., affine function)?

$$f(ax) = \alpha + (\beta(ax)) = \alpha + a\beta x,$$
$$af(x) = a\alpha + a\beta x,$$
$$\alpha + a\beta x \neq a\alpha + a\beta x.$$

This property doesn't hold either because $\alpha \neq a\alpha$. Again, the linear function satisfies the property, but the affine function (linear equation) does not.

The only difference between the two functions is the constant, α. Recall that α represents the value where the function crosses the vertical (y) axis. If it crosses at zero, then the two functions are equivalent. Thus, a linear function must cross the vertical axis at the origin (i.e., where x and y have a value of zero). You might recall that ratio level measurement requires a meaningful zero value, whereas interval level measurement does not, and that division and multiplication operations are valid on ratio level measures but not on interval level measures. Linear transformations require preservation of the order of the variables, the scale, and the zero, and only linear functions meet such criteria. Affine transformations preserve order and scale, but not the placement of zero.

Above we noted that political (and other social) scientists frequently refer to the linear equation $y = \alpha + \beta x$ as a linear function. *Technically*, this is inaccurate, but it is a rather fine mathematical point. The linear equation does produce a line, and a linear transformation with the affine function preserves order and scale, with the exception of the intercept. And that is all most political scientists are typically trying to indicate when they talk about linear functions and linear transformations. That said, there are some applications (e.g., time series analysis) where the proper definition of a linear function is important, and we raise the discussion here so as not to later confuse those who go on to study those issues in more detail.

3.2.3 Nonlinear Functions: Exponents, Logarithms, and Radicals

Technically speaking, nonlinear functions are all those that do not meet the two properties we just discussed. Practically speaking, nonlinear functions are all those that are neither linear nor affine: those functions that describe (the graph of) a curve that is not a line. For example, $y = \cos(x)$, in Figure 3.5, is a nonlinear function. Functions with exponent terms, including quadratics and other polynomials, are the most commonly used nonlinear functions in political science. Logarithms are another commonly used class of nonlinear functions, as are roots (or radicals). We briefly introduce the relationship among these functions and then turn our attention to graphing these functions and using them in algebra.

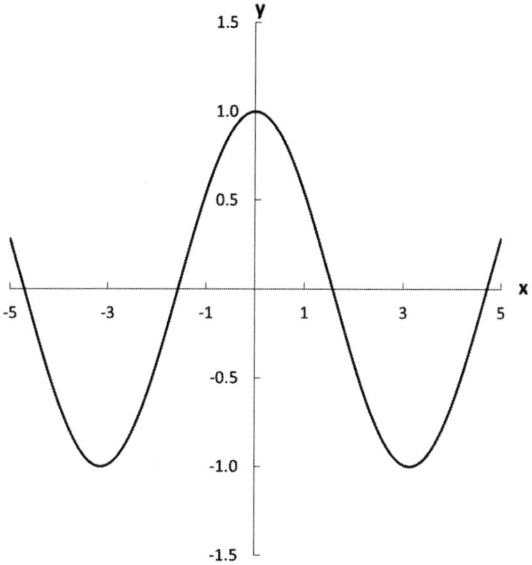

Figure 3.5: Graph of $y = \cos(x)$

Exponents, logarithms, and roots are related: one can transform any one such function into a representation of one of the others. In fact, in high school you may have focused on doing that. More specifically, when two of the following variables in the equation $b^n = x$ are known, one can solve for the unknown using

- Exponents to solve for x,
- Logarithms to solve for n,
- Radicals to solve for b.

That said, we will not focus on the relationship among the functions as political scientists do not frequently make use of those relationships.[13] Instead, we introduce each function and its notation, discuss their graphs, and then describe algebraic manipulations.

3.2.3.1 Exponents and the Exponential Function

As notation, **exponents** (aka power functions) are a shorthand for expressing the multiplication of a number by itself: $x^3 = x \times x \times x$. More generally, $x^n = x \times$

[13]Those interested in studying this might find the following Wikipedia entries useful: `http://en.wikipedia.org/wiki/Logarithm`, `http://en.wikipedia.org/wiki/Radical_(mathematics)#Mathematics`, `http://en.wikipedia.org/wiki/Exponential_function`.

$x \times x \ldots x$ (n times). This is all familiar, but you may be less familiar with other exponential notation: $x^{-n} = \frac{1}{x^n}, x^{\frac{1}{n}} = \sqrt[n]{x}$. In words, x to a negative power represents the fraction "1 divided by x^n" and x raised to a fraction represents a root of x, where the root is determined by the value in the denominator of the exponent. Perhaps an easier way to remember this is that a negative exponent indicates that one *divides* (rather than multiplies) the term by that many factors. Similarly, a fractional exponent indicates that one takes the nth *root* rather than multiplying the term n times. Mixed exponents work similarly. So $x^{\frac{2}{3}} = \sqrt[3]{x^2}$ and $x^{\frac{-3}{2}} = \frac{1}{\sqrt[2]{x^3}}$. Finally, $x^0 = 1$.[14]

Nonlinear functions with exponents are of interest to political scientists when we suspect that a variable x has an impact on y, but that the strength of the impact is different for different values of x. The best way to see this is to look at the graphs of some functions with exponents.

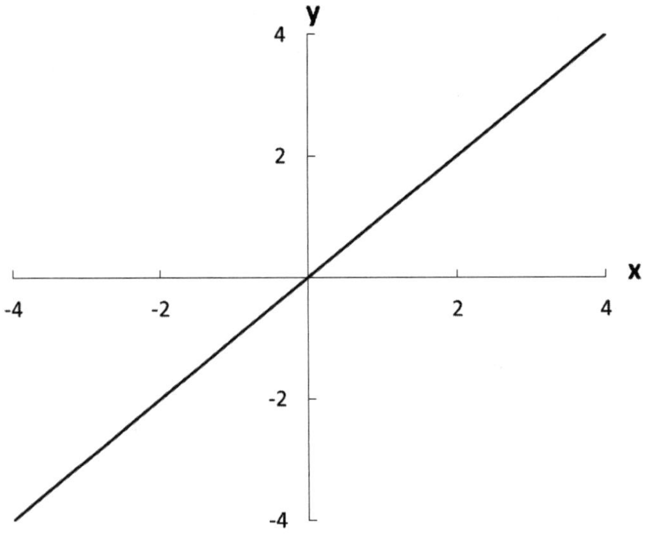

Figure 3.6: Graph of $y = x$

Consider the graphs of the functions $y = x$ and $y = x^2$, in Figures 3.6 and 3.7. The linear function produces a line with a constant slope: if we calculate the change in y due to a one-unit change in x, it does not matter what point on the x-axis we select; the change in y is the same.[15] However, the slope of the curve for $y = x^2$ is not constant: the impact of x on y changes as we move along the x-axis (i.e., consider different values of x). To be more concrete, a one-unit

[14]This holds for all $x \neq 0$, but people often treat $0^0 = 1$ as if it were true when they are simplifying equations.

[15]Another way to make this point is to observe that linear functions meet the scaling property.

increase from 0 to 1 produces a one-unit increase in y, but a one-unit increase from 2 to 3 produces a five-unit increase in y, and a one-unit increase from 5 to 6 produces an 11-unit increase in y. Thus, the impact of x on y increases over the range of x.

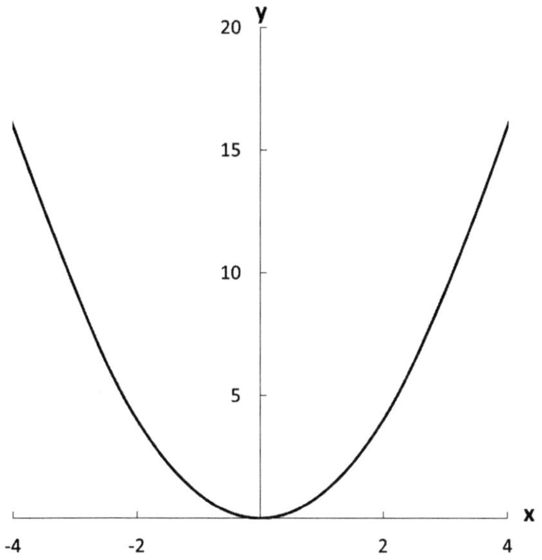

Figure 3.7: Graph of $y = x^2$

This has important implications for developing theory. If reflection, deduction, or inspiration leads one to conjecture that a causal relationship between two concepts is constant over the range of values for the causal concept, then a linear or affine relation represents that conjecture. However, if one suspects that the strength of the relationship varies across the values of the causal concept, then a nonlinear relation is needed. As we discuss below, exponential terms play an important role in quadratic and other polynomial functions.

We covered some of these above, but below is a list of the algebraic rules that govern the manipulation of exponents.

Multiplication: to calculate the product of two terms with the same base one takes the sum of the two exponents:

$$x^m \times x^n = x^{m+n}.$$

To see that this is so, set $m = 3$ and $n = 4$ and write it out:

$$x^3 \times x^4 = (x \cdot x \cdot x) \times (x \cdot x \cdot x \cdot x) = x \cdot x \cdot x \cdot x \cdot x \cdot x \cdot x = x^7.$$

This works when m and n are positive, negative, or zero. When the bases are different, you can simplify the expression *only when the exponents are the same.*

In this case, multiplication is distributive:

$$x^m \times z^m = (xz)^m.$$

To see why, set $n = 2$, and note that $x^2 \times z^2 = x \cdot x \cdot z \cdot z = (x \cdot z) \times (x \cdot z) = (xz)^2$.[16]

Last, when both the base and the exponent are different, you cannot simplify to a single term. Thus, e.g.,

$$x^m \times z^n \neq (xz)^{m+n}.$$

Assume that $m = 2$ and $n = 3$, and write the expressions out to see that this is so:

$$x^2 \times z^3 = (x \times x) \times (z \times z \cdot z) = z((x \times z) \times (x \times z)) \neq (xz)^5.$$

One cannot combine the terms fully.[17] To return to the point made above, if we assume that $m = n = 3$, then when we write it out we get:

$$x^3 \times z^3 = (x \cdot x \cdot x) \times (z \cdot z \cdot z) = (x \cdot z) \times (x \cdot z) \times (x \cdot z) = (xz)^3.$$

To determine the **power of a power**, one multiplies the exponents. For example,

$$(x^m)^n = x^{mn}.$$

To see that this is so, let's assign $m = 2$ and $n = 3$, and write out:

$$(x^2)^3 = x^2 \times x^2 \times x^2 = x \cdot x \cdot x \cdot x \cdot x \cdot x = x^6.$$

Division: to calculate the quotient of two terms with the same base and different powers, one takes the difference of the exponents:

$$\frac{x^m}{x^n} = x^{m-n}.$$

To see why this is so, recall that

$$\frac{1}{x^n} = x^{-n}.$$

We can therefore write out:

$$\frac{x^m}{x^n} = x^m x^{-n} = x^{m-n}.$$

---

[16]Note that this assumes that multiplication is commutative; hence this will not hold for matrix multiplication, as we'll see in Part IV of the book.

[17]If this illustration is not clear to you, then assign values to x and z (say, 2 and 3) and work it out. It will become clear that one can take the product when the exponents are equal and the bases are different, but one cannot take the product when both the exponents and bases are different. Note that we can simplify to a degree: $x^2 \times z^3 = (xz)^2 z$, but this is not usually helpful.

We can assign the values $m = 2$ and $n = 3$ and verify

$$\frac{x^2}{x^3} = \frac{x \cdot x}{x \cdot x \cdot x} = \frac{1}{x}$$

and

$$\frac{x^2}{x^3} = x^2 x^{-3} = x^{-1} = \frac{1}{x}.$$

When the bases are different, one can simplify only if the exponents are the same. When the exponents are the same, one raises the fraction to that power:

$$\frac{x^m}{z^m} = \left(\frac{x}{z}\right)^m.$$

Put differently, like multiplication, division is distributive when the bases are different and the exponents are the same.[18]

Recall that $x^0 = 1$. We can now demonstrate this by observing that $\frac{x^n}{x^n} = 1$. Observe that $\frac{x^n}{x^n} = x^{n-n} = x^0$. Since anything divided by itself equals one, it follows that $x^0 = 1$ (except when $x = 0$).

This covers x^a, but what about a^x? This is called an **exponential**. The one most commonly used sets $a = e$, where e is the base of the natural logarithm, or $e \approx 2.7183$ (to four decimal places). This is the **exponential function**, written as $y = \exp(x)$ or $y = e^x$.[19] We discuss the base of the natural logarithm, and its relation to the exponential function, below; in Figure 3.8 we graph the exponential function.

3.2.3.2 Quadratic Functions

Quadratic functions are nonlinear functions that describe a parabola. More specifically, if y is a quadratic function of x, then $y = \alpha + \beta_1 x + \beta_2 x^2$. In other words, quadratic functions describe a relationship where a variable (y) is a function of the sum of another variable (x) and its square (x^2).[20]

---

[18] In fact, other than having to remember not to divide by zero, multiplication and division have basically the same properties. The same is true for addition and subtraction. Thus one need remember only the properties of multiplication and addition.

[19] Both notations are common, and they are equivalent expressions.

[20] It is worth observing that many political scientists refer to a quadratic function as linear. For example, in a regression course you may encounter the claim that $y = \alpha + \beta_1 x + \beta_2 x^2 + e$ is a linear model. That is true: it is a linear *model.* When discussing regression models people frequently distinguish between models that are *linear in parameters* from those that are *linear in variables.* A regression model that contains a quadratic function (e.g., $y = \alpha + \beta_1 x + \beta_2 x^2 + e$) is linear in parameters but nonlinear in variables. Put differently, if we plot the relationship between x and y, the plot will be nonlinear: it is not linear in variables. But the parameters of the quadratic function have the properties of an affine function (to see this, set $z = x^2$ and rewrite the linear model as $y = \alpha + \beta_1 x + \beta_2 z$), and if we assume that $\alpha = 0$, then they have the properties of a linear function. Returning to models, the model $y = \alpha + \beta_1 x_1 + \beta_2 x_2 + e$ is linear in parameters and variables (as long as we assume that x_1 and x_2 are not nonlinear transformations of one another (e.g., $x_1 \neq x_2^n$)).

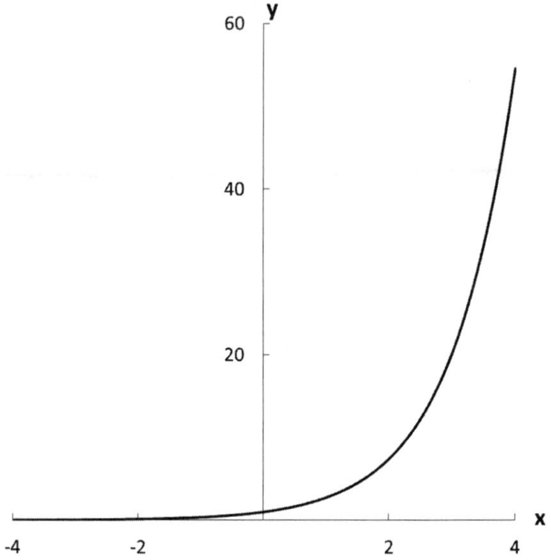

Figure 3.8: Graph of $y = e^x$

Note that since y is a function of only one variable, x, we can graph the function in two dimensions. If we set $\beta_2 < 0$, then we get a curve shaped like an inverse U (i.e., a concave parabola) as depicted in Figure 3.9. Switching the sign of β_2 produces a U-shaped curve (i.e., a convex parabola).

What sort of theoretical expectations might one want to sharpen by stating them as a quadratic relationship? Speaking generally, a quadratic function is quite useful for depicting relationships where we think the impact of an independent variable is positive (negative) for low values of the independent variable, flat for middle-range values, and negative (positive) for high values. Put differently, when one thinks that there is some (often unknown) threshold at which the relationship between two concepts (variables) switches (i.e., from positive to negative or from negative to positive), one might consider whether the quadratic can represent our conjecture.

For example, many scholars have hypothesized that rebellion will be low in countries that exert little to no government coercion *and* in countries that exhibit high levels of government coercion. Where will one find rebellion? This conjecture suggests that it will be highest among those countries that engage in mid-range levels of coercion (e.g., Muller and Seligson, 1987). If we let r represent rebellion and c represent coercion, then this conjecture can be represented as follows: $r = \alpha + \beta_1 c - \beta_2 c^2$.

Another example is the conjecture that the extent to which governments are transparent (i.e., noncorrupt) varies nonlinearly with the level of political competition. More specifically, over the range from authoritarian to democratic

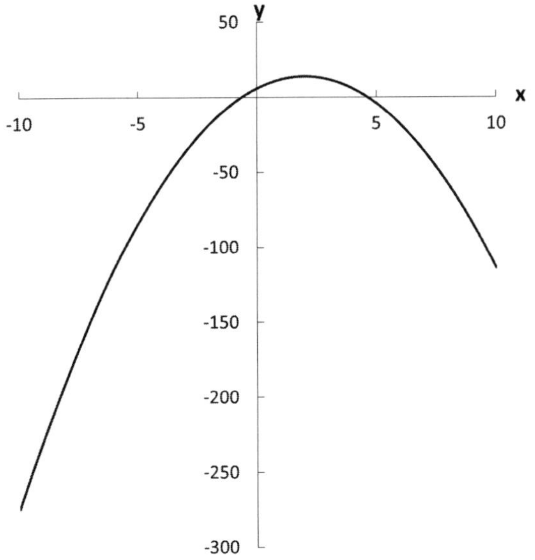

Figure 3.9: Graph of $y = 6 + 8x - 2x^2$

polities, transparency (e.g., the absence of bribery) is relatively common at both endpoints and least common in mixed polities that have a mix of autocratic and democratic institutions (e.g., Montinola and Jackman, 2002). If we allow t to stand for transparency and p for polity type, then we can represent that conjecture with the following quadratic equation: $t = \alpha + \beta_1 p + \beta_2 p^2$.

Finally, note that if we invert the concept we are trying to explain (i.e., flip the scaling of the dependent variable), we can represent the argument by flipping the signs on the quadratic (x^2) term. Thus, if we reconceptualize rebellion as quiescence, q, then we can write $q = \alpha + \beta_1 c + \beta_2 c^2$, and if we reconceptualize transparency as corruption, k, then we can write $k = \alpha + \beta_1 p - \beta_2 p^2$.

3.2.3.3 Higher-Order Polynomial Functions

Polynomial functions have the following general form: $y = \alpha + \beta_1 x + \beta_2 x^2 + \ldots + \beta_n x^n$, where n is an integer less than infinity. So both linear and quadratic functions are polynomials. Higher-order polynomials are those possessing powers of x greater than the reference. In this case, we are referring to the presence of cubed and higher terms. Like quadratics, higher-order polynomials are nonlinear: they describe curves, such as the cubic polynomial in Figure 3.10. More specifically, one can use them to explicitly represent the expectation that there are two or more thresholds over which the relationship between two concepts (variables) changes.

With the exception of the quadratic, polynomial functions are not very common in political science, though Mukherjee (2003) and Carter and Signorino

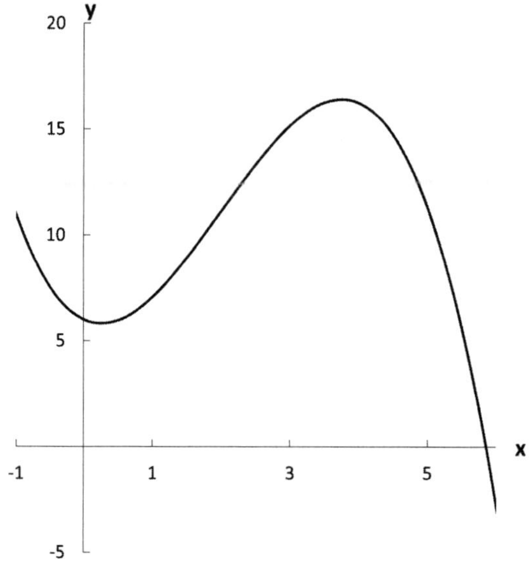

Figure 3.10: Graph of Cubic Polynomial

(2010) are exceptions. Mukherjee studies the relationship between the size of the majority party and central government expenditures in parliamentary democracies. A majority party is one that has at least 50% of the seats in the legislature and thus can govern without having to form a coalition with other parties. The basic underlying idea is that there are two different thresholds at work between the number of seats the majority party holds in the legislature and the size of government spending. First, as the number of seats held by the majority party rises from a bare majority (i.e., 51% of the legislature), spending declines, because it takes more and more legislators to defect and bring down the government.

Yet, while Mukherjee expects an initial negative relationship as the size of the majority party increases above a bare majority, he expects the relationship to quickly become positive (perhaps at around 56% of the legislative seats). Expenditures rise because the party has greater electoral safety and thus can take greater risks of alienating other parties' constituents by more greatly rewarding its own constituents. Yet he does not argue that this incentive to spend more remains as party size grows beyond the supermajority threshold (roughly 67% of the seats).

Instead, Mukherjee expects the relationship between majority party size and government expenditures to again turn negative (above the supermajority threshold) because the size of the population that the majority can tax without suffering electorally shrinks. That is, as majority party size rises beyond the supermajority threshold, the number of constituents that support other parties grows sufficiently small that it becomes increasingly difficult to write legislation that transfers income from those people to one's own constituents. He uses a cubic polynomial, $GovExp = \alpha - \beta_1(SizeMajParty) + \beta_2(SizeMajParty)^2 - \beta_3(SizeMajParty)^3$, to represent his verbal argument, and the results of his empirical analysis are consistent with his conjecture.

Carter and Signorino (2010) propose the use of a cubic polynomial to model time dependence in binary pooled cross-sectional time series data. Though it sounds complex, it is a fairly straightforward proposal. One takes the measure of time in one's data (perhaps the year) and, like Mukherjee, includes the three-termed polynomial in the regression equation. They show that if the dependent variable can take only two values (e.g., absence or presence of war) and the researcher has both cross-sectional data (e.g., all the countries in the world) measured over time (e.g., 1816–2005), then the cubic polynomial of time will control for what is called "temporal dependence" in the regression model.

More generally, then, polynomial functions are appealing because one can use them to make specific claims about threshold effects. That is, when theorizing leads one to expect that the relationship between two variables changes across the values of one of the variables, then a polynomial function might help one make a more specific (and more easily testable and falsifiable) claim.

3.2.3.4 Logarithms

Logarithms can be understood as the inverses of exponents (and vice versa). They can be used to transform an exponential function to a linear one, or a linear function to a nonlinear one in which the impact of one variable on another declines as the first variable rises in value. The logarithm (or log) tells you how many times to multiply its base a in order to get x, where a is a positive real number not equal to 1. If we denote the log with base a by $\log_a x$, then we have $a^{\log_a x} = x$ and $\log_a a^x = x$. Similarly, we can see that if $\log_a x = b$, then $a^{\log_a x} = a^b$, since the exponents are the same, and thus $x = a^b$. This lets us transition between logs and exponents readily.

Logs can be written in any base, though the most common are base 10 and the natural log. The base for the natural log is the $e \approx 2.7183$ from the exponential function. The concept of the base of a number is abstract and often confuses students. This owes in part to the commonality of the base 10 system in our lives. It is, after all, how we write numbers: we use 0 through 9, and at 10, 100, 1,000, and so on, we add a digit. You may also be familiar with binary from living in an age of computers, however. In binary one uses only 0 and 1, and at 2, 4, 8, and further powers of 2, one adds a digit. The base of a log is just an extension of this idea. We won't go into why e is one of the most common bases of logs used, though you are free to explore that topic on your own, of course. Rather, we'll just note that the natural log is *usually* identified with the notation ln, and log base 10 is *generally* denoted log, though some people use log to denote the natural logarithm. Throughout this book ln indicates natural log and log denotes log base 10.

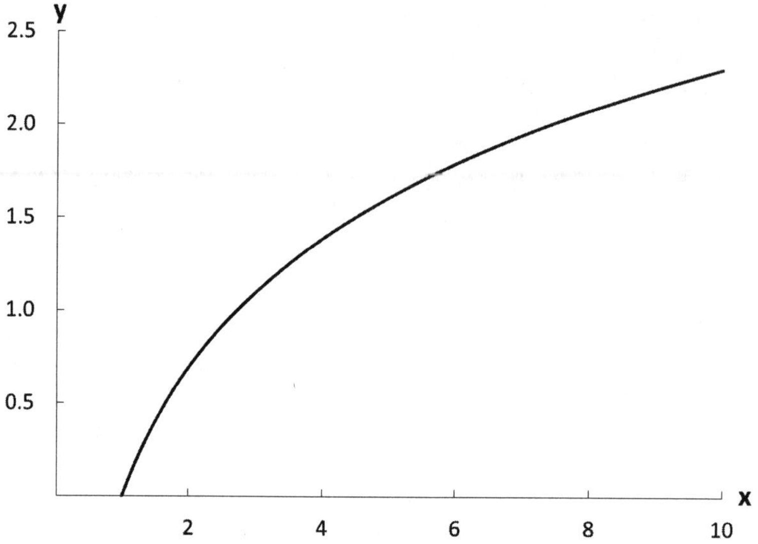

Figure 3.11: Graph of $y = \ln(x)$

Let's look at graphs of $y = \ln(x)$ and $y = \log(x)$ in Figures 3.11 and 3.12. Note that the impact of x on y diminishes as x increases, but it never becomes zero, and it never becomes negative. Theoretically, the log functions are very appealing precisely because of this property.[21] If you suspect, for example, that education increases the probability of voting in national elections, but that each additional year of education has a smaller impact on the probability of voting than the preceding year's, then the log functions are good candidates to represent that conjecture. Why? If p_v is "probability of voting" and ed is "years of education," then $p_v = \alpha + \beta ed$ specifies a linear relationship where an additional year of education has the same impact on the probability of voting regardless of how many years of education one has had. By contrast, $p_v = \alpha + \beta ed^2$ represents the claim that the impact of education on the probability of voting rises the more educated one becomes. Neither of these functional forms captures the verbal conjecture. But if we take the log of an integer variable such as "years of education," we transform the relationship between p_v and ed from a linear one to a nonlinear one where the impact of an additional year of education declines the more educated one becomes: $p_v = \alpha + \beta(\ln(ed))$.

There are several algebraic rules for logs that are important to know.[22] First,

[21] This is called *concavity*, and we will discuss it more in Part II of this book.
[22] The following holds for logarithms of any base, not just the natural log.

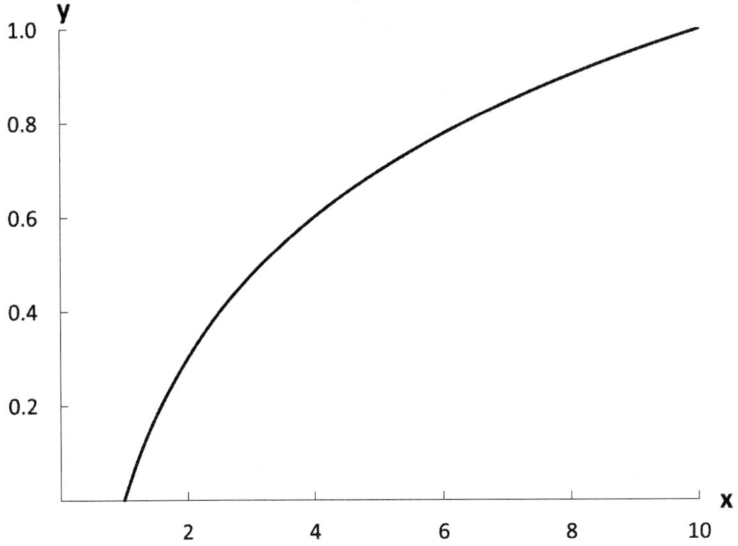

Figure 3.12: Graph of $y = \log(x)$

note that the log is not defined for numbers less than or equal to zero.[23] Further, $\ln(1) = 0$ (i.e., $\ln(x) = 0$ when $x = 1$), and $\ln(x) < 0$ when $0 < x < 1$.

Second, the log of a product is equal to the sum of the logs of each term, and the log of a ratio (or fraction) is the difference of the logs of each term:

$$\ln(x_1 \cdot x_2) = \ln(x_1) + \ln(x_2), \text{ for } x_1, x_2 > 0$$

and

$$\ln \frac{x_1}{x_2} = \ln(x_1) - \ln(x_2), \text{ for } x_1, x_2 > 0.$$

Note that addition and subtraction of logs *do not* distribute:

$$\ln(x_1 + x_2) \neq \ln(x_1) + \ln(x_2) \text{ , for } x_1, x_2 > 0,$$

and

$$\ln(x_1 - x_2) \neq \ln(x_1) - \ln(x_2), \text{ for } x_1, x_2 > 0.$$

These equations cannot be simplified further. Thus, if one takes the log of both sides of the equation $y = \alpha + \beta_1 x_1 + \beta_2 x_2$, the solution is *not* $\log y = \log \alpha + \log \beta_1 + \log x_1 + \log \beta_2 + \log x_2$ but $\log y = \log(\alpha + \beta_1 x_1 + \beta_2 x_2)$.

[23]This follows from the identity $a^{\log_a x} = x$. Assume $a > 0$, and that $x \leq 0$. Let $\log_a x = b$. Then we have $a^b \leq 0$ for $a > 0$, which is impossible, implying that b is undefined. Thus the log is defined only for $x > 0$. Other properties can also be derived from this identity and the rules on exponents we stated earlier.

Third, the log of a variable raised to a power is equal to the product of the exponent value and the log of the variable:

$$\ln(x^b) = b\ln(x), \text{ for } x > 0.$$

Finally, as $x > 0$ approaches 0 (so x is small), the log of $1+x$ is approximately equal to x:[24]

$$\ln(1+x) \approx x, \text{ for } x > 0 \text{ and } x \approx 0.$$

Political scientists generally use log functions to represent conjectures that anticipate a declining impact of some x over some y as x increases in value. For example, Powell (1981) studies the impact of electoral party systems on mass violence (as well as other forms of system performance). In the study, he controls for both the population size and per capita gross national product (GNP). The basic ideas are that (1) countries with larger populations will produce more riots and deaths from civil strife and (2) those with greater economic output per person will produce fewer riots and deaths from political violence. But Powell (and most social scientists) do not expect these relationships to be linear: an increase in population from 1,000,000 people to 2,000,000 people will have a greater impact on riots and deaths than will an increase in population from 100,000,000 to 101,000,000. Similarly, an increase from \$500 to \$1,500 GNP per capita is expected to have a greater impact on the number of deaths and riots a country will typically experience than an increase from \$18,000 to \$19,000 GNP per capita. That is, Powell hypothesizes that the positive and negative effects of population and economic output, respectively, on civil strife will decline as the value of population and economic output rises.[25] We can thus write Powell's expectations as: $CS = X + \ln(P) - \ln(G)$, where CS represents civil strife, X represents the party system variables that Powell considers, and P and G represent the control variables population and per capita GNP, respectively.[26] While a log function is only one of many one could use to convert those verbal claims to a specific mathematical statement, it is a common function that has often performed well in statistical tests.

Wallerstein (1989) provides another example. He explores the determinants of cross-national difference in labor unionization rates. One of the variables Wallerstein expects to have an effect is the size of the potential union membership (i.e., labor force). If we let U indicate unionization rate, L the size of the labor force, and X the other variables that he considers, we can represent his expectation as $U = \ln(L) + X$.[27] Why expect a nonlinear log relationship? Wallerstein explains that "using the log of the potential membership implies that the percentage increase, rather than the absolute increase, matters for

[24] This follows from a Taylor expansion of the log. We discuss this in Part II of this book.

[25] In other words, the marginal effect of these variables is decreasing. We discuss marginal effects at length in Part II of this book.

[26] Readers familiar with regression analysis in statistics might expect a representation like this: $CS = \alpha + \beta_1 X + \beta_2 \ln(P) - \beta_3 \ln(G) + \epsilon$.

[27] Using a regression representation, the argument is $U = \alpha + \beta_1 \ln(L) + \beta_2 X + \epsilon$.

union density" (p. 490). This argument stems from the equation for the difference in logs. If $\Delta \ln(L) = \ln(L_t) - \ln(L_{t-1})$ is the change in the natural log of the labor force variable, L, then $\Delta \ln(L) = \frac{\ln(L_t)}{\ln(L_{t-1})}$. This is a ratio rather than a difference in different values of the labor force.

The two most common usages of the log function are (1) to model the nonlinear expectation that the size of the effect of one variable on another declines as the second variable rises in value and (2) to model the expectation that the relative increase of a variable over time has a linear impact on another variable.

3.2.3.5 Radicals (or Roots)

Roots (sometimes called **radicals**) are those numbers represented by the radical symbol: $\sqrt[n]{\ }$. They are (almost) the inverse functions of x raised to the power n: $\sqrt[n]{x^n} = x = (\sqrt[n]{x})^n$ as long as n is odd or $x \geq= 0$.[28] Functions with radicals are nonlinear: $y = \sqrt[n]{x}$. Some roots are integers: $\sqrt[2]{9} = 3$. However, most are not: $\sqrt[2]{3} \approx 1.732050808$. Figure 3.13 graphs the function for $n = 3$ over the range $x = [1, 4]$.

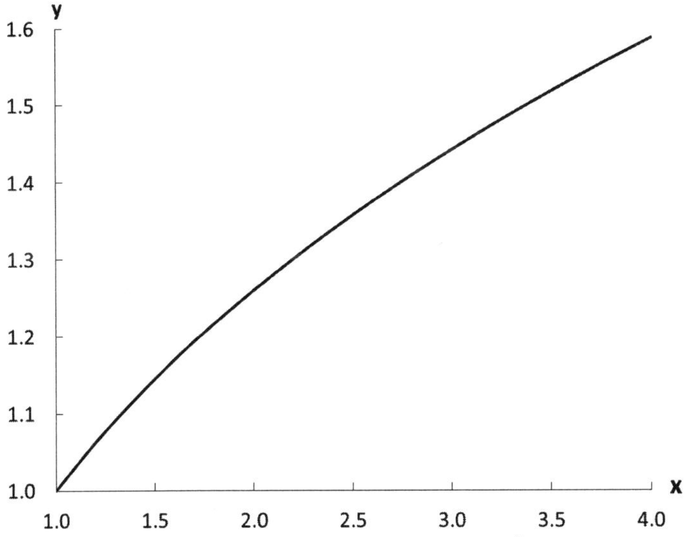

Figure 3.13: Graph of $y = x^{\frac{1}{3}}$

As noted above, radicals can also be expressed as fractional exponents: $\sqrt{x} = x^{\frac{1}{2}}$. We can express this more generally by observing that $\sqrt[n]{x^p} = (\sqrt[n]{x})^p = x^{\frac{p}{n}}$. When $n = 2$, we typically do not write the 2 in $\sqrt{x^p}$.

[28]Even roots (i.e., n is even) are undefined for negative values of x in the real numbers. They are defined in the complex number system using the definition of the imaginary number i, where $i = \sqrt[2]{-1}$.

Although roots do not play a large role in political science, one encounters them from time to time. For example, Gelman, Katz, and Bafumi (2004) explore a common assumption in the literature on the fairness (with respect to representation) of weighted voting systems such as the US Senate, where people living in states with smaller populations (e.g., Maine) have a greater influence on the votes cast in the Senate than people living in states with larger populations (e.g., Illinois). The conventional assumption is that all votes are equally likely (i.e., that voting is random), and a common indicator used to measure the "voting power" of an individual citizen is the Banzhaf index: $\frac{1}{\sqrt{N}}$. Gelman, et al. argue that this index "(and, more generally, the square-root rule) overestimates the probability of close elections in large jurisdictions" (p. 657). As an alternative indicator they recommend the fraction $\frac{1}{N}$.

To do algebra with roots one needs to memorize the following rules.

Addition and Subtraction

One cannot in general add or subtract two radicals. So:

$$\sqrt[n]{x} + \sqrt[n]{x} \neq \sqrt[n]{x+x} \text{ for } n > 1.$$

For example, $\sqrt{2} + \sqrt{2} = 2\sqrt{2} > 2 = \sqrt{4} = \sqrt{2+2}$.

Note that one cannot sum the roots, either:

$$\sqrt[n]{x} + \sqrt[n]{x} \neq \sqrt[n+n]{x+x} \text{ for } n > 1.$$

Observe that $\sqrt{9} + \sqrt{9} = 3 + 3 = 6 \neq \sqrt[4]{18}$ because $6^4 \neq 18$.

This is also so when the variables and roots are different, e.g.,

$$\sqrt[a]{x} + \sqrt[b]{y} \neq \sqrt[a+b]{x+y} \text{ for } a, b > 1.$$

To see this, note that $\sqrt[2]{9} + \sqrt[3]{8} = 3 + 2 = 5 \neq \sqrt[5]{17}$ because $5^5 \neq 17$.

The only exception is when one side would be zero, either because at least one of x or y is zero or because we are using subtraction and $x = y$.

Multiplication and Division

One can determine the product of two radicals *only* when they have the same order. In such a case, multiply the two variables (radicands) and collect the product under the root:

$$\sqrt[n]{x} \times \sqrt[n]{z} = \sqrt[n]{xz} \text{ for } n > 1.$$

But, e.g.,

$$\sqrt[a]{x} \times \sqrt[b]{z} \neq \sqrt[ab]{xz} \text{ for } a \neq b, \ a, b > 1.$$

To see that this is so, observe that $\sqrt{25} \times \sqrt{9} = 5 \times 3 = 15 = \sqrt{225} = \sqrt{25 \times 9}$ because $15^2 = 225$. However, $\sqrt{25} \times \sqrt[3]{8} = 5 \times 2 = 10 \neq \sqrt[6]{200}$ because $10^6 \neq 200$.

Finding the quotient of two radicals is similar; one can simplify the quotient of two radicals only when their order is the same:

$$\frac{\sqrt[n]{x}}{\sqrt[n]{z}} = \sqrt[n]{\frac{x}{z}} \text{ for } n > 1.$$

But, e.g.,

$$\frac{\sqrt[a]{x}}{\sqrt[b]{z}} \neq \sqrt[a+b]{\frac{x}{z}} \text{ for } a \neq b,\ a, b > 1.$$

3.2.3.6 *Other Functions*

Of course, this small array of functions is not the entirety of those used in political science. One commonly used is the *absolute value*, which we denote by $|x|$. In a single dimension it just means "remove the sign on the value." More formally, it can be represented as $|x| = \sqrt{x^2}$ in one dimension, where we take only the positive root, or $|x| = \sqrt{x_1^2 + x_2^2 + \ldots + x_n^2}$ in n dimensions, with $x = (x_1, x_2, \ldots, x_n)$. The absolute value is often used when one wants to keep the function positive (or negative with $-|x|$) over the entire range of x, or when one is interested in the distance between two points, which is $|a - b|$. Less commonly observed functions are rational functions (the ratio of two polynomials) and trigonometric functions (e.g., sine, cosine, tangent).[29]

Thus far all the functions we've defined have been the same over the entire domain. In other words, $f(x) = x^2$ doesn't change with the value of x. But we can also define functions *piecewise*, by which we mean simply "in pieces." These are useful, for example, when we expect an external intervention to alter the behavior of the relevant actors in a theory. There is nothing fancy to representing this sort of thing; we just write something like $f(x) = -(x-2)^2$ if $x \leq 2$ and $f(x) = \ln(x-2)$ if $x > 2$. This function states that below one's ideal point of 2, the function slopes downward at a faster rate than it slopes upward above one's ideal point.[30] One need only be careful to define the function across the entire domain, without missing some region. Piecewise functions are often expressed in the format

$$f(x) = \begin{cases} -(x-2)^2 & : x \leq 2, \\ \ln(x-2) & : x > 2. \end{cases}$$

3.2.3.7 *Why Should I Care?*

One encounters nonlinear functions both in formal theory and in statistics, as the examples sprinkled throughout this subsection demonstrate. We have already discussed the theoretical value of nonlinear functions: they provide us with a language to make very explicit statements about expected causal relations. And it turns out that exponential and log functions are useful for modeling and for transforming variables with highly skewed distributions, and that has both theoretical and statistical value, though this won't be very clear until we discuss distributions. Along these lines, the most used probability distribution, the normal distribution, is an exponential function composed with a quadratic.

[29] The trigonometric functions are rarely used in political science, but they can be important in situations in which they are used. Consequently, we include them in several of this book's discussions for reference but do not advise the first-time reader to worry about them.

[30] If this is not clear, draw the function. This is generally useful advice.

3.2.4 Why Should I Care?

Linear and nonlinear functions are nothing more than specific claims about the relationships among several variables, and thus can be very useful for making specific causal claims. Multivariate functions are especially useful as few political scientists suspect that much of politics can be usefully explained with bivariate hypotheses (i.e., conjectures that say only one concept is responsible for variation in another concept).

Having said that, let us briefly explain how one can move from verbal conjectures to writing down more specific functions. We will work with the probability of voting in a national election as an example. Suppose one suspects that the probability that a registered voter will cast a ballot will increase in response to (1) an individual's education level, (2) partisan identification, (3) income, (4) age, and (5) the closeness of the race.

Conceptualize education as a discrete count of the number of years of formal education, partisan identification as a distinction between those who identify with one of the two major parties and those who do not (we will assume this is a US election), income as continuous, age as a discrete count of years, and the closeness of the race as the gap between the Democratic and Republican candidates. We can represent the conjecture that the probability of voting is a function of each of these variables with the implicit function $p_v = f(ed, p, i, a, c)$, where p_v represents the probability of voting, ed represents education, p represents partisan identification, i represents income, a represents age, and c represents closeness. This equation is called the implicit functional form because it is not specific: we do not know whether the variables have positive, negative, linear, monotonic, or nonlinear effects on p_v. All the implicit function tells us is that they may have *some* effect. Hypotheses based on implicit functions are always more difficult to falsify than explicit ones that spell out the specific functional forms.

We might conjecture more strongly that the probability of casting a ballot is an affine function of each of these variables. That conjecture can be captured by the following function: $p_v = \alpha + \beta_1 ed + \beta_2 p + \beta_3 i + \beta_4 a + \beta_5 c$, where α is the intercept (i.e., the expected value of p_v when all of the explanatory variables have a value of 0) and the β_i parameters represent the strength of the impact that each explanatory variable has on voting probability.

Alternatively, we might conjecture that voting probability has a linear relation with some variables and a nonlinear relation with others. For example, one might argue that the impact of education is greatest at low levels (i.e., the difference between a fourth-grade and an eighth-grade education has a larger impact on voting probability than the difference between an eighth-grade and a twelfth-grade education, and the difference between completing a high school degree and completing a college degree has an even smaller impact). In addition, one might contend that greater levels of income have an even greater effect on voting probability. The following equation represents those conjectures: $p_v = \alpha + \beta_1 \ln(ed) + \beta_2 p + \beta_3 i^2 + \beta_4 a + \beta_5 c$.

The arguments presented in the preceding paragraphs are similar but distinct. By writing out an explicit functional form to represent the verbal arguments, one makes it very clear how the arguments are distinct (and how they are similar). Drawing graphs can often help one decide whether a given expected relationship comports with one's assumptions, intuition, or verbal argument. A functional representation of conjectures in equation form also makes very clear how one can be wrong—if statistical analysis shows that the parameters do not have the expected signs, for example—and that is another virtue of writing out the functional representation of a verbal argument.

Finally, one might conjecture an interactive relationship among some of the independent variables and probability of voting. There are many such possibilities, but, for example, one could believe that the higher one's level of education, the more the closeness of the race matters, as one will pay more attention to the media's reporting on the race. If this were true, we might expect a relationship like $p_v = \alpha + \beta_1 ed + \beta_2 p + \beta_3 i + \beta_4 a + \beta_5 c + \beta_6 (ed \cdot c)$. Or one might suspect that the relationship between each explanatory variable and the explained variable grows with the value of the explanatory variable, and that the strength of each relationship is conditional on the values of the other explanatory variables.[31] One can represent such an argument as follows $p_v = \alpha \cdot ed^{\beta_1} \cdot p^{\beta_2} \cdot i^{\beta_3} \cdot a^{\beta_4} \cdot c^{\beta_5}$. We can take advantage of the relationship between exponents and logs to rewrite that as $\ln p_v = \ln(\alpha) + \beta_1 \ln(ed) + \beta_2 (\ln p) + \beta_3 \ln(i) + \beta_4 \ln(a) + \beta_5 \ln(c)$.[32] Such a transformation is useful because while we cannot use common statistical routines to estimate the β parameters in the first representation, we can do so in the second representation. And while arguments that produce such a functional form are not usually observed in political science, they are in economics. It might be the case that few, if any, political processes are composed of concepts with such nonlinear, interactive relationships, but it might also be the case that few political scientists have explored those possibilities.

One can draw another illustration of the use of functions from game theory. In Chapter 1 we made reference to sets composing an actor's set of action. One can also create a set of strategic responses to all possible actions and all possible states of the world: a strategy set or strategy space. A strategy is a complete plan for playing a game (i.e., the choice an actor would make at each decision point the actor faces). So a strategy for player 2 might look like this: "if player 1 does x, then player 2 chooses a; if player 1 does y, then player 2 chooses b; etc." Strategies are functions (or correspondences): they map the relationship between the choices of the other players and the choices one makes at each opportunity. Strategies are sometimes represented as pairs of ordered sets rather than graphs or equations, but they are functions (or correspondences) nonetheless. Individual preferences in game theory also often take the form of

[31] We do not have a story to explain why such a conjecture is reasonable—it likely is not a reasonable conjecture. We offer it merely for illustrative purposes.

[32] If you found that too quick, observe that the first task is to take the log of all variables on both sides of the equation. The second step is to recall that $\ln(ed^b) = b \ln ed$.

functions, called utility functions. We provide an extended discussion of this example in the next section.

To reiterate the key point, one should develop a working familiarity with functional forms because they help one clarify the conjectures one is making. More specific causal claims are stronger because they are easier to falsify. Debates among scholars are also sharpened when there is greater clarity about the claims being advanced by the various factions. In short, good science becomes easier as clarity improves, and functions are a very basic and useful tool for adding clarity to one's conjectures.

3.3 PREFERENCE RELATIONS AND UTILITY FUNCTIONS

Game theory is a tool for understanding strategic interactions between political actors and developing theories about political behavior and the effect of institutions. The preferences of individual actors are foundational to game theory, as one cannot understand how one responds to incentives and others' actions without understanding what one actually wants. Typically, individual preferences are represented by functions, and the properties of these functions mirror the structure of one's preferences in the same manner that the form of the function described at the end of the last section matches one's theoretical expectations about the probability that one votes. In this section, we go into some detail as to why this is so, and how it all works, as an extended example of the usefulness of functions in political science. Before getting to functions, though, we need to return to relations.

3.3.1 Preference Relations

People frequently use a capital R to represent a relation, as follows: aRb, which is read "a is related to b." When applied to preferences, aRb is read "a is at least as good as b." If a and b were real numbers, this would translate to $a \geq b$; we return to this comparison below. There are also other analogues: aPb is "a is strictly preferred to b,"[33] or $a > b$ if both are numbers, and aIb is "I am indifferent between a and b,"[34] or $a = b$ if both are numbers. The study of these preference relations underlies decision theory, which, along with social choice theory, the study of group decision making, is often taught in parallel with or as a precursor to game theory. Many results from social choice theory are quite well known in political science. Black's median voter theorem, Arrow's (1950) impossibility theorem, and McKelvey's (1976) chaos theorem are notable examples you will likely be exposed to in other classes.

Our interest here is not in social choice heory, however, but rather in how to represent preferences with functions. To that end, we skip to a few important

[33] Formally, aPb if aRb but not bRa.

[34] Formally, aIb if aRb and also bRa.

propertiest that we often like preferences to have.[35] These are *completeness*, *transitivity*, *symmetry*, and *reflexivity*.

Completeness states that for any a and b, either aRb or bRa. In other words, all elements can be ordered pairwise, and there is no pair of elements for which one has no opinion. This is weaker than it may sound, as "no opinion" is distinct from indifference, which is allowed. What completeness disallows is the ability of someone to be unsure if she prefers a to b, b to a, or is indifferent between the two. For example, imagine a situation where a bureaucrat could (1) implement a new regulation (m), (2) implement the new regulation half-heartedly (h), or (3) ignore the new regulation (g). If the set is complete with respect to R, then one can have the preferences mRh, hRm, or both (mIh), but not neither. In other words, one can't say mPh and hPm simultaneously depending on mood, which is a formal way of denoting a lack of fixed opinion. Both the integers and the real numbers are complete relative to the normal ordering you are familiar with, given by the relations $>, \geq, =, \leq, <$. One never can be unsure whether $3 < 5$, for instance.

Transitivity states that if a is at least as good as b, and b is at least as good as c, then a is at least as good as c: if aRb and bRc, then aRc. The $>$, $\geq$, $<$, $\leq$, and $=$ relations are all transitive relations (e.g., if $a < b$ and $b < c$, then $a < c$) when applied to the integers or real numbers. To consider a political example, return to the set of implementation options for the bureaucrat: $\{m, h, g\}$. If she prefers ignoring the new regulation to implementing it half-heartedly, and also prefers implementing it half-heartedly to implementing it, then for her preferences to exhibit a transitive relation she would need to prefer ignoring it to implementing it.

Symmetry states that if aRb, then bRa for all a and b. In the realm of preference orderings, this implies complete indifference: everything is at least as good as everything else. The equality relation, $=$, is the only symmetric relation of $>, \geq, =, \leq$, and $<$ in the integers or real numbers: if $a = b$, then $b = a$.[36] Symmetric preference orderings are less common in the study of politics, though they do allow for a quite precise definition of the concept of "apathy," which otherwise might admit multiple interpretations. For instance, if our bureaucrat were indifferent between all three implementation options, then she would hold symmetric preferences. In this scenario, she would not only not care which option were chosen, but she would also be unlikely to put forth effort to affect the choice, assuming effort were at all costly.

Some people find **reflexivity** a bit of a brain bender. A relation on a set A is reflexive if for all $a \in A$, aRa is true. To illustrate, let's try the relations $>$, $\geq$, and $=$, and determine whether each is reflexive on the integers or real

[35] We state these in terms of preference relations, not more generally, as that is the only context in which we will have occasion to use them in this book. Note that these are normatively desirable properties, not properties observed to be true empirically. In fact, people violate these on a regular basis!

[36] While $a \geq b$ and $b \geq a$ might both be true (if a and b are equal), $a \geq b$ does not imply $b \geq a$ for all a and b.

numbers. To check the relation $>$, we replace the R in aRa with $>$ and see if it is true: $a \not> a$, so "greater than" is not a reflexive relation. However, $a \geq a$ and $a = a$ are both true, so "greater than or equal to" and "equal to" are reflexive relations. Now let's try a political science example. For our bureaucrat's preferences to exhibit reflexive order, each preference must be at least as good as itself: ignoring the new regulation must be at least as good as ignoring the new regulation, implementing the new regulation must be at least as good as implementing the new regulation, etc. We suspect you will agree that it would be odd indeed if someone's preferences were not reflexive.

3.3.2 Utility Functions

Complete and transitive individual preference is a fundamental assumption of rational choice theory and standard game theory, and is commonly assumed in the formal literature. It is true that people routinely violate this assumption in their everyday lives. However, the assumption buys us something very important—the ability to represent preferences with functions that take on real and integer values. To see why, let's return to the previous definitions. Integers and the real numbers are complete and transitive for all the usual ordering relations. Thus, if we want to represent our "at least as good as" relation with numbers, this relation had better have the same properties. With this assumption on individual (not group!) preference, we can translate the relation R on any set A to a function u on the same set. This u is called a *utility function* and assigns a value, typically a real number, to each element in A. So, for example, for a bureaucrat whose preferences are ordered $mRhRg$, we could assign $u(m) = 3, u(h) = 2$, and $u(g) = 1$.

This technique begins to pay dividends when the set of things one has a preference over is large, or infinite. For instance, while one could laboriously elaborate on preferences over dollar values of money ($100R99R98R\ldots$), it's far easier to define a utility function, $u(x) = x$, that represents those preferences. Varying the utility function alters what preferences are represented, in the same manner that varying the empirical model represents different theoretical ideas. A linear utility like $u(x) = ax$ for budgetary outlays, for example, would mean that each additional dollar is just as valued as the previous one. A quadratic utility like $u(x) = ax^2$, in contrast, would mean that each additional dollar is valued more than the one before it, an unlikely assumption in many cases (though see Niskanen, 1975, p. 619). In fact, for money, researchers typically assume that $u(x) = \ln(x)$, so that there are decreasing returns to increasing a bureaucratic budget.

This makes sense in the context of a single agency's preferences, but what about a Congress trying to distribute money over multiple agencies? Each congressperson might have some ideal budget number for each agency, with increases *and* decreases from that number being viewed negatively. In that case, we can use what is known as a quadratic loss function, $u(x) = -(x - z)^2$. If you graph this function, you will see it is a parabola that peaks at $x = z$, which is the point of highest utility, also known as an ideal point. This form of utility function is very common when modeling voting behavior (e.g., McCubbins, Noll, and Weingast, 1987).

3.3.3 Best Response Correspondences

Let's return to the example of the bureaucrat, but now assume there are two decision makers. One, Christine, has preferences $mPhPg$. She prefers to do it right, but also wants it done. The other, Bob, is lazy and has preferences $gPhIm$. He'd rather do nothing, but if it has to be done, he doesn't care which way it happens. Let's also assume that, for some unknown reason, the decision is made by asking Christine and Bob to write their choices on a piece of paper. If both agree, then that option wins. If only one writes m, then h happens. In this (odd) scenario, Christine will always write m. This is a dominant strategy for her, because it can secure her second-best option and possibly achieve her first-best option. Bob, on the other hand, is in a pickle. He can't get his best option given Christine's optimal action, and he is indifferent between h and m. Thus anything he does has the same outcome to him. His best response to Christine is any of the three options.

We can represent Christine's best response as a function. Let $S = \{m, h, g\}$, which is known as a player's strategy space. Then we can write the function $B_C(\cdot) = m$ for Christine, which means that her best option is to choose m regardless of what Bob does. To elaborate, $B_C(m) = B_C(h) = B_C(g) = m$. B_C here is called Christine's **best response function**. It takes as input Bob's strategies and returns the optimal action for Christine to take. It is a function because Christine has only one best response to each of Bob's actions.

Now consider Bob's best response to Christine's play of m. We can't represent this best response as a function, as it would have to return three values—m, h, or g—when presented with Christine's m. Instead, we can write Bob's **best response correspondence**. Formally, Bob's decisions are governed by the correspondence $B_B(m) = \{m, h, g\}$. In words, this means that Bob responds optimally to Christine's choice of m by choosing any of his options. We write such correspondences as $B_B(s_C) : S_C \to\to S_B$ where we have added subscripts for each player's name, and S_i and s_i are the strategy space and strategy choice for player i. Though we will not deal with correspondences much in this book, they will come up in your game theory classes.

3.3.4 Why Should I Care?

Whether or not they are your cup of tea, formal theories of political science are prevalent in the field and often referenced in empirical work to justify hypotheses. Being able to read them and understand their underlying assumptions are important skills. Further, formalizing theories can help sharpen your thinking. Finally, in the same manner that different utility functions represent different preferences, one can choose different underlying properties on preferences if one does not like, for instance, rational choice assumptions.

3.4 EXERCISES

1. For each pair of ordered sets, state whether it represents a function or a correspondence:

   a) {5, -2, 7}, {0, 9, -8}

   b) {3, 1, 2, 6, -10}, {5, 7, 1, 4, 9}

   c) {3, 7, -4, 12, 7}, {8, -12, 15, -2, 17}

2. Simplify $h(x) = g(f(x))$, where $f(x) = x^2 + 2$ and $g(x) = \sqrt{x - 4}$.

3. Simplify $h(x) = f(g(x))$ with the same f and g. Is it the same as your previous answer?

4. Find the inverse function of $f(x) = 5x - 2$.

5. Simplify $x^{-2} \times x^3$.

6. Simplify $(b \cdot b \cdot b) \times c^{-3}$.

7. Simplify $((qr)^\gamma)^\delta$.

8. Simplify $\sqrt{x} \times \sqrt[5]{x}$.

9. Simplify into one term $ln(3x) - 2ln(x + 2)$.

10. Visit the "Graphing Linear Functions" page at the Analyze Math website `http://www.analyzemath.com/Graphing/GraphingLinearFunction.html`. Read the examples and solve the two "matched problems."

11. Visit the Analyze Math website's "Slope Intercept Form of a Line" page at `http://www.analyzemath.com/Slope_Intercept_Line/Slope_Intercept_Line.html`. Print a copy of the page and then click on the Click to Start Here button to start the tutorial applet. Do numbers 2 through 8. What does this tutorial show?

12. Visit the "Quadratic Function(General Form)" page at Analyze Math: `http://www.analyzemath.com/quadraticg/quadraticg.htm`. Click on the Click Here to Start button and adjust parameters a, b, and c. What happens to the graph as you increase or decrease a? Note the changes when you increase b and c as well. Is there a value to which you can set one of the parameters to make the quadratic function a linear function?

13. Visit the "Graphs of Basic Functions" page at the Analyze Math site (`http://www.analyzemath.com/Graph-Basic-Functions/Graph-Basic-Functions.html`). Click on the Click Here to Start button and plot the graph of each function. After plotting each once, click the Y-Zoom Out button several times and plot each of the graphs again.

14. Visit "Polynomial Functional Graphs" at `http://id.mind.net/~zona/mmts/functionInstitute/polynomialFunctions/graphs/polynomialFunctionGraphs.html`. Plot polynomial functions of different orders, then adjust the parameters and observe how the graph changes in response to different values (use the Simple Data Grapher from the Math link on the main page). Write down a verbal conjecture about politics that you think might be captured by a specific polynomial function. Be sure to explain your thinking. Write down the function and print a copy of its graph.

15. Rewrite the following by taking the log of both sides. Is he result a linear (affine) function?
$y = \alpha + x_1^{\beta_1} + \beta_2 x_2 + \beta_3 x_3$.

16. Rewrite the following by taking the log of both sides. Is the result a linear (affine) function?
$y = \alpha \times x_1^{\beta_1} \times x_2^{\beta_2} \times x_3^{\beta_3}$.

17. Rewrite the following by taking the log of both sides. Is the result a linear (affine) function?
$y = \alpha \times x_1^{\beta_1} \times \frac{x_2^{\beta_2}}{x_3^{\beta_3}}$.

18. Is this problem done correctly? Yes or no.
Take the log of both sides of the following equation:
$y = x_1^{\beta} - x_2^{n} + x_3^{2}$.

19. Visit "The Universe Within" page on the website of Florida State University's magnet lab: `http://micro.magnet.fsu.edu/primer/java/scienceopticsu/powersof10/`. It is a visual display of the concept of scale—viewing the same object from different scales of measurement—as it begins with a view from 10^{+23} meters away and moves to 10^{-16} meters away. Besides being a cool visual, this page offers a graphic illustration of exponentiation.[37] Note especially what happens when the exponent shifts from positive to negative values.[38] If that does not make sense to you, review the discussion of exponents, specifically the arithmetic rules.

---

[37] We have purposely used political rather than physical examples throughout, but could not resist this one.

[38] Once it has run you may want to click on the Increase button to go back through it as it moves fairly quickly.

20. The graduate studies committee has asked the graduate students to provide the faculty with a list of three nominees to represent the students on the committee. After much discussion, three nominees are put forward, and you are asked to rank them, with the rank representing preference (i.e., 1 is most preferred, 2 is second best, and 3 is the third-best choice). The nominees are Beta, a seventh-year student who recently defended his prospectus; Gamma, a second-year student who is very bright but tends to dominate seminar discussion; and Alpha, a fourth-year year student who is preparing for her exams and is widely viewed as level-headed and realistic. Provide your pairwise preference rankings of each candidate. Check to see whether your rankings are transitive. If you have been assigned this problem for class, bring your ordering to class so that the class can determine whether it is transitive at the aggregate level under pairwise majority rule.

21. Recall the first question of Section 8.2 in Chapter 1. There we asked you to pick ideal spending points for three parties, as well as a status quo and a bill, and conjecture about whether or not it would pass. Now we want you to go further. Write a utility function for each party that is largest at that party's ideal point. How does that function decrease with distance from the ideal point? Try to draw a curve around each ideal point that gives the same utility to the corresponding party for every point on the curve. These are called *indifferences curves*, as the party is indifferent between all points on the curve. Draw these for all parties and see whether you can answer your earlier conjecture.

22. Propose and justify a quadratic utility function as representing the preferences of some political actor over something.

Chapter Four

Limits and Continuity, Sequences and Series, and More on Sets

We have now covered most of the building blocks we need to introduce the more complex topics in the remainder of the book, but there remain a few more important ideas to address. Specifically, we must describe properties of functions and sets related to *limit behavior*, the behavior of these mathematical constructs as we approach some point in or out of their domains. Those readers with a stronger math background will likely have seen many of these ideas before, though they may have not seen them presented in this fashion, and so could benefit from skimming this chapter. We expect many readers will not have spent a great deal of time on these topics prior to this course, and thus will benefit from a more thorough reading.

The first section discusses sequences and series, which are necessary background for the rest, along with being useful in their own right in political science. The second introduces the notion of a limit. The third uses both limits and sequences to explore some additional properties of sets necessary for understanding calculus. Finally, the fourth treats continuity, a fundamental property of functions that affects our ability to maximize them, and thus find optimal choices.

4.1 SEQUENCES AND SERIES

4.1.1 Sequences

A **sequence** is an ordered list of numbers. For example, $\{1, 2, 3, 4 \ldots\}$ is a sequence. So is $\{\frac{3}{10}, \frac{3}{100}, \frac{3}{1{,}000}, \frac{3}{10{,}000}, \ldots\}$. Sequences can be infinite (i.e., continue forever), as depicted above, or they can be finite: $\{2, 1, 3, 1, 4, 1, 5, 1\}$. All sequences are countable, in that one can assign a natural number to each element in the sequence. For this reason we typically identify an element in the sequence with the notation x_i, where i is an index corresponding to its place in the sequence. Sequences usually start at index 0 or 1. To avoid confusion on this point, we can write the whole sequence in shorthand as $\{x_i\}_{i=1}^{N}$. If N is finite, it is a finite sequence; if $N = \infty$, it is an infinite one.

Sequences are rarely just random numbers placed in a row, and usually are generated from some underlying process, particularly when the sequences are infinite. The elements in the first example above can be represented by $x_i = i$, and the elements in the second example by $x_i = \frac{3}{10^i}$. We can write these sequences as $\{i\}_{i=1}^{\infty}$ and $\{\frac{3}{10^i}\}_{i=1}^{\infty}$.

Subsequences are just parts of larger sequences. The sequence $\{2, 4, 6, \ldots\}$ is a **subsequence** of $\{1, 2, 3, 4 \ldots\}$, which one can represent as $x_i = 2i$.

4.1.2 Series

A **series** is the sum of a sequence. Put differently, a series is a sequence with addition operators between each of the elements (e.g., $\{1 + 2 + 3 + 4 + \ldots\}$ or $\{\frac{3}{10} + \frac{3}{100} + \frac{3}{1{,}000} + \frac{3}{10{,}000} + \ldots\}$. We can write this as $\sum_{i=1}^{N} x_i$, where $N = \infty$ for our two examples.

Finite series are easy to grasp because calculating the sum of a finite sequence is straightforward (we add the terms). One can also consider infinite series, and many times these are easy (e.g., the infinite sequence $\{1, 2, 3, 4, \ldots \infty\}$ sums to infinity). In other cases, though, they are more complex. We do not have a great need to sum many different series in political science, but there are some of importance, so it's worth discussing briefly how one can sum a series.

A common series that you will see time and again in game theory is $\sum_{t=0}^{\infty} \delta^t$, where $\delta < 1$. The story behind this series goes something like this. State A and state B are interacting in a repeated game (a game in which the actors have repeated chances to interact). In each period of the game, both states decide whether or not to cooperate (on defense policy, free-trade agreements, and the like) or to defect (by making private deals, raising trade barriers, and so on). After the states have decided simultaneously what to do, they receive some measure of utility, which we'll call their payoffs from the period. Since the game is repeated, this happens again and again, indefinitely. By definition, strategies must cover decisions by the states across all time. Choosing the optimal strategy requires that we know what the total payoff is across time, which means adding up all the individual periods' payoffs.

Before we can do so, we must ask ourselves how we should treat the payoffs for future periods. If they are as valuable to us as present payoffs are, then we'll just be adding something of roughly the same magnitude an infinite number of times, which would give an infinite total payoff. Is that reasonable? Game theorists generally answer no—future payoffs are worth less than present ones. Why? Well, the game might end at any time if, say, one state ceases to be or undergoes a radical shift in governance. If there is a chance that this might happen in any period, then we have to discount the payoffs for future periods that might not occur. Even if this is not the case, future payoffs may still be worth less. If I have \$5 now, I can put it in a bank (in theory) and receive some rate of return, getting back \$5(1+r) later. The \$5 I may get later is worth less than this \$5(1+r). Reversing things, compared to my \$5 now, the \$5 later is only worth $\$\frac{5}{1+r}$ or $\$5\delta$, where $\delta < 1$. This δ is known as a **discount rate** as it represents the rate at which the actors value future utility relative to present utility.

This can be repeated for every period onward, yielding the series $\sum_{t=0}^{\infty} \$5\delta^t$.

We can pull the \$5 out, so we just need to sum $\sum_{t=0}^{\infty} \delta^t$. It turns out this sum is $\frac{1}{1-\delta}$. To see this, note that $\sum_{t=0}^{\infty} \delta^t = \frac{1}{1-\delta}$ implies that $(\sum_{t=0}^{\infty} \delta^t)(1-\delta) = 1$ or $(1+\delta+\delta^2+\ldots)(1-\delta) = 1$. Multiplying out the LHS gives $1-\delta+\delta-\delta^2-\delta^2+\ldots$. All the infinite terms in δ cancel, yielding a sum of 1.

Sometimes scholars impose a finite number of rounds on an iterated (repeated) game instead, if they believe this best captures the structure of the political interactions they are studying. We can derive the sum of the finite sequence $\sum_{t=0}^{N} \delta^t$ from the infinite one by breaking up the infinite one into parts. First, note that $\sum_{t=0}^{\infty} \delta^t = \sum_{t=0}^{N} \delta^t + \sum_{t=N+1}^{\infty} \delta^t$. The second term on the RHS, since it goes on forever, is the same as $\delta^{N+1} \sum_{t=0}^{\infty} \delta^t$. Since $\sum_{t=0}^{\infty} \delta^t = \frac{1}{1-\delta}$, we can replace the infinite sums on both sides with $\frac{1}{1-\delta}$ to get $\frac{1}{1-\delta} = \sum_{t=0}^{N} \delta^t + \frac{\delta^{N+1}}{1-\delta}$. Isolating the finite sum yields our answer: $\sum_{t=0}^{N} \delta^t = \frac{1-\delta^{N+1}}{1-\delta}$.

4.1.3 Why Should I Care?

Some readers might be thinking, "This is pretty abstract. Did we really have to go through this?" In addition to providing some useful experience practicing manipulation of algebra and indices, this material will be useful for students in formal theory courses. You've already seen the utility of series in the extended example above; it is this math which underlies the importance of the "shadow of the future." However, sequences are also common even in games that are not repeated. Extensive form games, in which players alternate taking actions in some order, model sequential political interactions. These include bargaining scenarios and the legislative process of bill proposal, voting, and veto. In these cases the structure of the game has a fixed sequence. If you study game theory, you will be introduced to the terms *history*, *terminal history*, *subhistory*, and *proper subhistory*, and these are all defined in terms of mathematical sequences. Some histories are finite and others are infinite, but all histories express a sequence of events in the game.

For example, a game's history might consist of the House Speaker's choice of the open or closed rule, followed by the choice of a congressional committee to invest in specialized knowledge, the subsequent bill it proposes, and finally the floor of Congress's response to that proposal (Gilligan and Krehbiel, 1989). Or a game's history could consist of a series of alternating offers in an attempt to reach a bargain, with each player getting to make a new offer if she rejects the previous one. If an offer is accepted the game's history ends, but a rejection extends the history by an additional offer (and decision over it). This alternating-offer, infinite-horizon (because the game may never end, in theory) model of bargaining was proposed and solved by Rubinstein (1982) and has become an influential model of bargaining in political science (e.g., Baron and Ferejohn, 1989).[1] In Rubinstein's model, the player making the first offer seeks

[1] These games are also called "divide-the-dollar" games because they often focus on how to distribute one unit's worth of goods among the players.

to provide just enough of an incentive via the offer to make the other actor indifferent between taking the offer immediately and rejecting it in favor of making her own offer later. Taking the offer can be attractive because the discount factor is less than one; i.e., future periods' payoffs are not worth as much as the present period's, implying that waiting to accept an offer is costly for both players. In equilibrium, the first player's offer is just enough to take advantage of this cost, and the second player accepts the offer immediately, ending the game in the first period. Note that even though the game ends immediately, the potential future history of the game (i.e., the expected future sequence of actions) matters as it determines the payoff each player could expect to get should she reject the other player's offer.

Other games of note in political science have histories that are not truncated and so are truly infinite. The infinitely repeated prisoner's dilemma is an example of this. The prisoner's dilemma is an interaction in which mutual cooperation is beneficial to both players, but each player also has an incentive to deviate from this cooperation, defecting against his partner. When the game is played once—a one-shot game—mutual defection is the rule. The same is true for any finite history. But in an infinite history, the future gains to cooperative play and the lack of an end to the game prove sufficient to allow for mutual cooperation. Attention to future payoffs and their dependence on present actions is often called the "shadow of the future" in political science. Here sequences describe not only the equilibrium path of cooperation but also the punishment strategies each player uses to enforce cooperation. We provide an example of such a strategy below.

Game theory utilizes other sequences as well. For example, as you go on, you will encounter what is known as a *sequential equilibrium.* That concept is defined as the limit of a sequence.[2]

Last, one cannot study difference equation models without understanding sequences and series (and their limits), as the concept of a solution to such equations appeals to sequences and series (see Blalock, 1969, chap. 5, and Huckfeldt, Kohfeld, and Likens, 1982, chaps. 1–3).

4.2 LIMITS

A **limit** describes the behavior of a function, sequence, or series of numbers as it approaches a given value. That perhaps sounds rather strange and may

[2]Game theory has explicit mathematical foundations. In fact, even though game theory has most strongly influenced the field of economics, the scholars who created game theory were mathematicians. John Nash, who won a Nobel Prize in Economics for his foundational work in game theory, is a mathematician, not an economist, as were John von Neumann and Oskar Morgenstern, who are widely recognized as the founders of game theory. Though we refer only to a generic equilibrium in the text, there are actually several different (though closely related) equilibrium concepts in game theory, of which sequential equilibrium is one. However, you are more likely to encounter Nash equilibrium and subgame perfect equilibrium early in a first course in game theory.

produce questions about why one would want to know the behavior of a function, sequence, or series as it approaches a particular value. It turns out that knowing the limit can be quite useful. More specifically, the limit of a function can help us determine what the rate of change of the function is, and knowing the rate of change is useful in both statistics and formal models. Further, as noted above, the limit of a sequence or a series is important for the study of game theory and other formal models, and it is also appealed to in some areas of statistics.

4.2.1 Limits and Sequences

Limits are connected to sequences in a fundamental way. Consider our two example infinite sequences from the previous section, $\{i\}_{i=1}^{\infty}$ and $\{\frac{3}{10^i}\}_{i=1}^{\infty}$. We might want to know what is the "endpoint" of these sequences, despite knowing that the sequences continue without end. Does this idea have any meaning?

To answer this, we start with some definitions. A **limit of a sequence** $\{x_i\}$ is a number L such that $\lim_{i\to\infty} x_i = L$. The expression is read as "the limit of x_i as i approaches infinity is L." It means that as you traverse the sequence further and further (i.e., the index i gets bigger and bigger) the elements in the sequence get closer and closer to L. They may never equal L as long as i is finite, but they get arbitrarily close.[3]

Of course, this only makes sense if the elements *do* get closer to something. If they just oscillate between -1 and 1 forever, for instance, then there is no one value they approach. We say a sequence (or series, or function) **converges** if it has a finite limit, and **diverges** if it either has no limit or has a limit of $\pm\infty$.

Let's now return to our examples. The sequence $\{i\}_{i=1}^{\infty}$ gets larger and larger forever and approaches infinity, so it diverges. In contrast, the sequence $\{\frac{3}{10^i}\}_{i=1}^{\infty}$ gets smaller and smaller, approaching zero as $i \to \infty$. Thus, its limit is zero. Some useful limits to remember are $\lim_{i\to\infty} \delta^i = 0$ if $|\delta| < 1$,[4] and $\lim_{i\to\infty} \frac{1}{i^z} = 0$ if $z > 0$.

It is important to note that limits differ in general from the **extreme values** of a sequence. The extreme values of a set are the minimum and maximum values in that set. Some students mistakenly assume that the limit of a sequence is one or both of the extreme value(s) of that sequence. This may be true, but it may not. For example, in the sequence $\{1, 0, \frac{1}{2}, \frac{1}{2}, \ldots\}$ the limit is $\frac{1}{2}$, but the maximum is 1 and the minimum is 0.

4.2.1.1 Why Should I Care?

Though limits of series are more often used in formal theory than are limits of sequences, understanding the latter helps us to understand the strategic logic of repeated behavior. Because payoffs are discounted with a discount factor

[3]More formally, one can choose any tiny number $\epsilon > 0$ and find an N such that, for all $i > N$ (i.e., for all elements of the sequence further along than is N), $|L - x_i| < \epsilon$.

[4]To see this it might help to plug in a fraction for δ, say $\delta = \frac{1}{2}$. Then $\delta^2 = \frac{1}{4}$, $\delta^3 = \frac{1}{8}$, etc., with a limit of 0.

less than one, from the standpoint of the present time, payoffs accrued in the future get smaller and smaller the longer one must wait to receive them. In the limit, the size of the payoffs goes to zero, which is a necessary condition for an individual to want to trade off present gains for future losses. If the limit were *not* zero, total payoffs in the limit would effectively exceed payoffs in any finite time period, and no one would ever trade off present gains for permanent future losses. This would take away a key strategic insight of repeated behavior.

4.2.2 Limits and Series

The **limit of a series** is much like that of a sequence, except that one is looking for the sum of all elements in an infinite sequence rather than the "endpoint." If the series is $\sum_{i=1}^{N} x_i$, then the limit is $\lim_{N\to\infty}\sum_{i=1}^{N} x_i = S$. You've already seen the limits of two series. The limit $\lim_{N\to\infty}\sum_{i=1}^{N} i = \infty$, and so the series is divergent. The limit $\lim_{N\to\infty}\sum_{t=0}^{N} \delta^t = \frac{1}{1-\delta}$, so the series converges to $\frac{1}{1-\delta}$.

We can use this second example to connect sequences with series. Let $S_N = \sum_{t=0}^{N} \delta^t$ and $S_\infty = \lim_{N\to\infty}\sum_{t=0}^{N} \delta^t$. Construct the sequence S, where $S = \{S_i\}_{i=0}^{\infty}$; this is a sequence of partial sums. We calculated that $S_N = \frac{1-\delta^{N+1}}{1-\delta}$ in the previous section. Since $\delta < 1$, as $N \to \infty$, $\delta^{N+1} \to 0$, so $S_N \to \frac{1}{1-\delta} = S_\infty$. Thus the sequence of partial sums converges to S_∞.

What about our other example? That infinite series is $\lim_{N\to\infty}\sum_{i=1}^{N} \frac{3}{10^i}$. This is not immediately obvious, but as with many series we can write out some terms to try and see a pattern. In this case, using decimals makes it more clear. The sum is $.3 + .03 + .003 + .0003 + \ldots = .\bar{3} = \frac{1}{3}$.[5] Thus the series converges to $\frac{1}{3}$.

As a further illustration, consider the paradox of Zeno's runner. Zeno was a Greek philosopher who produced a number of apparent paradoxes. One of them involves a sequence and is thus of some interest for our purposes. Zeno asks us to consider a runner who is to complete a course from point A to point B. Call the distance between the points one unit (perhaps a mile). Imagine that the runner completes half the distance from A to B, and then completes half the remaining distance, and again half the remaining distance, and so on. If we think about the runner proceeding in this manner, does it not follow that the runner will never reach the endpoint, B?[6]

To answer this, start by noting that the sequence of distances the runner traverses is $\{\frac{1}{2}, \frac{1}{4}, \frac{1}{8}, \frac{1}{16} \ldots\}$. This can be represented as the element $x_i = \frac{1}{2^i}$. To see whether the runner reaches the goal, we want to calculate the sum of all these distances traveled, which is a series with this element. This is $S_N = \sum_{i=1}^{N} \frac{1}{2^i}$, where we have again used S_N to refer to a partial sum. We can write down the first few elements of a sequence of these partial sums to get a pattern:

[5] The bar over the .3 indicates that this number repeats in the decimal indefinitely.

[6] Zeno's paradox of the runner relies on the observation that an infinite number of points exists between any two points, and thus regardless of how many times the runner travels half the remaining distance, he will never reach point B. You will find a silly cartoon about it at `http://xkcd.com/1153/`.

$\{\frac{1}{2}, \frac{3}{4}, \frac{7}{8}, \ldots\}$. Each element gets closer and closer to 1. From this, we conjecture that the series converges to 1 in the limit.[7] Thus, in the limit, the runner reaches point B.[8]

4.2.2.1 Why Should I Care?

Let's return once more to the infinitely repeated prisoner's dilemma. To determine when it is beneficial to cooperate and when one would prefer to defect, one must consider the total payoff that one receives for cooperating and compare that to the total payoff one would receive if one were to defect and suffer the punishment. To compute these payoffs, one must add a sequence of infinite payoffs, which is to say, compute an infinite series.

Let's see how this works. Assume the payoff if both players cooperate in any period is 4. What total payoff does one get if both players cooperate forever? Well, in each period one gets the payoff 4, discounted by one more multiple of δ for every period that has passed. In other words, the total payoff is $4 + 4\delta + 4\delta^2 + \ldots$. Rewriting, this is the infinite series we discuss above, multiplied by 4: $\lim_{N\to\infty} \sum_{t=0}^{N} 4\delta^t = \frac{4}{1-\delta}$.

What if one defects? The answer depends on the punishment the other player will enact, which is represented by an infinite sequence of actions. There are many possible punishments, but we consider the simplest and most robust one: grim trigger. The grim trigger punishment (aka Nash reversion strategy) requires that one defect against one's opponent for all time after the opponent defects once. In other words, one defection leads to mutual defection forever, since there is no reason for the original defector to cooperate in the future given her opponent's permanent defection. The total payoff in this case is the payoff for the one-period defection in the presence of her opponent's cooperation (often called the temptation payoff), plus the infinite series of payoffs arising from mutual defection. Let the temptation payoff be 6, and the mutual defection payoff be 2. Then the total payoff is $6 + \lim_{N\to\infty} \sum_{t=1}^{N} 2\delta^t$. Note that the infinite sum here begins with $t = 1$, since the zeroth period is the one in which the first deviation from cooperation happens, giving a payoff of 6. But since the

[7] There are other ways to see this. You can draw a number line and mark a (generously sized) interval from 0 to 1 on it. Sector it by following the sequence of Zeno's runner: mark the halfway point, then mark the halfway point of the remainder, and so on. Do so until you are bored with it (or run out of room, as the case may be). What number do you get closer to each time you identify the next value? Would the number you are getting close to change if you continued the process beyond the number of times you did it? Nope. You are moving closer and closer to 1 no matter how many times you repeat the procedure, just as you are moving closer and closer to infinity as you add the positive integers or move closer and closer to $\frac{1}{3}$ as you add 3 divided by ever higher powers of 10.

[8] The reader might notice that this doesn't really answer the question, since who runs an infinite number of distances anyway? The trick, recognized by Aristotle, is that the time it takes to run the ever-smaller distances also decreases. This, however, leads to a discussion of infinitesimal change, a topic we cover in Part II.

sum is infinite, we can shift the series index by one if we pull out a δ, like so: $\lim_{N\to\infty}\sum_{t=1}^{N} 2\delta^t = \lim_{N\to\infty} 2\delta\sum_{t=1}^{N}\delta^{t-1} = \lim_{N\to\infty} 2\delta\sum_{t=0}^{N}\delta^t = \frac{2\delta}{1-\delta}$.[9]

Thus, we need to compare the cooperation payoff, $\frac{4}{1-\delta}$, to the payoff for defecting, $6+\frac{2\delta}{1-\delta}$. If the first exceeds the second, then cooperation is possible. This happens when $\frac{4}{1-\delta} \geq 6+\frac{2\delta}{1-\delta}$ or $4-2\delta \geq 6-6\delta$, which implies $\delta \geq \frac{1}{2}$. In other words, whenever the discount factor is sufficiently high, implying that people value future payoffs sufficiently much, cooperation can be maintained by the threat of a complete breakdown of cooperation.

4.2.3 Limits and Functions

Recall that a function is a mapping of the values in one set to the values in another set (or, equivalently, a graph of that mapping, or, equivalently again, an equation that describes that graph). If we want to know the value of a function for a given x, we can plug that x into the function and calculate its value. For example, if $f(x) = x^2$, then the value of $f(x)$ for $x = 2$ is 4 (since $2^2 = 4$). However, even though we can always find the value of the function at a point, as long as that point is in the function's domain, it's not always obvious how the values of the function *near* that point relate to the value at that point.

Why do we care about values close to some point $x = c$? There are several reasons, but the one of most importance to us relates to the roles of the limit in calculus, the topic of Part II of the book. Let's consider just one example. We are often interested in political science in the rate of change of a function at a point. We briefly mentioned this above when discussing the linear equation. If $y = \alpha + \beta x$, then the rate of change of y with x is given by β, and is constant for all x. This β is the important output of a statistical analysis, as it tells us the degree to which the concept represented by x affects the concept affected by y. Discerning such relationships, of course, is central to research in political science.

This still doesn't explain why we need limits, though. After all, the slope of the linear equation is constant. So let's try instead $y = \beta x^2$. It turns out, as you'll see in Part II, that the rate of change of y with x now is given by $2\beta x$, and so varies with the value of x. In this case, y is changing increasingly fast with x as x increases.

The way we figure out the value of the slope at any point is to look at the values of the function on both sides of a point $x = c$, but very close to that point. This difference tells us how much y changes for a small change in x. As we make this change smaller and smaller (i.e., take the limit as it goes to zero), we get a better and better approximation of the slope at that point, which is exact at the limit. Thus, limits of functions allow us to figure out the rate at which

[9] The key here is that the sum from $t = 1$ to infinity of δ^{t-1} is the same as the sum from $t = 0$ to infinity of δ^t. To see this, plug in the first few values of t and observe that the sum in each case is the same. Any difference in the sum with the shifted index would manifest at the end of the series, but it has no end, so there is no difference.

Table 4.1: Limit of $f(x) = x^2$ as $x \to 2$

x	$f(x)$
1.9	3.61
1.95	3.8025
1.98	3.9204
1.99	3.9601
2.0	4
2.01	4.0401
2.02	4.0804
2.05	4.2025
2.1	4.41

independent variables affect dependent variables, *even when the dependence of one on the other is complex.*

So, what is the **limit of a function**? For the function $y = f(x)$, a limit is the value of y that the function tends toward given arbitrarily small movements toward a specific value of x, say $x = c$. The limit either exists or does not exist for a given value of x, and if it does exist we can calculate it. Formally, much as for a sequence we can write $\lim_{x \to c} f(x) = L$.[10] One reads that notation as "the limit of f of x as x approaches c is L." Unlike a sequence, however, it's possible to approach a point (other than $\pm\infty$) from two different directions. If x approaches c from above (i.e., x decreases toward c), then we write $\lim_{x \to c^+} f(x) = L^+$. If x approaches c from below, we write $\lim_{x \to c^-} f(x) = L^-$. If the limits from above and below are equal, so that $L = L^+ = L^-$, then the function has a unique limit at c.[11]

Let's consider some examples, starting with a straightforward one: the limit of $f(x) = x^2$ as $x \to 2$. At $x = 2$, the function, as we have noted, takes value 4. What about near 2? When we are evaluating the limit of a function we want to know about *the function's behavior* at and near a given value, not the function's value. We want to ask: As we move a small distance away from $x = 2$ in either direction, does the function return a value that is only a small distance from $y = 4$? Let's look at a table to try to get a fix on this. Try the following x values: 1.9, 1.95, 1.98, 1.99 and 2.1, 2.05, 2.02, and 2.01. Table 4.1 reports both the x and y values.

Regardless of whether we begin at the top of the table or the bottom of the

[10] The similarities to the limit of a sequence do not end there. On the one hand, one can view a sequence as a function with a domain of the natural numbers; its limit is then just the limit of the function at infinity. On the other hand, one can construct a sequence with elements equal to $f(c + \frac{1}{n})$ or $f(c - \frac{1}{n})$. Its limit is the limit of the function from above or below, respectively.

[11] More formally, one can choose any tiny number $\epsilon > 0$ and find a $\delta > 0$ such that $|L - f(x)| < \epsilon$ whenever $0 < x - c < \delta$ for the limit from above, and whenever $0 < c - x < \delta$ for the limit from below.

table,[12] we can see that the closer we get to $x = 2$, the closer we get to $f(x) = 4$. In other words, the behavior of $f(x)$ around $x = 2$ is smooth: the function does not produce any surprising values in its sequence. If we graphed the function, it would not produce any surprising jumps in its graph around $x = 2$ (we do this in the exercises below). So we can conclude that the function $f(x) = x^2$ has a limit of 4 at $x = 2$.[13]

What often strikes students as odd about an example like this is that we can simply calculate that $f(x) = 4$ at $x = 2$, so what is all this business about saying that the limit of $f(x)$ at $x = 2$ is 4? Isn't it obvious that it is 4? Well, yes, the value of the function at $x = 2$ is indeed 4, and this is the same as the limit in this case, but information about the function at a point does not alone tell us whether $f(x)$ has a limit at $x = 2$. Put another way, the existence of a limit depends upon information about *the behavior of the function near a specific value of the variable(s) defined in the function*, and that is not the same as the value of the function at a specific value of the variable(s). Instead, we need to evaluate the behavior of the function near that point, as we did in Table 4.1.

Still don't believe us? Try this example:

$$f(x) = \begin{cases} x^2 & : x < 2, \\ (x-2)^2 & : x \geq 2. \end{cases}$$

The limit from below is still 4 at $x = 2$, since the function is the same as before. But now the limit from above is $0 \neq 4$. So this function has no limit at all. Is this a realistic scenario? Well, that depends on what the function is intended to represent, but certainly piecewise functions[14] see use in political science. For example, payoffs could decrease after some bureaucratic deadline is reached or after brinkmanship goes too far and leads to war rather than continued bargaining. Or a sudden change such as a shift in the party in power might lead to reduced circumstances for the party now out of power.

We don't need to invoke piecewise functions to find those that lack limits, though. Consider $f(x) = \frac{1}{x}$. As x approaches 0 from below, $f(x)$ gets ever more negative without limit. However, as x approaches 0 from above, $f(x)$ gets ever more positive without limit.[15] Thus, not only does this function have no finite limit at 0, it has no definable limit at all!

Before moving on, we consider some properties of limits. For any $f(x), g(x)$ that *both* have a well-defined limit at $x = c$, we have that (the last one as long

[12] If we think about this with respect to a number line, then we would say "regardless of whether we approach 2 from the left or the right."

[13] More formally, for any $\delta > 0$, $f(2 - \delta) \to f(2)$ and $f(2 + \delta) \to f(2)$ as $\delta \to 0$.

[14] A piecewise function is one that behaves differently depending on the value of x, and thus can be written in pieces. We introduce these in Chapter 3.

[15] If this is not clear, replicate the type of analysis in Table 4.1.

as $\lim_{x\to c} g(x) \neq 0$):

$$\begin{aligned}
\lim_{x\to c}(f(x)+g(x)) &= \lim_{x\to c} f(x) + \lim_{x\to c} g(x),\\
\lim_{x\to c}(f(x)-g(x)) &= \lim_{x\to c} f(x) - \lim_{x\to c} g(x),\\
\lim_{x\to c}(f(x)g(x)) &= (\lim_{x\to c} f(x))(\lim_{x\to c} g(x)),\\
\lim_{x\to c}(f(x)/g(x)) &= (\lim_{x\to c} f(x))/(\lim_{x\to c} g(x)).
\end{aligned}$$

These are helpful, but don't cover all cases. The most notable of these is the situation in which the value of the function is undefined at c. For example, let $f = \frac{x^2-4}{x-2}$. This function is undefined at $x = 2$. However, it has a limit of 4 at $x = 2$. How did we get this? One way is to factor the numerator into $(x-2)(x+2)$, and then divide the numerator and denominator by $(x-2)$. This is not allowed at $x = 2$, because of division by zero, but it is just fine at all points other than $x = 2$, which are the points we need to calculate to figure out the limit. Canceling the $(x - 2)$ leaves $x + 2$, which approaches 4 from above and below as $x \to 2$.

Even when the numerator and denominator of a rational function do not cancel, one can still use the same approach when the limit is $x \to \infty$ and the numerator and denominator of the function both go to ∞ as well. In this case, just pick the terms on both bottom and top that are the biggest for any finite value of x, and then cancel common terms as above.[16] For example, the limit of $f = \frac{x^3+2x^2}{3x^3+2x-1}$ as $x \to \infty$ is $\frac{1}{3}$, which is what you get when you consider only the terms of highest order (the biggest ones for all finite x) in both the numerator and denominator, and then cancel the x^3.

4.2.4 Why Should I Care?

We have already discussed the utility of the limit of a function in determining the rate at which independent variables affect dependent variables. Further, by discussing the importance of summing infinite series of payoffs in repeated games in game theory, we implicitly illustrated the benefits on limits of series, since an infinite series is just the limit of a finite series as the number of terms goes to infinity. There are many other examples like these. For instance, it is possible to develop a number of ideas relevant to the study of time series (e.g., permanent shocks with respect to unit roots, or impulse response functions), though these concepts are frequently presented without reference to the limit of a sequence.

Limits are also a building block in several other concepts that are frequently used in political science. The first is the maximum or minimum of a function. More specifically, the limit is a stepping stone to understanding derivatives, and derivatives are used to find maximum and minimum values of a function.

[16] A more general way to deal with limits like these (and one that makes this procedure make quite a bit more sense) invokes l'Hôpital's rule, which says that $\lim_{x\to c} \frac{f(x)}{g(x)} = \lim_{x\to c} \frac{f'(x)}{g'(x)}$ if $\lim_{x\to c} f(x) = \lim_{x\to c} g(x) = 0$ or $= \pm\infty$ and $\lim_{x\to c} \frac{f'(x)}{g'(x)}$ exists. Here $f'(x)$ and $g'(x)$ are first derivatives, which we'll cover in Part II.

When might one want to know the minimum or maximum of a function? Some theories assume that an actor wants to maximize something (e.g., their power, utility, time in office, etc.). If one writes down a function to describe the individual's power, utility, time in office, etc., then one will be interested in limits (and derivatives) to help one determine whether a maximum exists and what it is. In Chapter 8 of this book we put these tools to work to do precisely that.

In addition, statistical analysis often leads one to have interest in minima and maxima of functions. For instance, one might want to minimize the predicted error in a regression or maximize the likelihood of matching the data to the statistical model that the researcher specified. Again, the limit (and the derivative) are useful for that purpose.

Limits also arise in the definition of other important concepts in addition to those in calculus. We discuss two below: open and closed sets, and continuity of functions.

4.3 OPEN, CLOSED, COMPACT, AND CONVEX SETS

In Chapter 1 we discussed several properties of sets, but did not have the background to be as complete as we would have liked. Now we do and we introduce four new types of sets: **open**, **closed**, **compact**, and **convex**. All four will prove central to later developments in the book. We address them in the order stated.

Often one sees an open set defined as the opposite of a closed set, or vice versa. Since we are not overly concerned with formalism and want to highlight intuition, we will define both independently. An **open set** is one in which there is some distance (which may be arbitrarily small) that you may move in any direction within the set and stay in the set. We'll deal mostly with the concept of openness in reference to spaces that have a defined distance metric. In these cases we can be more intuitive. Picture a perfectly spherical ball of the same dimension as the space. So, if we're in normal three-dimensional space ($\mathbb{R}^3$), you should be picturing a sphere, while in two dimensions you should be imagining a circle. If you can find a small enough ball around any point such that the entirety of the ball remains in the space, then the space is open.[17]

The modal example for an open set is $(0, 1)$.[18] The point 1 is not in the set, but everything less than 1 (but greater than 0) is. Similarly, 0 is not in the set, but everything greater than 0 (but less than 1) is. So no matter which point less than 1 and greater than 0 you choose there is always some small enough distance that allows you to go a little to the left or a little to the right and

[17]Formally, define the distance between two points $x, y \in \mathbb{R}^n$ as $d(x, y) \geq 0$. Then a set $A \subset \mathbb{R}^n$ is open if for all $x \in A$, there exists an $\epsilon > 0$ such that all points $y \in A$ with $d(x, y) < \epsilon$ are in A too.

[18]Recall from Chapter 1 that this notation means the set of all real numbers between zero and one, exclusive, and that the use of curved brackets signifies openness.

stay in the set. The union of any number of open sets remains open, while the intersection of a finite number of open sets is also open.

Now consider the set $[0, 1]$. This is very similar, except for one important property: 0 and 1 are in the set. That means that one can't fit a ball around the points 0 or 1 and stay in the set, since *nothing* less than 0 or greater than 1 is in the set. So $[0, 1]$ isn't open. Instead, it is closed, though these terms are not opposites. For example, the empty set and the universal set are both open and closed, while the sets $(0, 1]$ and $[0, 1)$ are neither open nor closed. So how do we know $[0, 1]$ is closed? One definition of a closed set is that it is the complement of an open set, but this may not be that intuitive at this point. Instead we turn to limits.[19]

A **closed set** is one that contains all its limit points. What does that mean? It means that if you make a sequence with all its elements contained in a set A, then for a closed set the limit of this sequence must also be in A. More intuitively, start by choosing points within a set, such that the sequence of these points gets closer and closer to some other point. If a set is closed, then the point that this sequence approaches is also in the set. In other words, one can't leave a closed set by following a sequence that is otherwise within it to its limit. A set that is not closed, in contrast, can have the point the sequence approaches—the set's **limit point**—outside the set, even if all prior points in the sequence are in the set. In an open set, one can get closer and closer to a point, all the while staying in the set, but leave the set in the limit of the sequence.

Let's try this out with $[0, 1]$. We can choose a sequence with elements $x_i = \frac{1}{i}$ or $x_i = 1 - \frac{1}{i}$, depending on which limit we want to show. In both cases, for any finite value of i the corresponding element of the sequence is within A for either of $(0, 1)$ or $[0, 1]$. However, the limit of the first sequence is 0 and the limit of the second is 1, and neither of these points is in $(0, 1)$. They are both in $[0, 1]$, though. And any other sequence whose elements are all in $[0, 1]$ will also have a limit in $[0, 1]$. So $[0, 1]$ is closed. Now we can also make more sense of the other definition. The complement of a closed set is everything outside the closed set. In this case, the complement of $[0, 1]$ is $(-\infty, 0) \cup (1, \infty)$, which is the union of two open sets, which is open. The important thing is that the complement does not contain the points 0 and 1, since they are in $[0, 1]$. One can come ever closer to the boundary of a closed set from outside it without reaching it.

Any intersection of closed sets is closed, while the union of a finite number of closed sets is closed. One can get from an open (or other) set to a closed set by adding all the original set's limit points into the set. The closed set you get by doing this is called the closure of the original set. So $[0, 1]$ is the closure of $(0, 1)$.

A subset of $\mathbb{R}^n$ that is both closed and bounded—that is, the set contains all its limit points and can itself be contained within some finite boundary—is

---

[19]And thereby justify this section's inclusion in this chapter.

called **compact**.[20] Compact sets are in some sense self-contained: they don't go on forever, either by having no bound or by containing sequences that don't stay in the set in the limit. These properties help ensure that continuous functions defined over compact sets have nice properties, notably maxima and minima that are present in the set.

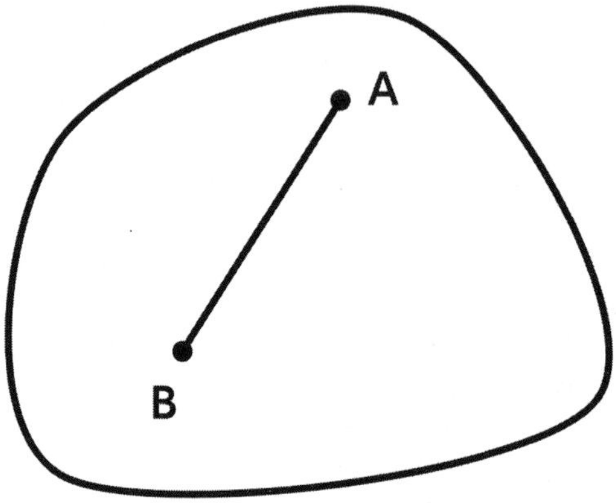

Figure 4.1: Convex Sets

Finally, a subset of $\mathbb{R}^n$ for which every pair of points in the set is joined by a straight line that is also in the set is called a **convex set**. Formally, a set is convex if for all points x and y in it and all $\lambda \in [0, 1]$,[21] the point $(1 - \lambda)x + \lambda y$ is also in the set. The **convex hull** of a set A is the set A plus all the points needed to make A convex. See Figures 4.1 and 4.2 for pictures of convex and nonconvex sets. Convex sets are useful when one is considering linear combinations of variables and wants to make sure that these combinations remain in the set. For example, a political party in a proportional representation system may get utility both from the coalition government policy and from the share of the cabinet it controls in the coalition government. If all linear combinations of policy and cabinet share are possible—i.e., if the set of possible government outcomes is convex—then the strategy the party might take in trying to form a government may be very different than if all combinations are not possible

[20] Recall the definition of bounded from Chapter 1. This is one of those points where we sacrifice formalism and generality for simplicity. Something bounded and self-contained seems compact in common language. The more general definition of compact—every open cover has a finite subcover—requires additional definitions and adds little intuition, in our opinion. One should be aware that the definition of compact as closed and bounded is a consequence of the Heine-Borel theorem and applies specifically to subsets of $\mathbb{R}^n$. However, these are by far the most common sets we will encounter in political science.

[21] Parameters that vary between zero and one in this way and multiply other variables are often called weights.

because the set of possible outcomes is not convex. A party might, for instance, seek out a much larger cabinet share if a larger share is the only way to secure a beneficial policy than if a preferred policy is available with a smaller share.

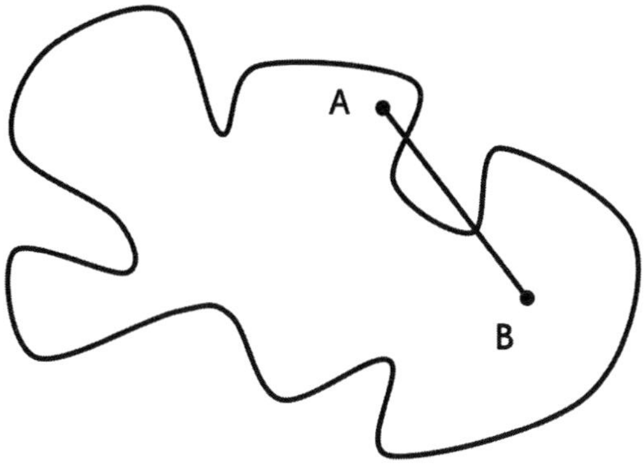

Figure 4.2: Nonconvex Sets

4.3.1 Why Should I Care?

This section may have seemed fairly abstract, but these set concepts will prove fairly central to the rest of this book and political science in general. Compactness in particular will appear time and again for the simple reason that it permits us to maximize (and minimize) functions over a domain, and maximizing (and minimizing) functions is, as we have noted, central to analysis in political science. We address this more in Part II, but an example can help us to see how this works. Consider the sets $A = (0, 1)$ and $B = [0, 1]$. B is compact; A is not. We ask: What's the maximum of the linear function $f(x) = x$ on each of these sets?

For B the answer is 1. One is the largest value in B, and so the biggest value that f can take on the domain B. But there is no corresponding biggest value of f on A because A *has no biggest value.* That's right, even though it is bounded by 1, there is no single largest value in it because one can always go just a bit closer to 1. Each movement closer to 1 increases f, but since the limit of a sequence approaching 1 is outside the set A, we can never increase f as much as possible, since we're limited to values within the domain A. So f has no maximum on the domain A. Isn't math fun?

4.4 CONTINUOUS FUNCTIONS

Intuitively, a **continuous function** is a function without sudden breaks in it. When you draw the graph of a continuous function, you never need to lift your pencil from the page. Put differently, the graph of a continuous function forms an unbroken curve, whereas the graph of a **discontinuous function** has at least one break in it. This break is usually called a jump, since the function jumps up or down quite a bit for a small change in x. We can define continuity in a couple of ways.

More formally, a continuous function is one for which an arbitrarily small change in x causes an arbitrarily small change in y for *all* values of x. That's why you never need to lift your pencil when drawing the graph of a continuous function: each time you move a little bit to the right, further along the x-axis, you need only move a little bit up or down along the y-axis as well.[22]

This is a fine definition (particularly the one in the footnote), but limits provide perhaps a more intuitive one. A function is continuous at a point if the limit of the function at that point exists and is equal to the value of the function at that point. In the language of math: $f(x)$ is continuous at $x = c$ if and only if $\lim_{x \to c} f(x) = f(c)$. A function that is continuous at all points c in its domain is called continuous. Why is this more intuitive (according to us)? Consider what this means. That the limit exists means that if I traverse the function toward c from above and from below, I get closer and closer to a point, and this point is the same on both sides. That means the two pieces of the function *should* connect at c. The only problem is that the function at c need not equal the limit at that point, as we have noted. So, the function is continuous only when the value of the function at c is equal to this limit; i.e., there is no gap in the function right at that point to force you to move your pencil. In a way, this concept is very similar to that of a closed set. A closed set contains the limit points of all sequences that lie within it. In a continuous function, the range of the function, loosely speaking, contains the limit of the function.

Let's look at some examples. Most of the functions described in Chapter 3 are continuous. So, affine (and linear) functions are continuous, quadratic and higher-order polynomials are continuous, logarithms are continuous over the region on which they are defined, and so on. In fact, nearly every function you've had occasion to consider has likely been continuous. But there are lots of functions that are not continuous. For instance, the function $f(x) = \frac{1}{x}$ is not continuous over the domain $\mathbb{R}$, i.e., the real number line, since no limit exists at zero. Put another way, there is no way you can get from $-\infty$ to $+\infty$ while crossing $x = 0$ without picking up your pencil!

The piecewise function described at the end of the section on limits is also

[22] More formally still: a function is continuous at some point $x = c$ if for every $\epsilon > 0$, there exists a $\delta > 0$ such that whenever $|x - c| < \delta$, $|f(x) - f(c)| < \epsilon$. In other words, points very near to c map to points very near to $f(c)$, and by going closer to c one can go as close as one wants to $f(c)$. A function that is continuous at all points c in its domain is called continuous.

not continuous. Why not? Consider the function again:

$$f(x) = \begin{cases} x^2 & : x < 2 \\ (x-2)^2 & : x \geq 2. \end{cases}$$

Its limit at $x = 2$ is 4 coming from the left and 0 coming from the right, so it's not continuous at $x = 2$. Could we change it to make it continuous? Well, yes. Since it's continuous at all points other than $x = 2$, all we need to do is either to shift the left piece down by 4, to $x^2 - 4$, or the right piece up by 4, to $(x-2)^2 + 4$. Then the value of each piece at $x = 2$ would be the same, as would be the limits from above and below. Of course, this would change the meaning of our function and so we might not want to do this.

These examples of discontinuity all relate to the lack of a limit at some point, but that's not necessary. Consider the function $y = \frac{x^2}{x}$. If you plot it over the range -5 to 5 it appears to be a straight line, and its limit is well defined at every point. However, division by zero is undefined, so the function is undefined at $x = 0$, as indicated by the open circle at $x = 0$ on the plot in Figure 4.3. This discontinuity is the easiest to remove; all we have to do is to define the function piecewise: $y = 0$ at $x = 0$ and $y = \frac{x^2}{x}$ everywhere else. This fixes the discontinuity and most likely does not change the meaning of the function either.

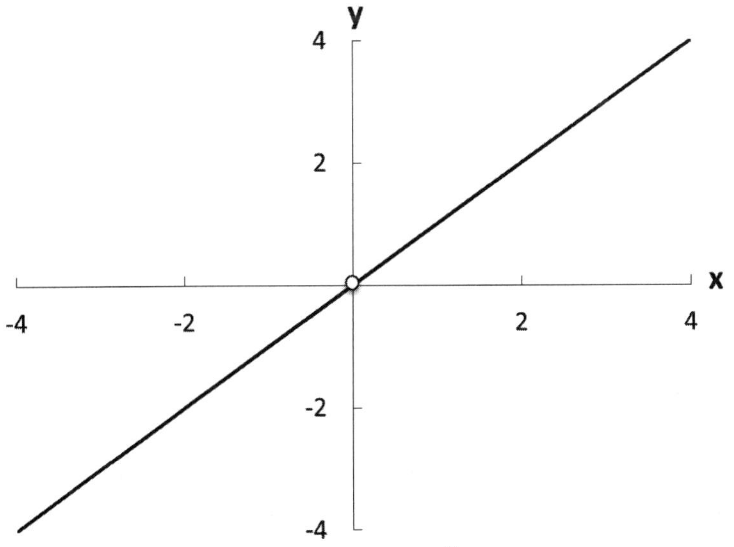

Figure 4.3: Graph of $y = \frac{x^2}{x}$, $x \in [-5, 5]$

4.4.1 Why Should I Care?

Continuity is perhaps one of the most central concepts you will encounter, for much the same reason that compactness proves useful: it helps us to maximize (or minimize) functions, and maximizing (or minimizing) functions is important in both game theory and statistics. To see how this works, consider one last time the piecewise function from above, but now let's say that the domain of x is only between 0 and 3. In other words, $x \in [0, 3]$, a compact subset of the real numbers. The expression $(x-2)^2$ never gets higher than 1 in this domain, so the maximum has to come from x^2. But the function only takes these values for $x < 2$. We can keep increasing the function by moving closer and closer to $x = 2$, but we never hit a maximum, just as when maximizing some functions on open sets.

We often assume continuous functions when representing preferences (i.e., a utility function) or the outcome of interactions between actors for this reason. This is particularly true when scholars work with functions that are only implicitly defined and are not given explicit functional forms. These implicitly defined functions are usually assumed continuous. Why might they make these assumptions? Well, the reason why they do not write down an explicit functional form is probably that they want their theories to be as general as possible. If they were to write down a specific functional form, then the theory would apply to that function, but not necessarily to others. So keeping it general expands the domain of their theories. Yet there are some functional forms that might produce an internal inconsistency in one's theory if they are not ruled out. For example, if one's theory relies on a result where one takes a derivative, then one will want to limit the theory to only those functions where it is possible to take a derivative. As we will learn in the next part of this book, the derivative is a type of limit. Thus, if a function is not continuous, there are points where it does not have a limit, and that means that the function is not differentiable (i.e., has no derivative) at those points. If one were not to limit one's theory to continuous functions, then one would be asserting that the theory is relevant to functions that have points over which the derivative is undefined, and then appealing to the derivative to substantiate one's result (or hypothesis). Doing so would be contradictory. Thus, to avoid contradiction, one begins with the assumption that the functions to which one's theory applies are continuous.

Though continuous functions are thus common, some solutions of problems require analysis of discontinuous functions. For example, in Chapter 3 we referred to best response correspondences in game theory, which describe what one player's optimal strategy (or strategies) is (are) given a particular choice in strategy by the other player. These are often discontinuous: one strategy may be optimal for some subset of the other player's strategy space, but a quite different one might be optimal for a different subset. For example, a particular distribution of the budget might be optimal if each party bargains in good faith, but if one of the parties asks for too much, the optimal response might abruptly drop to an offer of nothing. When one's best response to another actor's actions is discontinuous in those actions, one often needs an alternative technique to figure out the equilibrium of the interaction. One technique is monotone comparative statics, which allows you to relax assumptions on the continuity of best response functions (McCarty and Meirowitz, 2007; Ashworth and Bueno de Mesquita, 2005). So it is important to understand continuity and to be able to determine whether a function is continuous over all possible values of x or whether it is continuous only for some values of x.

4.5 EXERCISES

1. Draw a graph to show that the sequence $\{1, -1, 1, -1, 1, -1 \ldots\}$ is divergent.

2. Find the sum of the infinite series $\sum_{t=0}^{\infty}(\delta^t)^2$.

3. Show whether $f(x) = x + x^3$ has a limit at $x = 3$ and, if so, the value of the limit.

4. Show whether $f(x) = (x - 3)(x + 5)$ has a limit at $x = 4$ and, if so, the value of the limit.

5. Show whether $f(x) = \frac{3x^2-12}{x-2}$ has a limit at $x = 2$ and, if so, the value of the limit.

6. Show whether $f(x) = \frac{x^3-4}{x-2}$ has a limit at $x = 2$ and, if so, the value of the limit.

7. For each of the following sets, state whether they are (a) open, closed, both, or neither; (b) bounded; (c) compact; (d) convex:

   a) $[1, 3]$
   b) $(2, 5)$
   c) $[0, 6] \bigcup [10, 12]$
   d) $(2, 4) \bigcap [3, 4]$
   e) $[0, \infty)$

8. Is the function $f(x) = \frac{\ln(x)}{x}$ continuous for $x \in [2, \infty)$?

9. Is the function

$$f(x) = \begin{cases} x^3 - 3x + 4 & : x \leq 3, \\ x^2 & : x > 3, \end{cases}$$

continuous? If so, why? If not, what changes would make it continuous?

Part II

Calculus in One Dimension

Chapter Five

Introduction to Calculus and the Derivative

In our experience, calculus and all things calculus-related prove the most stressful of the topics in this book for those students who have not had prior calculus coursework. We conjecture that this is due to the foreignness of the subject. While probability and linear algebra certainly have some complex concepts one must internalize, much of the routine manipulations students perform in applying these concepts use operations they are used to: addition, multiplication, etc. In contrast, calculus introduces two entirely new operators, the derivative and the integral, each with its own set of rules. Further, these operators are often taught as a lengthy set of rules, leading to stressful rote memorization and little true understanding of what are relatively straightforward concepts, at least as used in most of political science.[1] To try to avoid this, we're going to take a little more time with the topic. In this chapter we will cover the basics of Calculus and the derivative in what we hope is an intuitive manner, saving the rules of its use for the next chapter. If you are working through this chapter as part of a course and are not sure of something, this is the time to ask questions—before you end up trying to take derivatives without having a clear understanding what they *are*. The first section below provides a brief overview of calculus. The second section introduces the derivative informally, and the third provides a formal definition and shows how it works with a few functions.

5.1 A BRIEF INTRODUCTION TO CALCULUS

For our purposes, the primary use of calculus is that it allows us to deal with continuity in a consistent and productive manner. This is likely a useless claim at this point, so let us explain. As we discussed in Chapter 4, a continuous function is one that we can draw without lifting pencil from paper. Intuitively, such a function has no gaps or jumps in it.[2] Such functions are great for lots of applications, as we have noted, but they lead to some problems when we're trying to understand the concept of change. Change within a discrete function is pretty clear: if $f(1) = 1$ and $f(2) = 4$, and if f is not defined between 1 and 2, then we know moving from 1 to 2 results in a change of 3 in the function. Further, that's pretty much all we can say about it.

[1] As with the other topics in the book, calculus has plenty of complexity to it. We'll just be avoiding nearly all of it.

[2] We're going to assume that unless explicitly stated otherwise functions are defined over subsets of the real numbers and in one dimension in this chapter.

What if the function actually is defined between these two points, though? Let $f(x) = x^2$, now and consider $x \in [1, 2]$. For this function, $f(1) = 1$ and $f(2) = 4$, and we can still say that moving from 1 to 2 results in a change of 3, but we can say a lot more. For example, $f(1) = 1$ and $f(1.5) = 2.25$, and moving from 1 to 1.5 results in a change of 1.25, which is less than half the change resulting from moving from 1 to 2. Or, $f(1) = 1$ and $f(1.1) = 1.21$, and moving from 1 to 1.1 results in a change of 0.21. In fact, there's no limit to how many of these changes we can write down. But there is a *limit* to these changes, in the sense we discussed in Chapter 4: we can keep making the second point closer and closer to 1 until, at the limit, it is the same as 1. What is the change in f at this limit? As we'll see shortly, this is the *derivative*, and the study of these objects constitutes *differential calculus*.

Differential calculus thus deals with the study of infinitesimally small changes in a function. As we'll see in Section 3 below, the derivative of the function $f(x) = 3x$ is 3, which is the slope (rate of change) of the line $y = 3x$. This example illustrates the way in which a derivative breaks down functions, removing information about their value at *any* point and providing just the value of the change at that point. In this case, the rate of change is 3 at all x, and this is all the information the derivative gives us. In other words, the derivative provides for us a graph of the marginal rate of change in any variable that we can represent as a continuous function of another variable, with respect to the variable of which it is a function. This is powerful stuff, and we'll use it extensively in Chapter 8 to find maxima and minima of functions, which prove to be terribly important in both computational statistics and formal theory.

But what if we had a derivative, and wanted to build back up a function from it? Well, we'd need to start at the smallest value of x about which we cared and add up all the infinitesimal changes to $f(x)$ that occurred as we increased x from that point. But how do we add infinitesimal things? The answer is the integral, or antiderivative.[3] The integral is a tool for adding infinitesimals, the same way a sum ($\sum$) is a tool for adding discrete quantities. The symbol for an integral ($\int$) even looks like an "S" to help you remember. As we'll see in Chapter 7, integrating the function $f'(x) = 3$ returns the function $f(x) = 3x + C$, where the C is a constant that has to be added because the derivative doesn't contain information about the value of the function at $x = 0$, so we don't know where to start adding infinitesimals.

We go into much more detail about both of these objects in the coming chapters, but before doing so it's fair to ask why we'd want to do so. Calculus is foundational to higher math, so there are many answers, but two will prove particularly important in this book and political science more generally. Consider first a continuous function with a maximum. Intuitively, this means that there are some values of x for which the function increases in value, but for some other values of x the function has to stop doing so; otherwise it wouldn't have

[3]Technically speaking, only the indefinite integral is the antiderivative. We'll talk more about this in Chapter 7.

a maximum. At the point it stops doing so, and possibly *only* at that point, it is no longer increasing, but not yet decreasing either. This means at that point its *instantaneous* rate of change is zero, so the derivative at that point is zero. We'll discuss what instantaneous rate of change means shortly, and the procedure for finding maxima (and minima) in Chapter 8, but the important point is that *derivatives help us maximize (and minimize) functions.*

We put off until Chapter 11 of Part III of this book the most common use of integrals: in probability, specifically continuous probability distributions. There the integral will allow us to understand statistical inference with continuous variables and to compute expected values and expected utilities, which are vital when considering uncertainty of any sort in game theory. Needless to say, being able to make inferences and deal with uncertainty is necessary for quantitative and formal political research, and learning the ways of the integral will certainly pay off down the road.

5.2 WHAT IS THE DERIVATIVE?

Before we get too far ahead of ourselves, though, let's start with the fundamentals. What is a derivative? As we discussed briefly above, the **derivative** is the instantaneous rate of change of a function. That's it. The notation in the next section might be a bit intimidating, but the underlying concept is very straightforward. And it turns out that it can be quite useful to know what the rate of change is. This is true both for constructing theories of politics and for developing statistical techniques to test hypotheses.

You might be thinking that we snuck something in with that word *instantaneous.* Rate of change is certainly straightforward, but what is the *instantaneous* rate of change? Perhaps that is where things get tricky. If things do, in fact, get tricky, then we suppose the word *instantaneous* is the locus of the tricky bit. But we are going to forestall that issue for the moment and focus on *rate of change.*

5.2.1 Discrete Change

Perhaps one is interested in budgetary politics, the rise and decline of political parties, or arms races between countries. All of these topics can be considered with respect to growth rates over time. For example, we might calculate the percentage change from time $t - 1$ (e.g., 2000) to time t (e.g., 2001) in a budgetary expenditure, the seats a political party wins in the legislature, or the nuclear weapons a country builds. Alternatively, we might calculate and then plot the first difference of the series. The **first difference** of a variable is the value of that variable at time t minus the value of that variable at time $t - 1$, and it is a measure of discrete change. Table 5.1, for example, lists the total number of heavy weapons held by China for several years, as well as the annual discrete change (labeled first difference).

Table 5.1: Aggregate Heavy Weapons, China

Year	Total	First Difference
1995	37,095	——
1996	35,747	−1,348
1997	36,910	1,163
1998	37,032	122
1999	36,494	−538
2000	31,435	−5,059
2001	34,281	2,846

Source: SIPRI (`http://www.sipri.org/databases`).

Once we have calculated the first difference, we can calculate the **percentage change** using the formula $\frac{(x_{t+1}-x_t)}{x_t} \times 100\%$, where the subscript t indicates the first observation and the subscript $t + 1$ indicates the second observation.

Why discuss differences and percentage change in a chapter on the derivative? Our purpose is to get you comfortable thinking about rates of change using math with which you are familiar: arithmetic. The derivative is nothing more than a refinement of these ideas. And you guessed it: that refinement involves the word *instantaneous*.

Discrete change, then, is the first difference between two observations. It is a measure of change in a variable across two *discrete* moments in time. It follows that the size of a first difference is going to vary across different temporal scales. For example, we might have a measure of the number of times that the US Department of State lodges a complaint with a foreign government. We could calculate the first difference of that variable across two years, two quarters (i.e., two chunks of three months each), two months, or two weeks. The size of those differences will vary across the scale of discrete time we consider.

5.2.2 Instantaneous Change

The difference $(x_{t+1} - x_t)$ has a limitation: it can only represent the rate of change over two discrete moments in time. What if we want to know the rate of change at a specific moment in time, not over a discrete interval? The difference cannot help us if we want to know the rate of change at a specific point of a function. However, as we discussed in the previous section, if we could evaluate the rate of change at a point on a function by taking the *limit* of the difference as the interval gets smaller and smaller, then that would tell us what the **instantaneous rate of change** was at the point (or moment in time). And that is precisely what the derivative does.

The derivative is defined for a function with respect to a specific variable. We begin with univariate functions but will discuss multivariate functions as well, very briefly at the end of this chapter and in more depth in Part V of the book. The derivative of $f(x)$ with respect to x tells us the instantaneous rate of change of the function at each point. Just as we can calculate the value of a function for a specific value of x we can calculate the value of the derivative of $f(x)$ for a specific value of x, but it is typically of more interest to have a general representation of the derivative of $f(x)$ over a range of values of x. And it turns out that there are rules for taking derivatives that make it possible for us to determine what that general representation is. We will cover these in the next chapter. First, we flesh out what we mean by instantaneous a bit more, and then, in the next section, formalize these notions in an intuitive fashion, making use of our study of limits in Chapter 4.

5.2.3 Secants and Tangents

Let's return to the linear example $f(x) = 3x$. To figure out the discrete rate of change between any two points on this line we look at the amount of change on the y-axis relative to a particular amount of change on the x-axis. In other words, we compute for two points:

$$m = \frac{f(x_2) - f(x_1)}{x_2 - x_1} = \frac{y_2 - y_1}{x_2 - x_1}.$$

This should look familiar, as it is the equation for the slope of a line. For $f(x) = 3x$ and $x_2 = 2$ and $x_1 = 1$, we get $m = \frac{6-2}{2-1} = 3$, so the slope is 3. You can try other values of x to convince yourself that the slope is the same for all values of x. The rate of change between two discrete points is just the slope of the line connecting those two points, known as a **secant**.[4] We can't see it on a plot of a straight line because it overlaps the line (which is why we haven't created a figure to show you the line). However, in Figure 5.1, which plots the function $f(x) = x^2$, we can see the secant drawn between $x = 1$ and $x = 2$—it is above the curve.[5] The discrete change between these two points is 3. Of note, this change is not constant over values of x. The discrete change between $x = 2$ and $x = 3$ is $\frac{9-4}{3-2} = 5 > 3$, so the slope of the secant line between the points corresponding to $x = 2$ and $x = 3$ on the function is $m = 5$. Trying other points should convince you that the secant is increasing as x gets bigger.

This example illustrates that we can get an idea of change just by subtraction, but even in the case of a quadratic function that is familiar we can see that change itself is more complicated. If the rate of change is itself constantly changing, then how can we really speak about change? That's where the derivative and the instantaneous rate of change come in. Consider first the linear function $f(x) = 3x$. Since the secant overlaps the function itself, we can compare the function at closer and closer points ($x = 1$ and $x = 1.5$, $x = 1$ and $x = 1.1$, etc.) and still get a slope of 3 for the secant line. This is true for all differences, and if

[4] A secant is a line that intersects two points on a curve.

[5] Later we will see that this means $f(x) = x^2$ is a convex function.

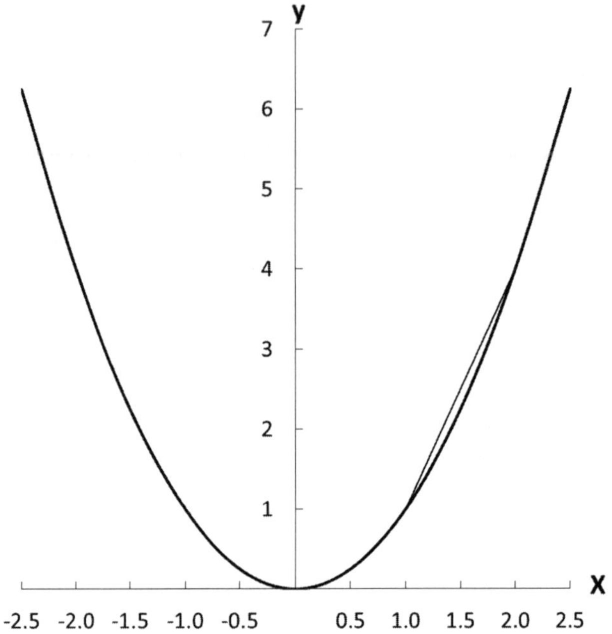

Figure 5.1: Graph of $y = x^2$ with Secant Line

we constructed a sequence of decreasing differences, such as $1, 0.1, 0.01, 0.001, \ldots$, represented by $x_n = \frac{1}{10^n}$, it would be true for every element in the sequence. It's not hard to see that it remains true in the limit as $n \to \infty$. Thus we can say that the instantaneous rate of change at $x = 1$ is also 3. A line that intersects the function at $x = 1$ and has a slope equal to the instantaneous rate of change of the function at $x = 1$ is known as the **tangent**[6] line of the function at the point $x = 1$. We could repeat this analysis to see that all tangent lines of the function have the same slope. In short, we can say the derivative of the function is 3, since this is the instantaneous rate of change at all points on the function.

Now turn one last time to $f(x) = x^2$. Since we already know that the slope of the secant lines change with different values of x, we should expect that changing the distance between points should also change the slopes of these lines. As we can see in Figure 5.2, it does. This plot adds secant lines between 1 and 1.5 and between 1 and 1.1 (which is barely visible) to the one between 1 and 2. As we can see, the slope of the secants decreases as the difference gets less. If this slope approaches some number as the difference approaches zero (i.e., in the limit), then we can talk about the instantaneous rate of change of the function at 1. Plugging numbers into the slope formula, we see that for the three secants we

[6]The tangent line is a line that just touches the curve at a given point.

drew, the slopes are $3, 2.5$, and 2.1, which seem to approach 2. Assuming this holds up under more formal analysis (as we'll see in the next section, it does), then the instantaneous rate of change of the function at $x = 1$ is 2. You can sort of see this graphically too in Figure 5.2: the slopes of the secants get flatter as they approach the tangent line, which has a slope of 2.

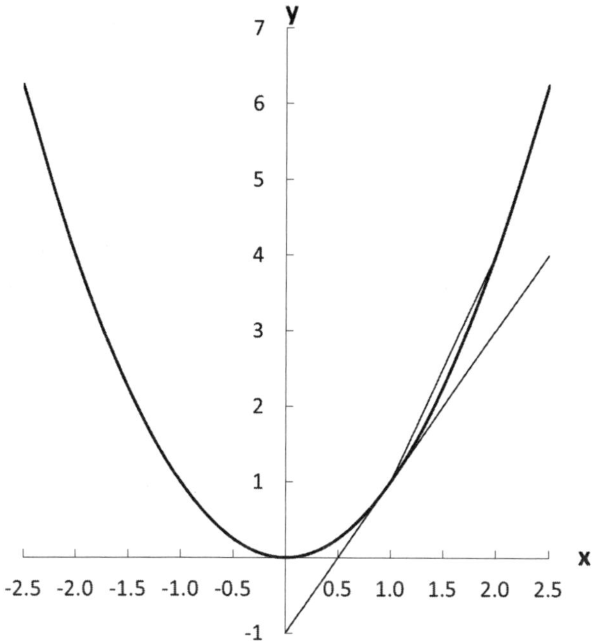

Figure 5.2: Graph of $y = x^2$ with Tangent Line

What about at other points? We could repeat this analysis over and over again, discovering that the slope of the tangent line is 2 at $x = 1$, 4 at $x = 2$, 6 at $x = 3$, and so on, but this is horribly inefficient, and will never give the full picture for the whole function because it's defined over the uncountable real numbers. We might guess that the slope of the tangent line at a point x is equal to $2x$ from the pattern above, and we'd be right in this case, but this would not work for harder problems. If we care about the instantaneous rate of change at all, and we hope we have convinced you that we do, we need a better technique.

5.3 THE DERIVATIVE, FORMALLY

At this point some readers might be readying themselves for the other shoe to drop. Though the graphical representation of secants and tangents might have been intuitive, surely one of the two fundamental pieces of calculus must be much more conceptually difficult. It turns out it's not. Before justifying this claim, however, we need a little notation.

5.3.1 Notation, Notation

Sir Isaac Newton, one of the fathers of calculus, represented a derivative this way: $\dot{y}$ is the derivative of $f(t)$ if $y = f(t)$. This is only used for derivatives with respect to time, however, and you will likely never see it in political science. Instead we turn to the notation of the other father of calculus, Baron Gottfried Wilhelm von Leibniz, who used the following notation to identify the derivative of a function: $\frac{d}{dx}f(x)$, which is read "the derivative of f of x with respect to x." One might say this more cumbersomely, but completely, as the instantaneous rate of change in f of x with respect to x.

If $y = f(x)$, then we can use Leibniz's notation to write the derivative of y with respect to x as $\frac{dy}{dx}$. Leibniz's notation sometimes throws students because it looks like a fraction. But it is not a fraction, so don't try to simplify it by eliminating the d's and concluding that $\frac{dy}{dx} = \frac{y}{x}$ (i.e., that the derivative is the ratio of y to x). That conclusion is *false.*

A benefit of Leibniz's notation is that it allows you to specify the variable with respect to which you are differentiating. This is the variable in the denominator. When this is clear, as it always is when there is only one variable, then we often use Lagrange's prime notation: $f'(x)$ (read "f prime x") is the derivative of $f(x)$. There is no difference between $\frac{d}{dx}f(x)$ and $f'(x)$; they mean the same thing. Finally, you may on occasion see Euler's differential operator D, indicating that the derivative is $Df(x)$. When there is potential confusion about the variable with respect to which the function is being differentiated, then we instead use $D_x f(x)$.

Formal notation isn't the only reference to derivatives that vary. You will sometimes see work that references **differentiation**. This is the process of taking a derivative, and it has the added bonus of a corresponding verb: to differentiate. That is, if one wants to calculate the derivative of $f(x)$, then one differentiates $f(x)$. Again, they mean the same thing.

5.3.2 Limits and Derivatives

Though it is reasonable to be confused at this point by the different ways of representing a derivative, believe us when we say this will pass. Notation is only a way of writing down a concept, and if the concept is clear, one will get used to the notation. We would assume that readers of this book do not have a great deal of trouble using "many" and "a lot," after all, notation that means more or less the same thing. Different notation exists simply for convenience in different applications.

With this notation in hand, let's return to the concept of the derivative, discussed in the previous section. The derivative of a function is the instantaneous rate of change at a point. As noted in the previous section, we can take the difference between two points and calculate the discrete rate of change (i.e., the rate of change between any two points). But the derivative is the continuous rate of change (i.e., the instantaneous rate of change at any given point). How do we calculate it?

We noted in the previous section that taking the difference between discrete points that are closer and closer together (i.e., two values of x that are increasingly close in value) resulted in determining this instantaneous rate of change, and that doing so is effectively equivalent to taking the limit as this difference goes to zero. It turns out that this verbal description is actually all we need to produce a formal definition of the derivative that relies on our prior definition of the limit of a function. As we calculate the discrete rate of change over smaller and smaller units, we get better and better approximations of the derivative. In other words, the discrete rate of change over very small units is an approximation of the continuous rate of change. Taking the limit as the small units approach zero makes the approximation exact.

To see how this works, start with the equation for the slope of the secant:

$$m = \frac{f(x_2) - f(x_1)}{x_2 - x_1}.$$

Now let $x_2 = x_1 + h$, where h is any real number. Rearranging, we see that $h = x_2 - x_1$ is the difference between the two discrete points. We can then write the slope as

$$m = \frac{f(x_1 + h) - f(x_1)}{h}.$$ [7]

Note that we have rearranged the equation for the slope of a line to more clearly express the rate of change in $f(x)$ over the interval h.

All that we have to do now is make the difference between points smaller and smaller until it approaches zero. That is, to get the derivative we take the limit as $h \to 0$:

$$\lim_{h \to 0} \frac{f(x + h) - f(x)}{h} = f'(x) = \frac{d}{dx}f(x) = \frac{dy}{dx}.$$

This equation gives the slope of the tangent line to the function at any point. Note that the derivative is also a function in its own right: it varies with x and returns the value $f'(x)$.

The only thing missing from this formal presentation is the existence of derivatives. Derivatives can only be calculated at points at which limits exist because derivatives are in a sense limits themselves, but it is also necessary for a function to be continuous at a point to have a derivative at that point. In other words (recalling the logic in the section on proofs in Chapter 1), a function differentiable at a point is also necessarily continuous at the point. However, continuity is not sufficient for differentiability. For example, continuous func-

[7] You often see Δx used instead of h. There is no difference; these variables have no intrinsic meaning.

tions with "kinks," such as the absolute value function $f(x) = |x|$ at $x = 0$,[8] or with vertical tangent lines at a point, such as $f(x) = x^{\frac{1}{3}}$ at $x = 0$, don't have derivatives at those points. However, these functions are differentiable at all points *other than* $x = 0$.

5.3.3 Some Examples

Let's go through some examples to make this concrete, starting with the linear function $f(x) = 3x$. We calculate its derivative by plugging into the definition:

$$\begin{aligned} f'(x) &= \lim_{h\to 0} \frac{f(x+h) - f(x)}{h} = \lim_{h\to 0} \frac{3(x+h) - 3x}{h} \\ &= \lim_{h\to 0} \frac{3x + 3h - 3x}{h} = \lim_{h\to 0} \frac{3h}{h} = 3. \end{aligned}$$

So the derivative is 3. We didn't even need to use the limit, because a linear function has a constant slope, equal to the slope of its tangent line (which overlaps the function, as noted in the previous section). In other words, only one line is tangent to any point on a linear equation, and it is the linear equation itself.

Now let's do a slightly more complicated one, $f(x) = x^2$. Again we plug into the definition:

$$\begin{aligned} f'(x) &= \lim_{h\to 0} \frac{f(x+h) - f(x)}{h} = \lim_{h\to 0} \frac{(x+h)^2 - x^2}{h} \\ &= \lim_{h\to 0} \frac{x^2 + 2xh + h^2 - x^2}{h} = \lim_{h\to 0} \frac{2xh + h^2}{h} = \lim_{h\to 0} 2x + h = 2x. \end{aligned}$$

We see that our earlier guess was correct: the derivative of $f(x) = x^2$ is $2x$. This means that the slope of the tangent line increases with x, and there is a different tangent line at each point.

Next, consider an even more complex derivative $f(x) = x^3 + x - 5$:

$$\begin{aligned} f'(x) &= \lim_{h\to 0} \frac{f(x+h) - f(x)}{h} \\ &= \lim_{h\to 0} \frac{(x+h)^3 + (x+h) - 5 - (x^3 + x - 5)}{h} \\ &= \lim_{h\to 0} \frac{x^3 + 3x^2h + 3xh^2 + h^3 + x + h - 5 - x^3 - x + 5}{h} \\ &= \lim_{h\to 0} \frac{3x^2h + 3xh^2 + h^3 + h}{h} \\ &= \lim_{h\to 0} 3x^2 + 3xh + h^2 + 1 = 3x^2 + 1. \end{aligned}$$

[8] Draw the function to convince yourself that it has a negative constant slope for negative x and a positive constant slope for positive x, so that the point at which they meet can't have a well-defined instantaneous rate of change.

There are several things to note here. First, the constant term, 5, played absolutely no part—it cancels in the definition. Since this should be true for all constants, we should expect their derivatives to be zero. This makes sense: constants don't change by definition, so their rate of change should be zero.[9] Second, we know the slope of the linear term, x, is 1, by our first example. It appears as 1 by itself in the answer, suggesting that we'll be able to treat added terms separately. Third, only the second leading term in the expansion of $(x + h)^3$ (the term in x^2) survives the limit. This will be true in general, as the second term has h only to the first power, allowing it to cancel with the denominator and leave no h to go to zero in the limit. This will make longer expansions easier, as we can ignore all the other terms.

Finally, a derivative that will prove particularly useful in the next chapter $f(x) = \frac{1}{x}$:

$$\begin{aligned} f'(x) &= \lim_{h \to 0} \frac{f(x+h) - f(x)}{h} \\ &= \lim_{h \to 0} \frac{\frac{1}{x+h} - \frac{1}{x}}{h} \\ &= \lim_{h \to 0} \frac{\frac{x-(x+h)}{x(x+h)}}{h} = \lim_{h \to 0} \frac{-1}{x(x+h)} = -\frac{1}{x^2}. \end{aligned}$$

5.3.4 Multivariate Functions and the Partial Derivative

In general, we are keeping to functions of one variable in this part of the book, saving multivariate calculus for Part V. However, we recognize that some students will not get to that part on a first read-through despite our exhortations in the preface to read the first chapter of Part V. Given the importance of partial derivatives to understanding statistical work and formal theory, we offer an exceedingly brief discussion of the partial derivative here and again entreat our readers to visit the first chapter of Part V for more.

Without providing any formal definitions, let's extend our functions to consider two variables.[10] Write these as $f(x, z)$. Examples would be $f(x, z) = 3x^2z + 2z$ or $y = \sqrt{x^2 + z^2}$. Say we want to know how y changes with x, holding z constant. This is a fundamental question in both statistics and formal theory. We can find this via what is known as the **partial derivative**. It is written $\frac{\partial}{\partial x} f(x, y)$, or sometimes simply ∂_x, and means "treat every variable other than x as a constant, and just take the derivative with respect to x." Consequently, partial derivatives are no more complex to take than derivatives of one-dimensional functions. For instance, if $f(x, z) = 3z^3 - 3z^2 + \sqrt{z} + x$, then $\frac{\partial}{\partial x} f(x, z) = 1$, since only the last term has an x in it and it is linear in x. In essence, relative

[9] This is why we had to add the constant C in the integral we considered in Section 1 of this chapter. The derivative eliminates all information about constants in a function, which removes information about translations.

[10] We show in Part V of the book that all of this extends to more than two variables.

to the partial derivative with respect to x, $f(x,z) = 3z^3 - 3z^2 + \sqrt{z} + x$ is no different from $f(x,z) = x$.

Partial derivatives appear commonly in early statistics courses, which is why we bring them up here. The context is usually of an empirical model such as $y = \beta_0 + \beta_1 x + \beta_2 z + \beta_3 xz$. Here we have both a direct dependence on x by itself, and an interaction term with z. If we want to know how x affects y, holding z constant, we need to take the partial derivative with respect to x. Since this treats z as a constant, this partial derivative is $\frac{\partial}{\partial x} f(x,z) = \beta_1 + \beta_3 z$. When interpreting the effect of x on y, this is the expression of interest, which varies with z.

For example, Mondak et al. (2010) hypothesize that the size of a person's political discussion network will influence the impact that her personality characteristics will have on her political attitudes. They show that the extent to which a person is extroverted or conscientious influences her political attitudes, but that the size of the impact of extroversion or conscientiousness varies with the number of people with whom she has conversations about politics. Similarly, Hopkins (2010) argues that the local and national context interact to produce Americans' political attitudes toward immigrants. He finds that Americans' interpretation of demographic change varies depending on the local increase in immigrant population. Both of these studies present results based on discrete changes in the value of the intervening variable (network size for Mondak et al., 2010 and local increase in immigrant population for Hopkins, 2010), and use a graph to display those results. Alternatively they could have calculated, and reported, the first derivative. In statistical work economists frequently calculate and report derivatives (e.g., Eeckhout, Persico, and Todd, 2010) and have even adopted the convention of referring to these derivatives as marginal effects. In their statistical work, political scientists often make reference to marginal effects, though we use the term loosely, and more often than not it is used to describe discrete change, rather than continuous change arising from the derivative. Familiarity with the derivative permits greater precision in presentation as well as greater flexibility in choice about what results are of most interest given one's research question and hypothesis.

5.4 SUMMARY

This is the take-home point: derivatives help us study rates of change in continuous functions. The derivative is the instantaneous rate of change in a function at a point. If you are comfortable with discrete change (i.e., the difference between two values), then you will be able to master the derivative: it is the change in a function at a point rather than over two points. The key is to become familiar with the notation. The derivative is useful because it permits us to make precise statements about changes in relationships, and social scientists are often interested in the change that one variable causes in another, including the case where we want to hold all other variables constant. In a multivariate function the partial derivative provides us with precisely that information.

You will encounter the opinion that bringing mathematics (and especially calculus) to bear on the study of politics is a misguided effort motivated by "physics envy." This is a canard. One employs the tools of calculus to study politics if one wishes to sharpen one's deductions and/or one's ability to draw inferences by using statistical inference. Those are the primary motives that have merit. If one feels that one has sufficient specificity in one's theory, then there is no reason to employ calculus to develop theory. Similarly, if one feels that one can draw inferences sufficiently well without appeal to inferential statistics, then there is no reason to appeal to calculus as it is used in statistics. However, to the extent that one is dissatisfied with the rigor of deduction in one's theory, one may want to appeal to mathematics generally and, in some cases, calculus in particular to strengthen one's ability to deduce the implications of sets of assumptions and conjectures. And if one wishes to employ statistical inference, mathematics is essential and calculus is often helpful for understanding the appropriate use of that tool.

Finally, we remind our readers that many students find calculus quite challenging at first. However, as long as one can recall that the underlying concepts are fairly straightforward, we believe that it is merely a matter of time and practice before calculus becomes a useful tool in one's belt. To that end, we offer a few online resources. Those students who have never studied calculus before will likely find Daniel Kleitman's Calculus for Beginners and Artists site helpful: `http://www-math.mit.edu/~djk/calculus_beginners/`. Dan Sloughter's online text is also generally considered quite useful: `http://math.furman.edu/~dcs/book/`. Students who have studied calculus but would like an online refresher might try Harvey Mudd College's Calculus Tutorial, which includes review and quizzes: `http://www.math.hmc.edu/calculus/tutorials/`. Finally, the following introduction to the fundamental theorem of calculus is useful and contains some helpful Java figures: `http://ugrad.math.ubc.ca/coursedoc/math101/notes/integration/ftc.html`.

5.5 EXERCISES

1. Visit Daniel Kleitman's "Derivative and Tangent Line Applet" page here `http://www-math.mit.edu/~djk/calculus_beginners/tools/tools04.html`. Select the $y = x^2$ function, and change the x min and max values to -5 and 5. Change the y min and max values to -5 and 135. Click the Plot Function box to generate a new graph. Now click the Show Derivative box, and drag the slide bar to change the values of x. What is the relationship between the tangent line shown and the function being drawn? Now click the Show Second Derivative box and move the slide bar again. What function is now being drawn, and what is its relationship to the tangent

lines? (We discuss the second derivative in Chapter 8. For now we just want you to look at it.) Repeat this for the function $y = x^3$.

2. Use the definition of the derivative to find the derivative of y with respect to x for the following:

   a) $y = 6$.
   b) $y = 3x^2$.
   c) $y = x^3 - 2x^2 - 1$.
   d) $y = x^4 + 5x$.
   e) $y = x^8$.
   f) $y = 4x^3 - x + 1$.
   g) $y = 2x^4 + x^2 - 1$.
   h) $y = 5x^5 + 4x^4 + 3x^3 + 2x^2 + x + 1$.
   i) $y = 7x^4 - 9x^3 + 5x + 117$.
   j) $y = 27x^3 + 5x^2 - x + 13$.

3. For each of the following, first sketch $f(x)$. Then draw a secant between the points on f corresponding to $x = 1$ and $x = 3$ and to $x = 1$ and $x = 2$. Using these as a guide, draw the tangent line at $x = 1$. Guess its slope. Now use the definition of the derivative to find $f'(x)$. Finally, sketch the derivative and compare it to the tangent at $x = 1$.

   a) $f(x) = 2x^2 + 7$.
   b) $f(x) = x^3 - x + 1$.

4. For each of the following, find the partial derivative with respect to x.

   a) $f(x, z) = 3zx + 2z$.
   b) $f(x, z) = x^2 + 2z^2$.
   c) $f(x, z) = 3z^2 - z + 1$.

5. For each of the following, find the partial derivative with respect to z.

   a) $f(x, z) = 3x + 3z + 3$.
   b) $f(x, z) = 9x^2 + 3z^2$.
   c) $f(x, z) = 5xz + 7xz^2 + 9x^z$.
   d) $f(x, y, z) = 3x + 4y + 5z + 6$.
   e) $f(x, y, z) = 11z + 3x^2y + 5x^2z + 7z^2y$.
   f) $f(x, y, z) = 4x^2y^2z^2 + 8xyz + 12xy + 14x$.
   g) $f(x, y, z) = 8xyz^2 + 10x^2y^2 + 12x^2y + 14x^2z^2$.

Chapter Six

The Rules of Differentiation

In the previous chapter we introduced the derivative as an operator that takes a function and returns another function made up of the instantaneous rate of change of the first function at each point. We also presented the definition of the derivative in terms of the limit of the discrete rate of change in a function across two points as the difference between the points went to zero. From this definition we could, with some algebra, compute derivatives of some polynomial functions. We'll need to calculate derivatives for more complex functions in political science, however. Rather than go back to the definition each time, it will help to have some rules for the differentiation of specific functions and types of functions we can call on. This chapter presents those rules.

A common method of presentation of this material provides a list of many seemingly disconnected rules. While that type of presentation provides clear rules and an easy memorization device, we feel that it can lead to the perception that calculus is somehow more complicated than it is, fraught with specificities that one must memorize. In reality, most of the commonly used rules stem from a handful of properties of the derivative operator, and most of these properties may be deduced from the definition. To help make these connections, we present the material accordingly.

In Section 1 we develop the rules typically called the sum rule, product rule, quotient rule, and chain rule largely from the definition of the derivative itself. In Section 2 we use these rules and the definition of the derivative to offer derivatives of many common and special functions used in political science. In Section 3 we relent and provide a summary accounting of the rules for differentiation for ready reference, along with a discussion as to when to use each of the rules. If you already know or don't care about calculus but need to calculate some derivatives, you can skip to that section. You can also skip to that section if you are finding the derivations of the rules too difficult to follow the first time through. We expect that many students who have not had calculus before will be in this boat, and might benefit from some experience manipulating derivatives before going back to see whence came the rules they used to perform the manipulations.

6.1 RULES FOR DIFFERENTIATION

The definition of the derivative provided in the previous chapter implies several properties of the derivative that are useful in computation. We alluded to some of them in the previous chapter's examples; now we present them formally.

6.1.1 The Derivative Is a Linear Operator

In Chapter 3 we defined a linear function as one that satisfies the properties of additivity and scaling. A linear operator satisfies pretty much the same definition. That is, an operator like $\frac{d}{dx}$ is linear if $\frac{d(f+g)}{dx} = \frac{df}{dx} + \frac{dg}{dx}$ and $\frac{d(cf)}{dx} = c\frac{df}{dx}$ for any two (differentiable!) functions f and g and any constant c. Is the derivative linear? Let's check the definition

$$\frac{df(x)}{dx} = f'(x) = \lim_{h\to 0} \frac{f(x+h) - f(x)}{h}.$$

First we try additivity

$$\begin{aligned} \frac{d(f+g)}{dx} &= \lim_{h\to 0} \frac{(f(x+h) + g(x+h)) - (f(x) + g(x))}{h} \\ &= \lim_{h\to 0} \frac{(f(x+h) - f(x)) + (g(x+h) - g(x))}{h} \\ &= \lim_{h\to 0} \left(\frac{f(x+h) - f(x)}{h} + \frac{g(x+h) - g(x)}{h} \right) \\ &= \frac{df}{dx} + \frac{dg}{dx}. \end{aligned}$$

That checks out, thanks to the fact that the limit of a sum is the sum of the limits. Scaling is easier, since the limit of a product is the product of the limits and the limit of a constant is a constant

$$\begin{aligned} \frac{d(cf)}{dx} &= \lim_{h\to 0} \frac{c(f(x+h)) - c(f(x))}{h} \\ &= \lim_{h\to 0} c \left(\frac{f(x+h) - f(x)}{h} \right) \\ &= c\left(\frac{df}{dx}\right). \end{aligned}$$

That was a little bit of algebra, but it got us something important: *the derivative is a linear operator.* This supersedes what are commonly referred to as the sum and difference rules: $(f+g)' = f' + g'$ and $(f-g)' = f' - g'$. Since the derivative is linear, we instead have the more general $(af + bg)' = af' + bg'$ for any two constants a and b and any two (differentiable) functions f and g. And, rather than having to memorize sum and difference rules, we just need to know that the derivative is a linear operator and know the characteristics of linearity, which come up in lots of contexts beyond just calculus.

Let's try some examples before moving on. Let $h(x) = x + x^2$. Call $f(x) = x$ and $g(x) = x^2$. Linearity then yields $h'(x) = f'(x) + g'(x)$. We can look up these derivatives in the previous chapter to get $\frac{dh(x)}{dx} = 1 + 2x$. We can add a constant a to this too without changing anything; if $h(x) = (x + a) + x^2$, then still $h'(x) = 1 + 2x$, because the derivative of a constant is zero, as we saw in the previous chapter. So don't let yourself get distracted by constants. Either they vanish in the derivative if they are added, or they get pulled out to multiply the relevant terms in the derivative. For example, if $f(x) = ax$, where a is a constant, then $f'(x) = a$ since the derivative of x is one. We can expand these ideas to any number of terms. For example, if $h(x) = x^3 + 6x^2 - 3x + 1$, then $h'(x) = 3x^2 + 12x - 3$. All we've done here is take the derivative of each term separately. Let's work out this last one to see the application of the rule a bit more carefully

$$\begin{aligned}
\frac{dh(x)}{dx} &= \frac{d(x^3 + 6x^2 - 3x + 1)}{dx} \\
&= \frac{d(x^3)}{dx} + \frac{d(6x^2)}{dx} + \frac{d(-3x)}{dx} + \frac{d(1)}{dx} \\
&= \frac{d(x^3)}{dx} + 6\frac{d(x^2)}{dx} - 3\frac{d(x)}{dx} + \frac{d(1)}{dx} \\
&= 3x^2 + 6(2x) - 3(1) + 0 \\
&= 3x^2 + 12x - 3.
\end{aligned}$$

6.1.2 Chain Rule

Believe it or not, the fact that the derivative is a linear operator takes us much of the way toward providing rules for differentiation. To go further, though, we need to know how to deal with composite functions. Recall from Chapter 3 that a composite function looks like $g(f(x))$. To see why we're bringing this up now, note that most complex functions can be written as composite functions. Take, for example, $h(x) = e^{2x^2}$. If we let $f(x) = 2x^2$ and $g(x) = e^x$, then $h(x) = g(f(x))$. Thus we can break complicated functions down into the composition of simpler functions. If we can devise a rule for differentiating composite functions, a rule known as the **chain rule**, then we can simplify the differentiation of complex functions immensely. Given this incentive, we begin this task.[1]

We let f and g be differentiable, and use the definition of the derivative to get $\frac{dg(f(x))}{dx}$. We start with

$$\frac{dg(f(x))}{dx} = \lim_{h \to 0} \frac{g(f(x + h)) - g(f(x))}{h}.$$

[1]We should note that the following "proof" for the chain rule is "hand-wavy" even by the standards of this book. But the idea is more or less correct.

Then we multiply and divide by $f(x+h) - f(x)$:[2]

$$\frac{dg(f(x))}{dx} = \lim_{h\to 0} \frac{g(f(x+h)) - g(f(x))}{f(x+h) - f(x)} \frac{f(x+h) - f(x)}{h}.$$

Since f is assumed differentiable, it is also continuous (see the previous chapter), so as $h \to 0$, $f(x+h) - f(x) \to 0$. Let's call $z = f(x+h) - f(x)$ and $u = f(x)$; rewriting the derivative with these substitutions and using the fact that the limit of a product is the product of a limit yields

$$\frac{dg(f(x))}{dx} = \lim_{z\to 0} \frac{g(u+z) - g(u)}{z} \lim_{h\to 0} \frac{f(x+h) - f(x)}{h}.$$

We're almost done. The second term on the right-hand side (RHS) is $f'(x)$. The first term is $g'(u) = g'(f(x))$, where we've plugged back in our substitution for x.[3] Finally, this gives us the chain rule:

$$\frac{dg(f(x))}{dx} = \frac{dg(u)}{du}\frac{du}{dx}, \text{ where } u = f(x).$$

In words, take the derivative of the outer function (g of x) evaluated at the inner function (f of x), then take the derivative of the inner function (f of x), and multiply the two terms. We can also write it as $(g(f(x)))' = g'(f(x))f'(x)$. The key concept here is that one can separate the derivative of a composite function into two derivatives: one of the outer function, evaluated at the inner function, and one of the inner function, evaluated at x.

This may have seemed complicated, but the chain rule is perhaps the most useful rule in differential calculus, and one of only a handful you should actually try to memorize.[4] Not only is it useful in its own regard, but it gives us many other rules as well.[5] But before getting into that, we start with some examples of its use.

Let's start with $h(x) = x^9$. We show below this is a surprisingly easy derivative to take, but right now you know only the definition of the derivative, and you might not feel like expanding $(x+h)^9$. So let's instead call $f(x) = x^3$, $g(x) = x^3$, and $h(x) = g(f(x))$; we check that $h(x) = (x^3)^3 = x^9$, recalling our rules on exponentiating exponents from Chapter 3. We've now seen several times that the derivative of x^3 is $3x^2$. So, plugging this into the chain rule, and letting $u = x^3$, we have $\frac{dg(f(x))}{dx} = \frac{dg(u)}{du}\frac{df(x)}{dx} = (3u^2)(3x^2) = 3(x^3)^2(3x^2) = 9x^8$. Not only is this much easier than expanding something to the ninth power, it

[2]For those keeping score at home, here's the "hand-wavy" part. The possibility of dividing by zero means a formal proof needs more care at this step than we offer.

[3]You'll see this technique a great deal in the next chapter, as the method of substitution in integration is closely related to the chain rule in differentiation.

[4]You will encounter many examples in political science publications, but for a specific one that uses the chain rule to find the long-run effects of a change in one variable in the context of a discrete difference equation, see Brandt and Williams (2001).

[5]It also proves immensely important in multivariate analysis, but we leave this until Part V of the book.

also lets us produce the derivative for a polynomial of any positive power we want, just by working up from ones we know already.

Of course, the two functions composed need not be equal. Let's return to our first example in this section, $h(x) = e^{2x^2}$. Again, we let $f(x) = 2x^2$ and $g(x) = e^x$, so $h(x) = g(f(x))$. We show below that $\frac{de^x}{dx} = e^x$, and we can use the fact that the derivative is a linear operator to find that the derivative of $2x^2$ is $2\frac{dx^2}{dx} = 2(2x) = 4x$. Given this, we can apply the chain rule to get $h'(x) = g'(u)f'(x) = e^u(4x) = 4xe^{2x^2}$, where in the last step we plugged back in for u.

One more example: let $h(x) = (2x - a)^2$. We'll set $g(x) = x^2$ and $f(x) = 2x - a$, so that $h(x) = g(f(x))$. Then $h'(x) = (g(f(x)))' = g'(f(x))f'(x) = 2(2x - a)(2) = 8x - 4a$, where in the second-to-last expression we evaluated $g'(x)$ at $f(x)$.

The major step in using the chain rule is to figure out how to assign $g(x)$ and $f(x)$. We'll talk more about how to do this in Section 3 below, but in general, you want each function to be one that you can differentiate easily. The tricky part in using the chain rule is to make sure that you substitute the inner function, $f(x)$, for the x in $g'(x)$. One way to make sure you do this right is to set $u = f(x)$ and then differentiate $g(u)$ with respect to u. This is what we'll usually do, and what we did in our first two examples. All you have to remember in this case is to plug back in for u at the end.

However, you won't always see it done that way, so it's good to get used to using the chain rule without this substitution, instead directly using $(g(f(x)))' = g'(f(x))f'(x)$ as we did in the last example. That is, you first calculate $g'(x)$ as if the function were not nested at all, and then you evaluate it at $f(x)$, which means, practically, replacing each x in $g'(x)$ with $f(x)$.[6] This is different from multiplying $g'(x)$ and $f(x)$.

There are many more examples like these, and we provide some at the end of the chapter. For now, however, we turn to using the chain rule to derive other rules. We start by considering the inverse function of $f(x)$, $f^{-1}(x)$. Recall that for an inverse function, $f(f^{-1}(x)) = x$. We can take the derivative of both sides, which is allowed since neither derivative is undefined or infinite. The RHS derivative is 1. To find the LHS, we use the chain rule to get

$$\frac{df(f^{-1}(x))}{dx} = \frac{df(u)}{du}\frac{df^{-1}(x)}{dx}, \text{where } u = f^{-1}(x).$$

The LHS of this equation is equal to one, as it is $\frac{dx}{dx}$. Assuming the derivative of the function exists and is not zero, the first term on the RHS is $f'(f^{-1}(x))$,

[6] If this is weird to you, think of evaluating a derivative at a point a. If we just had $g'(x)$, we could evaluate it at a to get $g'(a)$. (Remember that $g'(x)$ is a function, just as $g(x)$ is.) So this means replacing each x in $g'(x)$ with a. But instead we have a nested function, and we have to evaluate $g'(x)$ at the value that $f(x)$ maps a into: $f(a)$. So instead of replacing each x in $g'(x)$ with a, we replace each with $f(a)$. Both a and $f(a)$ are just values, so there's nothing much different here either way. This is true for all x in the domain, though, so we have to replace each x with a corresponding $f(x)$.

and we can divide by it. Putting these two facts together allows us to rearrange the equation to get the inverse function rule

$$\frac{df^{-1}(x)}{dx} = \frac{1}{f'(f^{-1}(x))}.$$

So we can get the derivative of the inverse of the function from the derivative of the function itself. Using this rule is no different from using the chain rule, so we leave examples until the exercises at the end of the chapter.

6.1.3 Products and Quotients

We can now deal with sums and differences of functions, as well as composite functions and functions multiplied by constants. The only remaining rule of immediate use is that for products and quotients of functions. Luckily, only the first need be memorized, and this too can be derived from the definition of the derivative if you happen to forget it.

We should note before this derivation that in contrast to sums and differences of functions, one does not simply differentiate the terms of a product. There are two ways to go about the task. The easier route is not always available, but if one can, do the multiplication first and then take the derivative of the resulting product. For example, if $y = f(x) \times g(x) = (2x + 3) \times (x^2 - 15)$, then we can take the product, yielding $2x^3 + 3x^2 - 30x - 45$. Differentiating yields $6x^2 + 6x - 30$.

Unfortunately, one cannot always multiply out before differentiating. In such cases the product rule is needed. If we have two functions, $f(x)$ and $g(x)$, then the derivative of their product is

$$\frac{d(f(x)g(x))}{dx} = \frac{df(x)}{dx}g(x) + f(x)\frac{dg(x)}{dx}.$$

In words, the **product rule** states that the derivative of the product is a sum of product terms. In each such term, one and only one of the constituent functions is differentiated. Or, if you like, the derivative of the product of two terms is the derivative of the first term times the second term, plus the first term times the derivative of the second. Clear as mud, eh? We try to clarify by first deriving it, and then providing an example.

Let f and g be differentiable functions. The definition of the derivative gives

$$\frac{d(f(x)g(x))}{dx} = (fg)' = \lim_{h\to 0} \frac{f(x+h)g(x+h) - f(x)g(x)}{h}.$$

We next add and subtract the term $f(x)g(x+h)$ within the RHS to yield

$$(fg)' = \lim_{h\to 0} \frac{f(x+h)g(x+h) - f(x)g(x+h) + f(x)g(x+h) - f(x)g(x)}{h}.$$

Grouping terms and making use of the properties of the limit gives

$$\begin{aligned}(fg)' &= \lim_{h\to 0} \frac{f(x+h)g(x+h) - f(x)g(x+h)}{h} \\ &\quad + \lim_{h\to 0} \frac{f(x)g(x+h) - f(x)g(x)}{h}.\end{aligned}$$

Pulling out the common terms in each case and making use again of the properties of the limit gives

$$\begin{aligned}(fg)' &= \left(\lim_{h\to 0} \frac{f(x+h) - f(x)}{h}\right)\left(\lim_{h\to 0} g(x+h)\right) \\ &\quad + \left(\lim_{h\to 0} f(x)\right)\left(\lim_{h\to 0} \frac{g(x+h) - g(x)}{h}\right).\end{aligned}$$

Finally, taking the limits gives us the rule:

$$(fg)' = f'g + fg'.$$

Now for the example. Consider the functions above: $y = f(x)g(x) = (2x + 3)(x^2 - 15)$. Use the product rule

$$\begin{aligned}\frac{dy}{dx} &= \frac{d(2x+3)}{dx}(x^2 - 15) + (2x+3)\frac{d(x^2-15)}{dx} \\ &= (2)(x^2 - 15) + (2x+3)(2x) \\ &= (2x^2 - 30) + (4x^2 + 6x) \\ &= 6x^2 + 6x - 30.\end{aligned}$$

We get the same answer regardless of which way we go about it, which suggests that we have done it properly.

Instead of considering $y = f(x)g(x)$, we might also need to calculate $y = \frac{f(x)}{g(x)}$. There is a **quotient rule** for this that people often memorize (note that we must have $g \neq 0$)

$$\frac{d}{dx}\frac{f(x)}{g(x)} = \frac{\frac{df(x)}{dx}g(x) - f(x)\frac{dg(x)}{dx}}{g(x)^2}.$$

You are free to memorize this as well, and it's not so hard. The numerator is just like the product rule but with a negative sign instead of a positive one between the terms, and the denominator is just the square of the second term. But the quotient rule is just an application of the product rule combined with the chain rule. Let us explain. The only difference between the quotient and the product under consideration is that we are dividing by g instead of multiplying by it. So let's go back to multiplying by letting $h(x) = \frac{1}{g(x)}$. Now $y = f(x)h(x)$ as before. We already know that $y' = f'h + fh'$ from the product rule, so all we need is h'. We can think of h as a composite function: $h(x) = k(g(x))$, where $k(x) = \frac{1}{x}$. From the examples in the last chapter, we know that $k'(u) = -\frac{1}{u^2}$.

By the chain rule, then, $h'(x) = -\frac{1}{(g(x))^2}g'(x)$. We can plug this back into the result from the product rule to get $y' = \frac{f'}{g} - \frac{fg'}{g^2}$. Multiplying numerator and denominator of the first term by g and combining terms yields the quotient rule $y' = \frac{f'g - fg'}{g^2}$.

At this point you might be saying, "Of *course* I'm going to memorize the quotient rule! Who wants to re-derive all that every time?" That is a good point, but it misses the intent of this derivation. The fact that one can get the quotient rule from the product rule and the chain rule is indicative of the fact that one can compute derivatives using just the product rule and the chain rule; one does not need the quotient rule at all. Whether or not you want to bother memorizing it, then, depends on your facility with algebraic manipulation and fractions.

Let's try an example. Consider a situation where we are interested in the quotient $y = \frac{f(x)}{g(x)}$. The derivative of a ratio would be an example. Let $f(x) = (3x - 7)$ and $g(x) = (x^3 + 6)$. We need to find $\frac{d}{dx}\left(\frac{3x-7}{x^3+6}\right)$.

First we use the quotient rule to differentiate y

$$\begin{aligned}\frac{dy}{dx} &= \frac{\frac{d(3x-7)}{dx}(x^3+6) - (3x-7)\frac{d(x^3+6)}{dx}}{(x^3+6)^2} \\ &= \frac{(x^3+6)(3) - (3x-7)(3x^2)}{36 + 12x^3 + x^6} \\ &= \frac{(3x^3+18) - (9x^3 - 21x^2)}{36 + 12x^3 + x^6} \\ &= \frac{-6x^3 + 21x^2 + 18}{36 + 12x^3 + x^6}.\end{aligned}$$

Now we use the product and chain rules. Let $h(x) = \frac{1}{x^3+6}$. Then $h' = -\frac{1}{(x^3+6)^2}(3x^2)$ from the chain rule and $y' = f'h + fh'$ from the product rule. Since $f' = 3$, we can plug in to get

$$\begin{aligned}\frac{dy}{dx} &= \frac{3}{x^3+6} - \frac{(3x-7)(3x^2)}{(x^3+6)^2} \\ &= \frac{(x^3+6)(3) - (3x-7)(3x^2)}{36 + 12x^3 + x^6} \\ &= \frac{(3x^3+18) - (9x^3 - 21x^2)}{36 + 12x^3 + x^6} \\ &= \frac{-6x^3 + 21x^2 + 18}{36 + 12x^3 + x^6}.\end{aligned}$$

As you can see, the second method requires a bit more algebra and keeping track of a few more terms, but it doesn't require memorizing the quotient rule. They both give the same answer.

6.2 DERIVATIVES OF FUNCTIONS

To sum up the previous section, we have three useful rules of differentiation: the chain rule, the product rule, and the fact that the derivative is a linear operator. Other rules, such as the quotient, sum, difference, and inverse function rules, can all be derived from these three. Further, all three can be derived (more or less) from the definition of the derivative. We weren't pulling a fast one when we said there's not a lot to this: the derivative is the limit of discrete change in a function as the distance over which the change is measured goes to zero, and the rules of differentiation follow from this definition. If you get the concept of a derivative and can do a little algebra, you can get the rules.

This isn't quite fair, though, as we haven't told you how to differentiate specific functions yet, beyond those polynomials for which we could easily use the definition of the derivative. To remedy this, in this section we present derivatives of some common and special functions. Again, though, there will not be many of these to memorize.

6.2.1 Polynomials and Powers

By now you have most likely noticed that nearly all the functions you've dealt with can be written as a sum, difference, product, or quotient of terms like ax^n, where a and n are constants. For example, if $n = 0$, then we have $ax^0 = a$, which is just a constant. If $n = 1$, then we have $ax^1 = ax$, which is a linear function. If $n = 2$ we have ax^2, which is a quadratic function. If we add all three of these together we get a polynomial that describes a parabola. If $n = -1$, we get $ax^{-1} = \frac{a}{x}$. Finally, if $n = \frac{1}{2}$, we get $ax^{\frac{1}{2}} = a\sqrt{x}$, and if $n = \frac{1}{3}$, we get $ax^{\frac{1}{3}} = a\sqrt[3]{x}$.

We know we can treat each term in a sum or difference separately, thanks to the fact that the derivative is a linear operator. For this same reason we can pull the constant a out of the derivative. We also know how to deal with product, rational, and composite functions thanks to the product, quotient, and chain rules, respectively. Thus, if we can just differentiate anything of the form x^n, where n is a rational number, we can address a large proportion of derivatives in political science.

It turns out that it is easy to take this derivative. You might even have guessed the pattern from earlier examples such as $(x^3)' = 3x^2$ and $(x^9)' = 9x^8$. It's just

$$\frac{dx^n}{dx} = nx^{n-1}.$$

In words, move the value of the exponent to the front of the variable and subtract one from the exponent. This is true for any rational n and real x, other than times when x^n would be poorly defined.[7] This is generally referred to as the **power rule**.

[7]For example, if x were negative and n were one-half, or if both n and x were zero.

How did we get this? For n a positive integer there's a proof that is worth a moment of your time. Note that the first two terms in $(x+h)^n$ are x^n and nhx^{n-1}. All other terms have higher powers in h. As we noted in the previous chapter, all terms with powers in h in the numerator of the definition of the derivative will vanish in the limit, so all we need to care about are these first two terms. So that leaves

$$\frac{dx^n}{dx} = \lim_{h\to 0} \frac{(x+h)^n - x^n}{h} = \lim_{h\to 0} \frac{x^n + nhx^{n-1} - x^n}{h} = nx^{n-1},$$

where in the last step we canceled the x^n in the numerator and the h from top and bottom, leaving no h to worry about in the limit.

For more general n, the easiest way to see the rule is to use a slightly different definition of the derivative. Consider the derivative at a point c. As x approaches c, the difference $c-x$ goes to zero, just as h does. So let $h = c - x$. Then the derivative definition becomes

$$f'(c) = \lim_{x\to c} \frac{f(x) - f(c)}{x - c}.$$

If $f(x) = x^n$, then the numerator is $x^n - c^n$. We can factor out $(x-c)$ to get

$$(x-c)(x^{n-1} + cx^{n-2} + \ldots + c^{n-2}x + c^{n-1}).$$

After canceling the $(x-c)$ from numerator and denominator, taking the limit, and adding all the n identical terms, we get

$$(x^n)'|_c = \lim_{x\to c} x^{n-1} + cx^{n-2} + \ldots + c^{n-2}x + c^{n-1} = nc^{n-1}.$$

The $|_c$ here means evaluate the derivative at $x = c$. Since this is true for any point c, we have $\frac{dx^n}{dx} = nx^{n-1}$.

The formula $\frac{dx^n}{dx} = nx^{n-1}$ is very flexible, and we advise you to memorize this relatively simple relation rather than try to reproduce it. Let's try some examples.

First, let's consider the simplest case, a constant. Let $y = f(x) = 5$. What is the derivative of this function? Our formula says that for $y = 5 = 5x^0$, $y' = 0(5x^{-1}) = 0$. Does this make sense? Recall that the derivative is the instantaneous rate of change of the function. Does y change? No, not in this case. Thus, the rate of change should be zero, as should the derivative of any constant. Some people call this the **constant rule**, but it is a simple application of our formula for x raised to a power.[8]

Now try a linear function, $y = f(x) = ax$. Since $x = x^1$ and $1x^{1-1} = 1x^0 = 1$, our formula implies that $f'(x) = \frac{d(ax)}{dx} = a$. The derivative of a linear function is a constant, which is just the slope of the line.

---

[8] As usual, we're ignoring some technical details, in this case the value at $x = 0$. But the constant rule we offer is true regardless, and this provides the intuition.

Higher exponents are more straightforward. Change the function to $y = g(x) = x^2$. Pulling down the exponent and replacing it with an exponent one less, as our formula requires, yields $2x^1 = 2x$. And we've already seen that if $y = x^3$, then $y' = 3x^2$, and if $y = x^9$, then $y' = 9x^8$.

We can combine this formula with the rules from the previous section to calculate the derivatives of rather difficult-looking functions. For example, what is the derivative of $y = 10x^3 + x^2 - 3x$? The answer is $30x^2 + 2x - 3$. Not too bad, eh?[9] And it only required linearity and the power rule. Or if $f(x) = (2x + 5)$ and $g = (x^3 + 2x + 1)$, you can multiply them out to get $y = f(x)g(x)$ and differentiate this, but you could also use the product rule: $y' = f'g + fg' = (2)(x^3 + 2x + 1) + (2x + 5)(3x^2 + 2)$. Finally, you can differentiate functions like $y = (x - c)^3$ by using the chain rule. Assign $g(x) = x^3$ and $f(x) = x - c$. Then $y' = g'(f(x))f'(x) = 3(x - c)^2$, since $f' = 1$. (This is in general true; because constants differentiate to zero, owing to the chain rule one can treat $(x \pm c)$ as x when differentiating. Note also that c can be left as a constant here; you need not substitute a real number to differentiate.) The point is, all polynomials are now differentiable by you via the application of only a handful of rules, all derived from the same definition.

Recall also that one can use exponents to represent fractions and radicals (or roots). For example, $x^{-n} = \frac{1}{x^n}$ and $x^{\frac{1}{n}} = \sqrt[n]{x}$. How does one apply the power rule to these functions? There are no tricks involved—just apply the rule. Let's try x^{-2}. Applying the rule yields $-2x^{-3}$. The case of a fractional exponent is the same. Consider the derivative of the function $y = x^{\frac{1}{3}}$. To apply the rule we move $\frac{1}{3}$ to the front of x and subtract one from $\frac{1}{3}$, yielding $\frac{1}{3}x^{-\frac{2}{3}}$. Again, we get a great deal of flexibility from a very simple rule.

6.2.2 Exponentials

As in Chapter 3, we've covered x^a, but now need to cover a^x. We start with the most commonly used exponential, e^x. The derivative of e^x is even simpler than that of x^n. In fact, it's just e^x, the same function with which we started. That is,

$$\frac{de^x}{dx} = e^x.$$

How can this be? To derive it we will need the definition of the derivative, as always, but we will also need to express the exponential function in a different way. It turns out that there are many ways to define the exponential function, all equivalent. You may see two common ones:

$$\begin{aligned} e^x &= \lim_{n \to \infty} \left(1 + \frac{x}{n}\right)^n, \text{ or} \\ &= \sum_{n=0}^{\infty} \frac{x^n}{n!} = 1 + x + \frac{x^2}{2} + \frac{x^3}{6} + \ldots. \end{aligned}$$

[9] Don't worry if you don't yet see this quickly. You will, though it'll come faster if you take each opportunity like this to practice and check your work against ours.

We use the second definition for our purposes.

Now to the proof:

$$\frac{de^x}{dx} = \lim_{h \to 0} \frac{e^{x+h} - e^x}{h} = \lim_{h \to 0} \frac{e^x e^h - e^x}{h} = \lim_{h \to 0} \frac{e^x(e^h - 1)}{h} = e^x \lim_{h \to 0} \frac{e^h - 1}{h}.$$

We've got the e^x now; we just need to show that the rest of the term on the RHS equals 1. This is where we use our definition of e^x:

$$\lim_{h \to 0} \frac{e^h - 1}{h} = \lim_{h \to 0} \frac{-1 + 1 + h + \frac{h^2}{2} + \frac{h^3}{6} + \ldots}{h} = 1,$$

where we've used the fact that all terms of order h^2 and higher in the numerator vanish in the limit. This completes the proof.

To get the more general case of a^x for any a, we can use the properties of the exponential and logarithm to write it as $a^x = e^{x \ln(a)}$.[10] Next we use the chain rule. We let $g = e^x$ and $f = x \ln(a)$, so $\frac{da^x}{dx} = (e^{x \ln(a)})(\ln(a))$. Or, after rewriting the first term on the RHS one more time

$$\frac{da^x}{dx} = a^x(\ln(a)).$$

This is the **exponential rule**.

Believe it or not, between the power rule and this exponential rule, plus the three rules in the previous section, we've covered nearly all functions used regularly in political science. With these, even really complex functions are straightforward to differentiate (though they do require you to be careful with the algebra to avoid mistakes!).

For example, let $y = e^{x^2}$. We have one function (e^x) composed with another (x^2), so the chain rule is in order. Let $g(x) = e^x$ and $f(x) = x^2$. Then $y = g(f(x))$ and $y' = g'(f(x))f'(x) = (e^{x^2})(2x)$. This shows the nice thing about exponentials: they always return the original function, multiplied by the derivative of the term in the exponent. So the derivative of the really complicated function $y = e^{x^4 - 3x^2 + 1}$ is just $y' = (e^{x^4 - 3x^2 + 1})(4x^3 - 6x)$.

6.2.3 Logarithms

The only other function you'll encounter with any regularity is the logarithm. Recall from Chapter 3 that it is the inverse of the exponential function. So, for the natural log, $e^{\ln(x)} = x$. We can use the inverse function rule from the previous section directly, or we can just differentiate both sides and use the chain rule again. We'll do the latter. Let $g = e^x$ and $f = \ln(x)$. Then by the chain rule the LHS is $\frac{d\ln(x)}{dx} e^{\ln(x)} = \frac{d\ln(x)}{dx} x$, where we've again used the definition of an inverse function. The derivative of the RHS is just one. Thus, we get $\frac{d\ln(x)}{dx} x = 1$, or

$$\frac{d\ln(x)}{dx} = \frac{1}{x}.$$

[10]This is true because $e^{x \ln(a)} = e^{\ln(a^x)} = a^x$.

Usually people just memorize this, but as you can see, the derivation is pretty quick.

What about the more general logarithm $\log_a x$? Well, this is the inverse function of a^x, so $a^{\log_a(x)} = x$, and we can do the same proof over again. We leave it to you do this as an exercise, but we'll give you the answer

$$\frac{d\log_a(x)}{dx} = \frac{1}{x\ln(a)}.$$

While derivatives of the log don't have quite the nice properties of the exponential, they do have two things going for them. One, rather than just return the entire complicated exponential, they return one over the object of the log times the derivative of the object of the log, via the chain rule. So, $\ln(x)$ becomes $\frac{1}{x}$ under differentiation, since the derivative of the object of the log, x, is one. For a more complex example, let $y = \ln(x^5 - 2x^2 + 12)$. Then $y' = \frac{1}{x^5-2x^2+12}(5x^4 - 4x)$ via the chain rule. We can also write this as $y' = \frac{5x^4-4x}{x^5-2x^2+12}$, which illuminates the second thing derivatives of the log have going for them: they give you the *relative rate of change*, and are often used to this effect in the literature.

To see how this works, let $y = \ln(f(x))$ for some function $f(x)$. Then $y' = \frac{1}{f(x)}f'(x) = \frac{f'(x)}{f(x)}$. In words, the derivative of the log of a function is the derivative of the function divided by the function itself, giving the rate of change of the function relative to the value of the function.

6.2.4 Other Functions

This covers most of the functions that we see in political science. We mentioned a few other types of functions in Chapter 3, however, so we discuss their derivatives briefly here, though we'll not include proofs.

The derivatives of trigonometric functions turn them into each other. So $(\sin(x))' = \cos(x)$, $(\cos(x))' = -\sin(x)$, and $(\tan(x))' = 1 + \tan^2(x)$.

Piecewise functions lack derivatives at points at which there are discontinuous jumps or kinks. The derivative of the rest of the function can itself be treated piecewise, omitting the troublesome points. For example, consider the function we presented in Chapter 3

$$f(x) = \begin{cases} -(x-2)^2 & : x \leq 2, \\ \ln(x-2) & : x > 2. \end{cases}$$

We can write its derivative as

$$f'(x) = \begin{cases} -2(x-2) & : x < 2, \\ \frac{1}{x-2} & : x > 2. \end{cases}$$

Note that the point $x = 2$ has no derivative, though it is defined for $f(x)$.

6.3 WHAT THE RULES ARE, AND WHEN TO USE THEM

As promised, in this section we list the **rules for differentiation** discussed in this chapter. In all these cases, f and g are assumed to be differentiable functions, and a is a constant. Table 6.1 lists the rules, and we hope that if you read the previous two sections as well, you have a pretty good idea of their origins. We then conclude with a brief discussion about how to use them.

Table 6.1: List of Rules of Differentiation

Sum rule	$(f(x)+g(x))' = f'(x)+g'(x)$
Difference rule	$(f(x)-g(x))' = f'(x)-g'(x)$
Multiply by constant rule	$f'(ax) = af'(x)$
Product rule	$(f(x)g(x))' = f'(x)g(x)+f(x)g'(x)$
Quotient rule	$\left(\frac{f(x)}{g(x)}\right)' = \frac{f'(x)g(x)-f(x)g'(x)}{(g(x))^2}$
Chain rule	$(g(f(x))' = g'(f(x))f'(x)$
Inverse function rule	$(f^{-1}(x))' = \frac{1}{f'(f^{-1}(x))}$
Constant rule	$(a)' = 0$
Power rule	$(x^n)' = nx^{n-1}$
Exponential rule 1	$(e^x)' = e^x$
Exponential rule 2	$(a^x)' = a^x(\ln(a))$
Logarithm rule 1	$(\ln(x))' = \frac{1}{x}$
Logarithm rule 2	$(\log_a(x))' = \frac{1}{x(\ln(a))}$
Trigonometric rules	$(\sin(x))' = \cos(x)$
	$(\cos(x))' = -\sin(x)$
	$(\tan(x))' = 1+\tan^2(x)$
Piecewise rules	Treat each piece separately

There is one basic tactic of differentiation: use the rules on combining functions as sums, products, or composites to reduce the problem to the smallest pieces you know how to manage, differentiate these pieces, and then build up the answer again according to the rules. We work through one complex example here to show you what we mean; others can be found in the exercises.

Let $y = \frac{(5\ln(x+3))e^{3x^3-10x}}{5x^2+2}$. This is messy and there's not much we can do about that, but its derivative is solvable if we're methodical and use what we've learned. First we get rid of the quotient. So call $f(x) = (5\ln(x+3))e^{3x^3-10x}$ and $g(x) = 5x^2+2$. We've seen enough polynomials by now to know that $g'(x) = 10x$, so that's broken down enough.[11]

The integral f is still pretty complicated, though, so let's set $u(x) = 5\ln(x+3)$ and $v(x) = e^{3x^3-10x}$. We could break these down further if we liked, but we've

[11] If this is not true for you, then break it up further $g(x) = a(x)b(x)+c(x)$, where $a(x) = 5, b(x) = x^2$, and $c(x) = 2$.

seen examples of each type before, so we can take their derivatives: $u'(x) = \frac{5}{x+3}$ and $v'(x) = (9x^2 - 10)e^{3x^3-10x}$.[12]

Now we have all the components, so we can put it back together. We've used the quotient and product rules (and also the chain rule, if you've followed the footnotes), so

$$y' = \frac{f'g - fg'}{g^2} = \frac{(uv)'g - fg'}{g^2} = \frac{(u'v + uv')g - fg'}{g^2}.$$

All that's left is to plug the functions f, g, u, v into this to get our answer

$$\begin{aligned} y' &= (5x^2+2)^{-2}\left[\left(\frac{5e^{3x^3-10x}}{x+3} + 5\ln(x+3)(9x^2-10)e^{3x^3-10x}\right)\right. \\ &\quad \left. \times(5x^2+2) - (5\ln(x+3))e^{3x^3-10x}(10x)\right]. \end{aligned}$$

This is long and a big mess, true (it's not really worth the effort to simplify), but the important point again is that, via a handful of rules, one can differentiate very complex functions by first breaking them down into manageable parts and then building the derivative back up again. At first these parts will be small, likely smaller than in our example, but with practice they'll get bigger and differentiation will get faster and less worrisome. And practice is key: there's nothing special about any of this beyond becoming accustomed to the techniques, and we hope we've convinced you that the techniques can be very useful.

6.4 EXERCISES

1. Find the derivative of y with respect to x for the following, using the rules in this chapter:

   - $y = 6$.
   - $y = 3x^2$.
   - $y = x^3 - 2x^2 - 1$.
   - $y = x^4 + 5x$.
   - $y = x^8$.
   - $y = x^{-3}$.
   - $y = ax^3 + 6$.
   - $y = 12x^{\frac{1}{2}} + c$.

[12] Again, if this is not clear, break them up further. In particular, let $d(x) = e^x$ and $k(x) = 3x^3 - 10x$ in $v(x)$ so that you can use the chain rule.

- $y = ax^n - 1$.
- $y = f(x) + g(x) = (3x - 2) + (c - 4x^3)$.
- $y = f(x) \cdot g(x) = (13x + 2x^3) \cdot (x^5 - 4x + r)$.
- $y = (x - 3)^3$.
- $y = (\frac{x^2+1}{x+1})$.
- $y = 5x^7 + 7x^4 + 3x^2$.
- $y = 5x^8 + 10x^7 - 5x^6 - 5x^5 + 3x^4 + 7x^3 - 2x^2 + x - 1,123$.
- $y = x^3 + x^2 + 1$.
- $y = x^4 - x^3 + x^2 - x + 1$.
- $y = (3x^2 + 4)(2x^3 + 3x + 5)$.
- $y = (5x^3 + 4x^2 + 3x + 2)(7x^5 + 6x^4 + 5x^3 + 4x^2)$.
- $y = (3x^2 + 4) + (2x^3 + 3x + 5)$.
- $y = (5x^5 + 3x^3 + x + 1) - (4x^4 + 2x^2 + 2)$.
- $y = (x + 5)^2$.
- $y = (x^2 + x + 2)^2$.
- $y = \left(\frac{x^2+1}{x+1}\right)^2$.
- $y = \frac{x^3+x^2+x+1}{x^2+x+1}$.

2. Using the rules in this chapter (i.e., you don't need to go back to the definition), differentiate the following:

   a) $f(x) = a_n x^n + a_{n-1}x^{n-1} \ldots + a_0$. Try also expressing the derivative as a series.
   b) $f(x) = (x^3 + 2)\ln(x^4 - 5x + 3)$.
   c) $f(x) = \frac{(x^2-4)}{x^5-x^3+x}$.
   d) $f(x) = e^{x-\ln(x)+5}$.
   e) $f(x) = xg(x) - 7x^2$, where $g(x) = e^x \ln(x)$.
   f) $f(x) = a^x x^2 - b^{x^2}$.
   g) $f(x) = e^{5x}$.
   h) $f(x) = e^{5x^2+x+3}$.
   i) $f(x) = 3e^{2x}$.
   j) $f(x) = \frac{1}{2}e^{\frac{x}{2}}$.
   k) $f(x) = e^{\ln(2x)}$.
   l) $f(x) = e^{g(x)}$, where $g(x) = 7x^3 + 5x^2 - 3x + \ln(x) - 7$.
   m) $f(x) = x^2 g(x) + 6x^2$, where $g(x) = \log_a(x) + x^7$.
   n) $f(x) = ax^2 g(x) + 9x^4$, where $g(x) = e^{\ln(x)+2x^2}$.

3. Show that $\frac{d\log_a(x)}{dx} = \frac{1}{x(\ln(a))}$.

Chapter Seven

The Integral

Recall from Chapter 5 that our primary use of calculus will come in allowing us to deal with continuity usefully. The derivative provides us with the instantaneous change in a continuous function at each point. The derivative, then, permits us to graph the marginal rate of change in any variable that we can represent as a continuous function of another variable. In the next chapter we make extensive use of the derivative to find maxima and minima.

But what if we care less about change than about the *net effect of change*? Say we had some continuous function that represented the marginal change in voter turnout with respect to some aggregate measure of education, and we wanted to know the total level of voter turnout for all levels of education. To get this, we'd need to start at some point, say, where aggregate education is equal to zero, and then add up all the changes in turnout as education increased. This sounds straightforward enough, if time-consuming, save for one little factor: the function we have describes the instantaneous marginal change in turnout as education varies continuously! Somehow we have to add *continuously*, and our summation operator, $\sum$, is not going to cut it. This is where the integral enters the picture.

As we see in more detail below, the integral is like the limit of a sum. In our example, the integral of the marginal change function essentially adds up an infinite number of infinitesimal changes in turnout to produce the required total turnout function. This function specifies the level of turnout at any amount of education.

Though the integral is perhaps most naturally thought of in terms of summing changes, it is not limited to this. For example, other uses of the integral include calculations of areas and volumes. The most common use of the integral in political science relates to its connection to probability density functions, which describe the chance of any particular value of a continuous function occurring. We'll meet these in Part III. For now, we assert that the integral is needed to find cumulative distribution functions (the area under the curve of the probability density function), necessary for statistical inference, and to compute expected values and utilities, necessary for game theory.

We can distinguish between definite and indefinite integrals, and in Section 1 we provide an intuitive description of the definite integral that explicates the concept of a limit of sums. In Section 2 we discuss the antiderivative, or indefinite integral, and offer the fundamental theorem of calculus. In Section 3 we show how integrals can be computed, using the integral's status as the antiderivative in order to derive rules of integration. These rules are summarized in Section 4.

7.1 THE DEFINITE INTEGRAL AS A LIMIT OF SUMS

Take a moment to flip back to the plots of $f(x) = x^2$ in Chapter 5. There we were concerned with discerning the instantaneous rate of change in the function at any point. To get this, we drew secants between neighboring points, and then shrunk the distance between them until the secant became a tangent to the curve.

What if instead we were concerned with the area under the curve? That is, we want to know what the area is between the curve and the x-axis, from some point x_1 to another point x_2. As we alluded to earlier in this chapter, calculating such areas is central to statistical inference, as we'll see in Part III. If we let $x_1 = 1$ and $x_2 = 2$ and use $f(x) = x^2$, then we can visualize this area by dropping lines down from $f(x)$ at the two endpoints of the secant in the figure in Chapter 5. We show this in Figure 7.1, in which the shaded area is the area of interest.

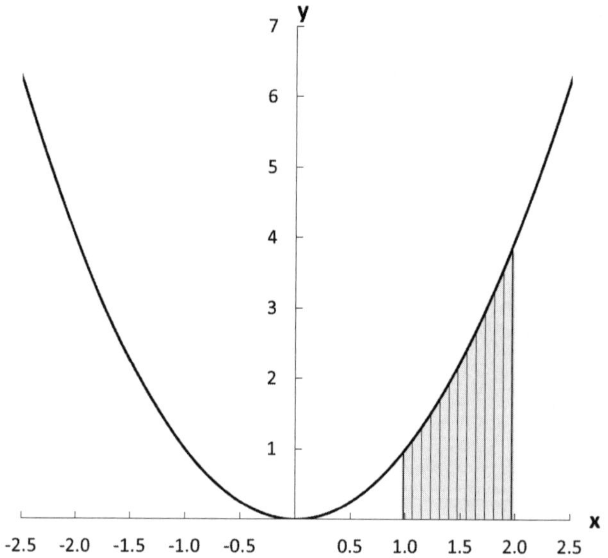

Figure 7.1: Area under $y = x^2$ from $x = 1$ to $x = 2$

How might we go about calculating this area? The intuition behind finding the area under a curve is pretty straightforward and is developed by thinking about finding the area within geometric shapes for which the formula for the area is known. If one were to take a curve and draw rectangles under the curve, then one could approximate the area under the curve by calculating the sum of the areas of the rectangles. That is, we can calculate the area of a rectangle, and we can sum those areas. Figure 7.2 does this, drawing rectangles with finite width within the area shaded in the previous figure. As we can see, the area contained within these rectangles does get close to matching the shaded area in the previous figure; however, it undercounts by the triangular regions on top of each rectangle.[1]

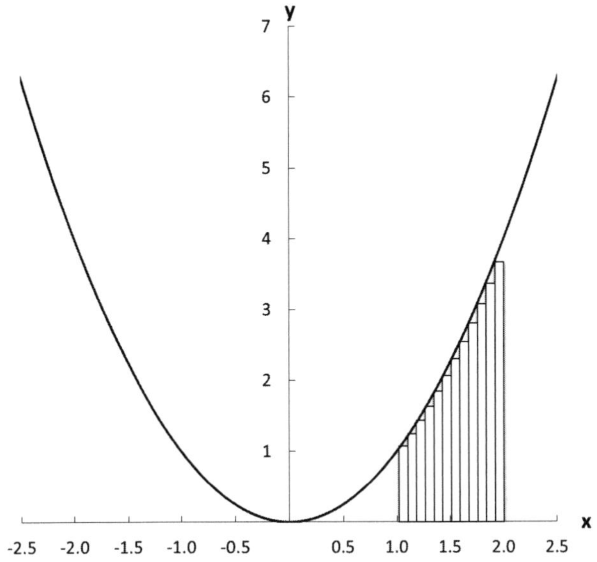

Figure 7.2: Area under $y = x^2$ from $x = 1$ to $x = 2$ with Rectangles

There are ways to do better than rectangles, using a similar method. We could add a triangle to the top of each rectangle, thus minimizing the underestimation. These would form trapezoids. They might overestimate or underestimate the area under the curve, depending on how they are drawn, but they would produce a closer approximation of the area. Or we could use the approximation developed by British mathematician Thomas Simpson, who proposed using tiny parabolas to fill in the gaps left by trapezoids (see Simpson's Rule). This effectively produces a quadratic interpolation, which should produce an exact solution for our case, since we have a quadratic function, $f(x) = x^2$, but will in general still over- or underestimate the area for more complex functions.

There is a way around these problems, however, and it is to draw our shapes very thinly: as they become more and more narrow, the divergence between the sum of their areas and the area under the curve will shrink. The German mathematician Georg Friedrich Bernhard Riemann proposed thinking about the area contained in the rectangles as they approach the limit of zero width. At the limit the area estimate will converge to the true value, and we can use these

[1] We could also have drawn the rectangles to overcount if the top of the rectangles had been drawn from the larger of the two points rather than the smaller.

rectangles not as an approximation of the area under the curve but as the actual area. The rectangle rule, then, can be used to calculate the area under a curve, and you might encounter reference to Riemann integrals or Riemann sums as a result.

How does this work? If we let the width be denoted Δx, then the area of each rectangle is $f(x_i)\Delta x$, where $f(x_i)$ is the value of the function at each (evenly spaced) point we have chosen. The total area might then be written as $\sum_i f(x_i)\Delta x$. We can take the limit of this sum as $\Delta x \to 0$ to produce the true area. We call this limit the **definite integral**, and write it in general as $\int_a^b f(x)dx$.[2]

The integral symbol, $\int$, looks like an S and is meant to, to remind you that it is in essence a sum. The dx in the integral tells you the variable of integration. It is exactly analogous to the dx in the derivative. The $f(x)$, or more generally, the expression multiplying the dx, is known as the **integrand**. The a and b are the bounds (or limits) of integration. They tell you the value of x at which to start and the value at which to stop. Unlike the derivative, which can be represented using a variety of notation, this is the single, common, agreed-upon notation for the integral. For our case, x^2 is the integrand, $a = 1$ and $b = 2$ are the bounds of integration, and the definite integral is $\int_1^2 x^2dx$.

We have now given a name to our limit of a sum, but what do we do with it? How does one add up an infinite number of infinitesimally small areas? To do this we must first specify the nature of the connection between the integral and the derivative, captured in what is known as the fundamental theorem of calculus. We turn to this now.

7.2 INDEFINITE INTEGRALS AND THE FUNDAMENTAL THEOREM OF CALCULUS

In Chapter 3 we discussed the notion of an inverse of a function. This object is a function $f^{-1}(x)$ such that when composed with $f(x)$ it yields the identity function, x. In symbols, $f^{-1}(f(x)) = x$. In Chapter 1 we discussed additive and multiplicative inverses but did not link them to inverse functions. Though these concepts are not the same, they do have some similarities. The difference operator $(-)$ is in a rough sense the inverse of the sum operator $(+)$: if you add 3 and then subtract 3, you end up where you started. Similarly, if you multiply by some non-zero number and then divide by this same number, you end up where you started. So, in a sense, $\times$ and $\div$ are inverses of each other. We'll see that the derivative and the **antiderivative** (aka indefinite integral) are similarly related.

[2]Strictly speaking this is a Riemann integral, which is very useful but limited in some important ways that we will not discuss in this book. A more flexible integral that is founded on measure theory is the Lebesgue integral. Though that is not the approach this book will take, one can formulate probability theory using measure theory, and statistics or math departments often offer courses that focus on doing just that. Other varieties of integral exist as well.

7.2.1 Antiderivatives and the Indefinite Integral

The derivative is an operator that takes one function and returns another that describes the instantaneous rate of change of the first at each point. This is certainly more complex than addition or subtraction, but it doesn't change the fact that the derivative also has an inverse. We call this inverse the antiderivative. Consider some function $f(x)$. The antiderivative of $f(x)$ is denoted $F(x)$. That is, it's the capital of the letter used to represent the function.[3]

Because they are inverse operations, differentiation and antidifferentiation, applied in sequence, should take you back to where you started. In other words, $\frac{dF(x)}{dx} = f(x)$. Let's see how this works. Start with $f(x) = 1$. What is its antiderivative? Well, we know that the derivative of x is one. So $F(x) = x$ is one antiderivative.

It's not quite that easy, though. Take $F(x) = x + 10$. The derivative of that is also one, as is the derivative of $F(x) = x + 1,000$. In general, because the derivative of a constant is zero, there are lots (an infinite number, actually) of antiderivatives that all produce the same $f(x)$. Luckily, they're all of the same form: $F(x) = x + C$. We call C the constant of integration, and it can be any constant value (i.e., any value that does not depend on x).

We can find other antiderivatives in the same way. So, if $f(x) = x$, then we try to figure out what function $F(x)$, when differentiated, yields x. Well, we know that the derivative of x^2 is $2x$, so that's pretty close. We also know that the derivative is linear, so we can divide x^2 by 2 to get a derivative of x. Thus, the antiderivative of x is $\frac{1}{2}x^2 + C$. (Don't forget the C!)

For a slightly more complicated example, consider $\frac{1}{x}$. There's no polynomial that when differentiated produces this; check for yourself. What about other functions? While e^x certainly doesn't work, its inverse function, $\ln(x)$, does. Since $(\ln(x))' = \frac{1}{x}$, the antiderivative of $\frac{1}{x}$ is $\ln|x| + C$, where the absolute value in the argument of the logarithm arises to account for the fact that the logarithm isn't defined over negative real numbers.

Other antiderivatives can be found in the same fashion. At this point you might be asking why we care about antiderivatives, and it's a fair question. As a way of answering, first note another name for the antiderivative, the *indefinite integral.* We write an indefinite integral as $\int f(x)dx = F(x)$, so it looks the same as a definite integral without the bounds on the integral. The difference is that the **definite integral** returns a value, the area under the curve, while the **indefinite integral** returns a function[4] that, when differentiated, reproduces the integrand. In symbols, $\frac{d \int f(x)dx}{dx} = f(x)$.

---

[3]You will see this notation a great deal in Part III, as we typically denote probability distribution functions (PDFs) by $f(x)$ or $g(x)$ and their cumulative distribution functions (CDFs) by $F(x)$ or $G(x)$, respectively. The CDF is the antiderivative of the PDF, or, conversely, the PDF is the derivative of the CDF.

[4]Or, more precisely, a set of functions given the "indefiniteness" of having an unknown constant C in the function.

7.2.2 The Fundamental Theorem of Calculus

This still doesn't tell us why we care about taking indefinite integrals, though; it just tells us why the antiderivative is in a chapter on integration. To get at this question, we offer the grand-sounding **fundamental theorem of calculus**:

$$\int_a^b f(x)dx = F(b) - F(a).$$

In other words, the definite integral of a function from a to b is equal to the antiderivative of that function evaluated at b minus the same evaluated at a. The theorem is fundamental because it bridges (or links) differential and integral calculus. We can make the connection even more clear with a little notation $\int_a^b f(x)dx = F(x)|_a^b$, where the vertical line means "evaluate the antiderivative $F(x)$ at b, and subtract the antiderivative evaluated at a."

With this theorem, to calculate the area under the curve—a value that will prove very important for statistical inference—all we need to know is the indefinite integral of the function. We don't even have to worry about the constant C: since it appears in both indefinite integrals on the RHS, it cancels when they are subtracted.[5]

Given this, let's return to the example of the figures above. There $f(x) = x^2$. What's the antiderivative of this? Well, we know that the derivative of x^3 is $3x^2$ from the previous chapter, so an x^3 will be involved. If we further divide by 3 we'll get x^2 without the 3 in front. Thus, $F(x) = \frac{1}{3}x^3$. Consequently, by the fundamental theorem of calculus, $\int_1^2 x^2 dx = F(2) - F(1) = \frac{1}{3}(2)^3 - \frac{1}{3}(1)^3 = \frac{8}{3} - \frac{1}{3} = \frac{7}{3}$, and that is the area under the curve of x^2 between 1 and 2.

Let's try another example. Consider some function $f(x) = 1 + 2x + x^2$. Suppose we want to know the area under the curve over the range from $x = 0$ to $x = 3$. This area is the shaded region in Figure 7.3.

We first need to compute the indefinite integral. Recalling that the derivative is linear, we note that x differentiates to 1, x^2 differentiates to $2x$, and, as we just saw, $\frac{1}{3}x^3$ differentiates to x^2. Putting them together yields $F(x) = x + x^2 + \frac{1}{3}x^3$. Evaluated at $a = 0$, this is 0. At $b = 3$, this is 21. As $21 - 0 = 21$, so $\int_0^3 (1 + 2x + x^2)dx = 21$. Put differently, the sum of the changes in the function $F(x) = x + x^2 + \frac{1}{3}x^3$ (changes given by the derivative of $F(x)$, $f(x) = 1 + 2x + x^2$) from $x = 0$ to $x = 3$ is 21.

7.2.3 Why Should I Care?

One will encounter a number of statistical tables that report integral values. For example, z-scores—which report the area between the mean of a distribution and a point a selected distance from that mean under the standard normal curve—are definite integrals. Recall that definite integrals are values: the sum of the area under the curve between two points. Thus, when you look up z-score

[5] Hence the appellation *definite*: there is no uncertainty here.

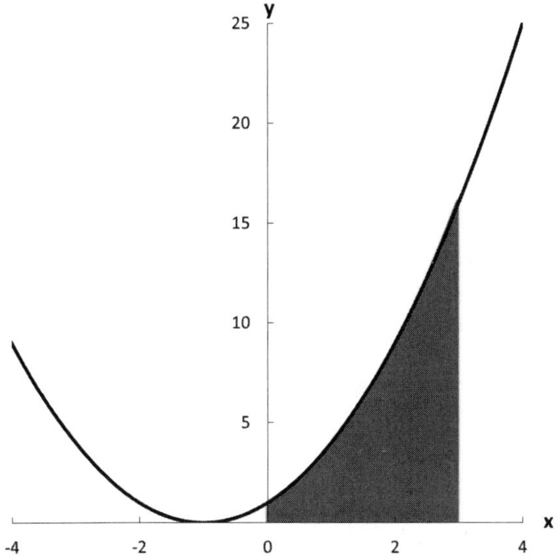

Figure 7.3: Shaded Area under $y = 1 + 2x + x^2$

values in a table in the appendix of a statistics text you are looking up the value of the difference between the antiderivative when $x = 0$ (the mean in a standard normal distribution) and the antiderivative at the point z (i.e., $\frac{X_i - \bar{X}}{s}$, where X_i is a given value of x, $\bar{X}$ is the sample mean, and s is the standard deviation of the sample). That value represents the area under the curve between those points. Conceptually, a z-score is the number of standard deviation units an individual observation is from the sample mean, and the area under the curve between it and zero relates to the chance of its being drawn from a standard normal distribution randomly.[6] Integral values are also important for studying continuous probability distributions in general, as we discuss in Chapter 11 of this book.

Formal theories also make appeals to integrals. Though some outcomes over which individuals may have an interest are certain, it is more often true that there is some level of uncertainty in the outcome, either because one is not sure what others with whom one is interacting will do, or because the payoffs themselves are variable. We can describe the chance that any particular outcome occurs by a probability distribution over outcomes. The expected value of the game (or expected utility) is the definite integral over all possible outcomes (or possible values of one's utility arising from different outcomes), weighted by the chance that each outcome occurs, i.e., by the probability distribution over outcomes.

[6] If you are unfamiliar with z-scores, make a note to return to this discussion when you are introduced to z-scores in your statistics coursework.

We cover continuous probability distributions in Chapter 11, where we provide an example that makes use of the integral of a probability distribution function to compute an expected utility. Here we provide a less intensive example by sticking to a decision theoretic context (i.e., one with a single decision maker). Assume that a bureaucrat must allocate money over two public works projects, one fixing a bridge and one building a new road. The payoff for the road might be known, but the payoff for the bridge is stochastic, since it depends on the likelihood that the bridge will collapse if it is not fixed. In other words, the return on investment in the case of the bridge is uncertain. To make things simpler, let the bureaucrat's decision be dichotomous: either money gets allocated or it doesn't, and money can only be allocated to one project. How would we figure out which project gets the money?

To model the bureaucrat's optimal decision we need to know the expected payout for each option. We show in Chapter 11 that this is computed by taking the expectation of the payout (or utility) function over the probability distribution of possible payouts. Here we simplify. Let the payoff for the road be 2, and the payoff for the bridge be uniformly distributed between -2 (if the bridge wouldn't have collapsed anyway) and 8 (if the bridge were in immediate danger of collapse). A uniform distribution places equal probability on every outcome. A rational actor will put money into the project with the greater payoff, but which is greater?

In Chapter 11 we define the uniform distribution formally. Here we assert that the expected value of a variable distributed according to the uniform distribution provided in the problem is the definite integral $\frac{1}{8-(-2)}\int_{-2}^{8} x dx = \frac{1}{10}\frac{1}{2}(x^2)|_{-2}^{8} = \frac{1}{20}(64-(4)) = 3$.[7] This gives us the payoff we need. Since $3 > 2$, the bureaucrat should fix the bridge.

7.3 COMPUTING INTEGRALS

We now know how to find the area under the curve in theory, but thus far doing so in practice has mostly involved trying to guess integrals from derivatives. Here we present rules for integration akin to those for differentiation. In fact, they are more than merely akin—they follow closely from the rules for differentiation, and we discuss them more or less in the same order, though we discuss integrals of functions first in order to get examples for the more general rules later. Since most of these rules apply for both definite and indefinite integrals, in our rules we refer generically to "integrals" and use their indefinite form, except for rules related to the bounds on the definite integral.

We start by noting that the fundamental theorem of calculus implies a couple of useful properties of the integral. Since $\int_a^b f(x)dx = F(b) - F(a)$, it must also

---

[7]Note that this is the midpoint of the range of the distribution. This is not an accident; since the uniform distribution places equal weight on all outcomes, the expected outcome is the one in the middle. Once you've read Chapter 11, you can come back here and see that out assertion was correct.

be true that $\int_b^a f(x)dx = F(a) - F(b)$. In words, if you flip the bounds of the integral, you switch the sign of the answer. In short, $\int_a^b f(x)dx = -\int_b^a f(x)dx$. Along these same lines, since $F(a) - F(a) = 0$, $\int_a^a f(x)dx = 0$. We can also split up the bounds of the integral. If $c \in [a,b]$, then $\int_a^b f(x)dx = \int_a^c f(x)dx + \int_c^b f(x)dx$. Finally, note that the bounds need not be constants—they may be functions as well. A common way to write a cumulative distribution function, for example, is $F(x) = \int_{-\infty}^x f(t)dt$. All that must be true is that the bounds on the integrals should not contain the variable as that over which you are integrating. In the case of the cumulative distribution function just provided, that means we shouldn't have a t in the bounds of the integral because we are integrating over it.

7.3.1 Polynomials and Powers

Recall from the previous chapter that $\frac{dx^n}{dx} = nx^{n-1}$. And recall from the previous sections of this chapter that differentiating the antiderivative must produce the original function. In this case, we'll call the original function $f(x) = x^n$. If $F(x)$ is the antiderivative, we need $\frac{dF(x)}{dx} = x^n$. How do we get this? The easiest way is to use what we know already. If we call $g(x) = nx^{n-1}$, then the definition of the antiderivative implies that $G(x) = x^n$. We can get $G(x)$ from $g(x)$ by first increasing the exponent in $g(x)$ by 1, and then dividing by this new exponent. Doing this transforms the $n-1$ in the exponent of $g(x)$ to n, and then eliminates the n out front. Since this is true for any n, we can use it as our general rule for powers of x. In symbols, $\int x^n dx = \frac{x^{n+1}}{n+1} + C$ if $n \neq -1$. Differentiating the RHS yields x^n, which is the original function in the integral on the LHS, so our rule checks out.

Note a few fine details, though. First, always remember the constant of integration, C! The indefinite integral is specified only up to a constant. Second, this holds only for $n \neq -1$ for the reasons we discussed in the example above: the derivative of x^0 is 0, not x^{-1}, so the antiderivative of x^{-1} can't be a multiple of x^0. As shown in that example, for this case we have $\int x^{-1}dx = \ln|x| + C$.[8]

Without the general rules for integration the examples we offer in this chapter must be kept pretty simple. But we can integrate a wide range of functions with this rule alone. For example, $\int x^3 dx = \frac{x^4}{4} + C$, $\int \frac{1}{x^5}dx = \frac{-1}{4x^4}$, and $\int \sqrt{x}dx = \frac{2}{3}x^{\frac{3}{2}}$.

7.3.2 Exponentials

In the previous chapter we saw that $\frac{de^x}{dx} = e^x$. Since there is no difference between the function and its derivative, it must be the case that the the antideriva-

[8] If you are worried about losing track of these rules, remember that we summarize them all in Section 4. Also note that, while integrals are in general more complicated to compute than derivatives, you will also most likely be doing fewer of them in typical political science applications.

tive is the same as the function as well, since otherwise differentiating it would produce something different from the function. Consequently, $\int e^x dx = e^x + C$. The more general a^x behaves similarly. Since $\frac{a^x}{dx} = (\ln(a))a^x$, the antiderivative must also have an a^x in it for the same reason as for e^x. That leaves only $\ln(a)$. Since we need the antiderivative to differentiate to a^x, and since differentiating multiplies by $\ln(a)$, dividing by this same factor should do the trick. So, $\int a^x dx = \frac{a^x}{\ln(a)} + C$. As there are literally no examples we could do right now that would be interesting here, we move on to the next section.

7.3.3 Logarithms

While integrals of exponentials are common in political science, in large part because the probability distribution function for the normal distribution can be represented by an exponential function, as we'll see in Chapter 11, integrals of logarithms are less so. Consequently we will spend almost no time on them and offer no examples. However, in the interest of completeness and benefit to those in game theory whose utility functions are logarithms, we offer the integral of a logarithm.

To get the antiderivative of $f(x) = \ln(x)$, we must guess what function differentiated produces a natural log. The problem is that we haven't seen any of those, so we're going to have to get creative. We might not be able to use a basic expression, but if we include a log in a product, then at least one of the terms in the derivative of the product will still have a log in it. Then all that will be left is to eliminate the other term. Let's try the easiest product, $x\ln(x)$. The derivative of this, by the product rule, is $\ln(x) + \frac{x}{x} = 1 + \ln(x)$. That's close; all we need to do is get rid of the one. Subtracting x from $x\ln(x)$ will do this since its derivative is one. Thus, $\int \ln(x)dx = x\ln(x) - x + C$. For the more general $\log_a$ we have to deal with the $\ln(a)$ that appears when it is differentiated. Thus we need to divide both terms by $\ln(a)$, so that $\int \log_a(x)dx = \frac{x\ln(x)-x}{\ln(a)} + C$.

7.3.4 Other Functions

Unlike with the derivative, we cannot be so casual with integrating more complex functions or piecewise functions. In general, for our purposes we'll want the functions we're integrating to be continuous (and sometimes differentiable too). Piecewise functions can be handled, however, particularly for definite integrals, by splitting up the integral using the rules on its bounds given above. Consider the example of the previous chapter

$$f(x) = \begin{cases} -(x-2)^2 & : x \le 2, \\ \ln(x-2) & : x > 2. \end{cases}$$

We could integrate this from $a \le 2$ to $b > 2$ by splitting it up: $\int_a^b f(x)dx = \int_a^2 (-(x-2)^2)dx + \int_2^b \ln(x-2)dx$.

There are also integrals for trigonometric functions that may sometimes be useful. We list these for completeness: $\int \sin(x)dx = -\cos(x) + C$, $\int \cos(x)dx = \sin(x) + C$, $\int \tan(x)dx = -\ln(|\cos(x)|) + C$.

7.3.5 The Integral Is *Also* a Linear Operator

We saw that the derivative is a linear operator in the previous chapter. This means that $(af + bg)' = af' + bg'$ for functions f and g and constants a and b. The fact that the derivative is linear means that when one differentiates an antiderivative to obtain the original function, each constituent expression in the antiderivative corresponds to one expression in the function. So the integral will also be a linear operator.[9]

More formally, to say that the integral is also a linear operator is to say that $\int (af(x) + bg(x))dx = a \int f(x)dx + b \int g(x)dx$. To show this we need to use the definition of the antiderivative. The proof follows from this definition and the linearity of the derivative, which we showed in the previous chapter.

By the definition of an antiderivative, $y(x) = \int \frac{dy}{dx} dx$ and $z(x) = \int \frac{dz}{dx} dx$. We multiply both sides of the first equation by $a \neq 0$ and both sides of the second by $b \neq 0$, and then add them to get equation (7.1):

$$ay + bz = a \int \frac{dy}{dx} dx + b \int \frac{dz}{dx} dx. \tag{7.1}$$

Now start with the fact that the derivative is a linear operator to get $\frac{d(ay+bz)}{dx} = a\frac{dy}{dx} + b\frac{dz}{dx}$. Next integrate both sides to get

$$\int \frac{d(ay + bz)}{dx} dx = \int \left(a\frac{dy}{dx} + b\frac{dz}{dx}\right) dx.$$

Finally, use the definition of the antiderivative on the LHS to get equation (7.2):

$$ay + bz = \int \left(a\frac{dy}{dx} + b\frac{dz}{dx}\right) dx. \tag{7.2}$$

Compare (7.1) and (7.2). They have the same LHS, so their RHS must also be equal. Therefore, $\int \left(a\frac{dy}{dx} + b\frac{dz}{dx}\right) dx = a \int \frac{dy}{dx} dx + b \int \frac{dz}{dx} dx$. Since the derivatives of y and z are just functions, let $f = \frac{dy}{dx}$ and $g = \frac{dz}{dx}$. Making this substitution yields $\int (af(x) + bg(x))dx = a \int f(x)dx + b \int g(x)dx$, which is what we wanted.

Like each of the three general rules of integration we offer, this rule is used to simplify integrals, turning them into something that looks like the simpler functions above. We saw this when we computed the integral of $f(x) = 1 + 2x + x^2$: we treated each term in the sum separately. Examples utilizing the linearity of the integral are of this form. We offer a few here.

---

[9]We are ignoring issues with the constant of integration here.

All polynomials can be tackled with this rule. For instance, consider $f(x) = 4x^5 + 2x^2 + 5$. Linearity implies we can treat each term separately. The integral $\int x^5 dx = \frac{x^6}{6} + C$, so $4 \int x^5 dx = \frac{2x^6}{3} + C$. The integral $\int x^2 dx = \frac{x^3}{3} + C$, so $2 \int x^2 dx = \frac{2x^3}{3} + C$. The integral $\int x^0 dx = x + C$ so $5 \int x^0 dx = 5x + C$. Combining all these yields the answer: $\int (4x^5 + 2x^2 + 5)dx = \frac{2x^6}{3} + \frac{x^3}{3} + 5x + C$, where we've combined all the arbitrary constants into one equally arbitrary constant. In a similar fashion, $\int (10x^6 - 4x^4 + \frac{1}{x^2})dx = \frac{10x^7}{7} - \frac{4x^5}{5} - \frac{1}{x} + C$, where all we've done is to consider each term separately and use the rule for powers of x. Nor are we limited to polynomials: $\int (5x^2 + e^x)dx = \frac{5x^3}{3} + e^x + C$.[10]

7.3.6 Integration by Substitution

We start describing the technique of substitution by taking a closer look at the notation in the integral $\int f(x)dx$. What is x here? Could we replace it by, say, u, to get $\int f(u)du$? The answer to this is yes. The variable of integration itself has no meaning. This is no different from saying the sum $\sum_{i=1}^{N} x_i$ is the same as the sum $\sum_{k=1}^{N} x_k$. The variables k and i are just indices in the sums, and x and u are just "infinitesimal indices" in the integral.

We bring this up because the fungibility of the variable of integration signals that we should have the ability to change things within the integral, as long as we are careful. Clearly, we can exchange the *name* of the variable, turning an x into a u. But what if we turn x into an entirely new function? For example, let $x = g(u)$. What then?

Well, if $x = g(u)$, then $f(x) = f(g(u))$, which is a composite function. Integration by substitution says that

$$\int_a^b f(g(u))g'(u)du = \int_{g(a)}^{g(b)} f(x)dx.^{11}$$

In words, we can change variables to an entirely different function if it helps us to compute the integral. We'll see why we'd want to do such a thing in the examples below in this subsection, but first we show why this equality holds.

We use a combination of the definition of an antiderivative, the fundamental theorem of calculus, and the chain rule from differentiation.[12] First, let $x = g(u)$, so that $f(x) = f(g(u))$. Next, define the antiderivative $F(x)$ in the (now) usual fashion. We can compose this with $g(u)$ as well: $F(g(u))$. The chain rule says that $(F(g(u)))' = F'(g(u))g'(u) = f(g(u))g'(u)$, where in the last step we have

[10] We went through these examples pretty quickly, and advise you to work them out on your own with more care.

[11] We could also have written $\int_{g^{-1}(a)}^{g^{-1}(b)} f(g(u))g'(u)du = \int_a^b f(x)dx$, depending on whether we wanted to define the bounds of the integral as measuring differences in u values or x values. The way given in the text makes it slightly simpler to prove.

[12] As you should be used to by now, we skip some steps that don't enhance intuition. In this case, we won't show that the integrals on each side of the substitution rule exist, just that they are equal, as the rule states.

used the fact that the derivative of the antiderivative is the original function $f(x)$.[13]

The fundamental theorem of calculus says that $\int_a^b (F(g(u)))'du = F(g(b)) - F(g(a))$. We can plug in $f(g(u))g'(u)$ to the LHS integral to get $\int_a^b f(g(u))g'(u)du = F(g(b)) - F(g(a))$. We can again use the fundamental theorem of calculus on the RHS to give us $F(g(b)) - F(g(a)) = \int_{g(a)}^{g(b)} f(x)dx$. Plugging this into the RHS of the previous equality gives our desired result, $\int_a^b f(g(u))g'(u)du = \int_{g(a)}^{g(b)} f(x)dx$.

This likely seemed a bit abstract, so let's break substitution down into a useful technique, and then try it out with some examples.[14] Integration by substitution is attempted whenever the integral contains a composite function that one cannot integrate easily. In this sense it is used similarly to the chain rule in differentiation. The big difference is that whereas the chain rule will provide the derivative in most cases, substitution will often fail. Integrals, in general, are less amenable than derivatives, and do not always produce straightforward answers. Sometimes they do, though, so it's worth trying substitution. A couple of examples that are quite important to probability theory will help.

First consider the function $f(x) = \frac{1}{2\pi}xe^{-\frac{x^2}{2}}$. This function is the probability distribution function of a standard normal distribution, multiplied by x. Its integral is $\int \left(\frac{1}{2\pi}xe^{-\frac{x^2}{2}}\right)dx$, which happens to be the expected value of x in a standard normal distribution. We'll see this more in Chapter 11. This integral looks complicated, but it turns out that it reduces quite easily via substitution. The composite function here is $e^{-\frac{x^2}{2}}$, which we don't know how to integrate. But we do know how to integrate e^u, so let's set $u = g(x) = -\frac{x^2}{2}$ and see what happens. We can use the power rule to see that $g'(x) = -x$, which means we can rewrite the integral as $\int \left(\frac{-1}{2\pi}g'(x)e^{g(x)}\right)dx$. Integration by substitution implies that this integral is the same as $\int \left(\frac{-1}{2\pi}e^u\right)du$, which just equals $\frac{-1}{2\pi}e^u + C = \frac{-1}{2\pi}e^{-\frac{x^2}{2}} + C$.[15] So we've completed our integral.

Now instead consider the function $f(x) = \frac{1}{2\pi}e^{-\frac{x^2}{2}}$, which is the probability distribution function of a standard normal distribution. The integral of this is $\int \left(\frac{1}{2\pi}e^{-\frac{x^2}{2}}\right)dx$, and if we took the definite integral from $-\infty$ to ∞ we would get 1, which is true for all probability distribution functions. But how would we take that integral? We can't use our substitution trick even though we still have the same composite function, because there is no longer an x there to satisfy the need for a $g'(x)$ in the integrand. So we're stuck.[16]

[13] If you don't see how the chain rule applies here, set $v = g(u)$. Then $(F(g(u)))' = \frac{F(v)}{du}\frac{dv}{du} = \frac{F(v)}{dv}\frac{dg(u)}{du} = F'(v)g'(u) = F'(g(u))g'(u)$.

[14] This technique is rough, but it works.

[15] If this is confusing, switch u and x in this example to see that we used integration by substitution here.

[16] It turns out that one can do this integral with tools from complex analysis, but this is considerably beyond the scope of this book.

The moral is that for our purposes, substitution is feasible to try, and as we show, it is not terribly challenging to determine whether or not it will work. First you identify a composite function $f(g(x))$. Then you substitute $u = g(x)$ for the inner function. If there is some multiple of $g'(x)$ in the integrand, you can use substitution. If there is not, you cannot, or at least not so easily.[17] To wit: $\int 3x^2 e^{x^3} dx = \int e^u du = e^{x^3} + C$, since $u = g(x) = x^3$ and $g'(x) = 3x^2$, and $\int x^2 e^{x^3} dx = \frac{1}{3} e^{x^3} + C$ because x^2 is a multiple of $3x^2 = g'(x)$,[18] but neither $\int 3(x^2+1)e^{x^3} dx$ nor $\int 3xe^{x^3} dx$, for example, is integrable in this way.

Another way to see how to perform integration by substitution returns to our discussion at the beginning of this subsection about changing variables of integration. Rather than view the change from x to $u = g(x)$ as merely substituting in a function, one can view it as a change in variables from x to u, where $g(x)$ specifies the relation between them. One can use this relation to link dx to du as well: $du = \frac{dg(x)}{dx} dx$, so $dx = (g'(x))^{-1} du$.[19] Then $\int g'(x) f(g(x)) dx = \int g'(x) f(u) (g'(x))^{-1} du = \int f(u) du$.[20] This way of going about substitution is equivalent to our first method but may be easier procedurally, as it makes clearer that the goal is to eliminate any independent x in the integrand when substituting. For example, to compute $\int x^4 e^{x^5} dx$, we set $u = g(x) = x^5$, so $dx = (g'(x))^{-1} du = \frac{1}{5x^4} dx$, and thus $\int x^4 e^{x^5} dx = \int x^4 e^{x^5} \frac{1}{5x^4} du = \frac{1}{5} \int e^u du$. This is just $\frac{1}{5} e^u + C = \frac{1}{5} e^{x^5} + C$.

Before moving on, let us inject a note on the definite integral. These examples all used indefinite integrals but would have been equally valid as definite integrals (recall that the distinction is between a limited range of x and a function valid over the full range of x). They would have differed only in the presence of the bounds on the integral in the case of definite integrals. With substitution, however, one must take care with the bounds. Specifically, one must convert the bounds as the substitution rule would dictate. For example, let's say we had $\int_1^3 3x^2 e^{x^3} dx$, i.e, our recent example made into a definite integral. With $u = g(x) = x^3$, the integral becomes $\int_{g(1)}^{g(3)} e^u du = \int_1^{27} e^u du = e^{27} - e^1$. If this seems confusing, note that one can always check this by putting the original

[17] This is of course a simplification, and there are many ways to be clever about this technique. As you become more practiced with integration, you will have plenty of opportunities for such cleverness. Trigonometric identities can figure heavily. We are just providing a guideline that will work in many common situations in political science.

[18] The $\frac{1}{3}$ in $\frac{1}{3} e^{x^3} + C$ arises because you have to multiply by $1 = \frac{3}{3}$ to get the $3x^2$ you need to use substitution in this case.

[19] An easy way to remember this is that $\frac{du}{dx} = \frac{dg(x)}{dx}$; just "multiply" both sides by the dx from the LHS to get the required equation. Don't do this in general, though; this is just a mnemonic!

[20] We could write this more generally: for any function $f(x)$, $\int f(x) dx = \int f(x)(g'(x))^{-1} du = \int f(u) du$ if $u = g(x)$. Of course, if $(g'(x))^{-1} f(x)$ does not simplify into a nice $f(u)$—which it won't if one can't get rid of any dependence of $f(u)$ on x—then substitution won't work. This simplification occurs when the $(g'(x))^{-1}$ cancels with a part of $f(x)$. The more specific way we have written substitution in the main text makes the need for cancellation clearer.

variable x back in the answer, and then using the original bounds. For our case, $e^u = e^{x^3}$, so $\int_1^3 3x^2 e^{x^3} dx = e^u|_1^{27} = e^{x^3}|_1^3 = e^{3^3} - e^{1^3} = e^{27} - e^1$, as before.

7.3.7 Integration by Parts

The third and final general rule we offer is integration by parts. Its analogue in differentiation is the product rule, and it is used to integrate products of functions. First we'll state the rule, then explain and prove it, and finally offer some examples for how to use it.

Integration by parts states that

$$\int f(x)g'(x)dx = f(x)g(x) - \int f'(x)g(x)dx.$$ [21]

Basically, the rule allows you to change around your functions. If you can't integrate $f(x)g'(x)$ but you can integrate $g'(x)$ and $f'(x)g(x)$, it's very useful. Practically, this occurs when $g'(x)$ is something like e^x that does not get more complex when you integrate it, while $f(x)$ is something like x that simplifies when you differentiate it. We'll see this shortly in the examples. First, the proof, which is straightforward.

The product rule states that $(f(x)g(x))' = f'(x)g(x) + f(x)g'(x)$. Integrating both sides and using the definition of the antiderivative on the LHS and the fact that the derivative is linear on the RHS yields $f(x)g(x) = \int f'(x)g(x)dx + \int f(x)g'(x)dx$. Moving $\int f'(x)g(x)dx$ to the other side of the equation produces the rule.

Now for why we'd use it. Let's start with $\int xe^x dx$. We can't do this directly, nor can we use substitution, because the only compound function available is xe^x and we don't have the derivative of this present so as to allow substitution. So we'll try integration by parts. This is promising, as we can integrate e^x in a way that does not complicate it further, and the derivative of x is just 1. We set $f(x) = x$, $g'(x) = e^x$, and the rule gives us $\int xe^x dx = xe^x - \int e^x dx = xe^x - e^x + C$. We could in theory do this multiple times for more complicated functions. For example, $\int x^2 e^x dx = x^2 e^x - \int 2xe^x dx + C$. We can then use our previous example of integration by parts to complete the problem: $\int x^2 e^x dx = x^2 e^x - 2(xe^x - e^x) + C$.

Most examples you'll see in political science that use integration by parts are of this form. For instance, the example we just did is related to taking the expected value of a function, or finding the mean, given an exponential distribution, which you'll meet in Chapter 11. Other cases you might see, though far less often, would involve a power of x, such as x^2 or x^3, and an integrable root of x, such as $(x+1)^{-\frac{3}{2}}$. For example, $\int x(x+1)^{-\frac{3}{2}} dx$. To solve it, let $f(x) = x$, and $g'(x) = (x+1)^{-\frac{3}{2}}$. We need to know $f'(x)$ and $g(x)$ to use integration by

[21] You will often see this as $\int udv = uv - \int vdu$, where $u = f(x)$, $v = g(x)$, and $du = f'(x)dx$ and $dv = g'(x)dx$ are differentials. These mean the same thing, so we'll stick with the one that is more familiar.

parts. Note that $f'(x) = 1$ and $g(x) = \int (x+1)^{-\frac{3}{2}} dx$. This requires substitution to solve, but a relatively simple one. Set $u = h(x) = x + 1$. Then $h'(x) = 1$, which is automatically in the integral already. So we rewrite the integral as $g(u) = \int u^{-\frac{3}{2}} du$. By the power rule, this is $-2u^{-\frac{1}{2}} + C = -2(x+1)^{-\frac{1}{2}} + C$, after plugging back in for u. Now we plug into the integration by parts rule: $\int x(x+1)^{-\frac{3}{2}} dx = -2x(x+1)^{-\frac{1}{2}} + 2\int (x+1)^{-\frac{1}{2}} dx$. The last integral can be solved by again using the substitution $v = (x+1)$, so that $\int (x+1)^{-\frac{1}{2}} dx = \int v^{-\frac{1}{2}} dv = 2\sqrt{v} + C = 2\sqrt{x+1} + C$. Plugging this back into the integration by parts rule yields the answer: $\int x(x+1)^{-\frac{3}{2}} dx = -2x(x+1)^{-\frac{1}{2}} + 4\sqrt{x+1} + C$.[22]

As with substitution, we have presented integration by parts for the indefinite integral. It is not difficult to switch to the definite integral, though:

$$\int_a^b f(x)g'(x)dx = (f(x)g(x))|_a^b - \int_a^b f'(x)g(x)dx.$$

7.3.8 Why Should I Care?

Though you will largely rely on tables and statistical software to compute integrals of cumulative distribution functions, you will have cause to compute integrals should you choose to delve further into the underpinnings of statistical methodology. This will become clearer in Part III of the book. Further, you will have ample opportunity to take integrals in formal theory when computing either expected values or expected utilities, which are foundational concepts.

7.4 RULES OF INTEGRATION

For convenience, we provide a summary here of the **rules of integration** we have discussed in this chapter. Though not necessary, to make this easier we assume that f and g are both differentiable and integrable functions, and a, b, and C are constants. Procedurally, it usually helps in computing the integrals of more complex functions to first check to see if the linear rule is sufficient; second, attempt substitution; and third, turn to integration by parts if neither of the previous two rules helps. You should also be aware that, unlike with differentiation, there will be times when none of these three techniques yields progress toward an answer. There are integrals you will simply not be able to do, either because they require specialized techniques (as with $e^{-\frac{x}{2}}$) or because they simply are not integrable. (You won't find any of those in our exercises, though.) Finally, though we would expect that in many cases you'd be able to reference a list like that in Table 7.1 should you need to compute an integral by hand, we note that despite the length of the list in Table 7.1, all the rules provided may be derived (more or less) from the corresponding rules for differentiation. Thus, as noted before, learning calculus need not be a matter of memorizing a large number of disconnected rules.

[22] Don't worry if this isn't immediately clear. It's a relatively complicated integral. That said, for this reason it is a good one to practice on.

Table 7.1: List of Rules of Integration

Fundamental theorem of calculus	$\int_a^b f(x)dx = F(b) - F(a)$
Rules for bounds	$\int_a^b f(x)dx = -\int_b^a f(x)dx$ $\int_a^a f(x)dx = 0$ $\int_a^b f(x)dx = \int_a^c f(x)dx + \int_c^b f(x)dx$ for $c \in [a, b]$
Linear rule	$\int (af(x) + bg(x))dx = a\int f(x)dx + b\int g(x)dx$
Integration by substitution	$\int_a^b f(g(u))g'(u)du = \int_{g(a)}^{g(b)} f(x)dx$
Integration by parts	$\int f(x)g'(x)dx = f(x)g(x) - \int f'(x)g(x)dx$
Power rule 1	$\int x^n dx = \frac{x^{n+1}}{n+1} + C$ if $n \neq -1$
Power rule 2	$\int x^{-1}dx = \ln\|x\| + C$
Exponential rule 1	$\int e^x dx = e^x + C$
Exponential rule 2	$\int a^x dx = \frac{a^x}{\ln(a)} + C$
Logarithm rule 1	$\int \ln(x)dx = x\ln(x) - x + C$
Logarithm rule 2	$\int \log_a(x)dx = \frac{x\ln(x)-x}{\ln(a)} + C$
Trigonometric rules	$\int \sin(x)dx = -\cos(x) + C$ $\int \cos(x)dx = \sin(x) + C$ $\int \tan(x)dx = -\ln(\|\cos(x)\|) + C$
Piecewise rules	Split definite integral into corresponding pieces

7.5 SUMMARY

We introduced the integral as a method for calculating the area under a curve. We also defined the indefinite integral as the antiderivative and the definite integral as the antiderivative evaluated over a range of values. The key points are the integral's relationship to the derivative and understanding why it is a tool we can use to measure the area under curves.

We also provided a variety of rules used to compute integrals. We note that, in general, computing derivatives is considerably easier than computing integrals. Although we have provided tools to compute most derivatives you'll encounter, you may quite well find you need to compute integrals that simply are not amenable to computation using the tools you have in your toolkit. For example, you may need to compute the expected utility over a normal distribution, or even simply the integral of the normal distribution's probability distribution function itself. If you should find yourselves in such a scenario, help is readily available. Searching on the phrase "list of integrals" on the Internet should get you far to start, as many of the integrals that you might want to do have been fully or partially computed by others, either via more advanced analytic techniques or by numerical analysis.

7.6 EXERCISES

1. Visit `http://math.furman.edu/~dcs/java/NumericalIntegration.html` and approximate the area under the curve using different rules as described on the page. What does this applet demonstrate?

2. Visit `http://math.furman.edu/~dcs/java/ftc.html`, and drag the red dot (as described on the page). What does this applet demonstrate?

3. Integrate the following derivatives to find y:

   a) $\frac{dy}{dx} = 4x + 3.$
   b) $\frac{dy}{dx} = 3x^2.$
   c) $\frac{dy}{dx} = -2x + 3 - 4x^3.$
   d) $\frac{dy}{dx} = -1.$
   e) $\frac{dy}{dx} = -3 + 4x.$
   f) $\frac{dy}{dx} = 5x^4 - x - 4.$
   g) $\frac{dy}{dx} = 4x^4 + 3x^2.$
   h) $\frac{dy}{dx} = 5x^5.$
   i) $\frac{dy}{dx} = 4x^4 + 3x^3 + 2x^2 + x + 1.$
   j) $\frac{dy}{dx} = 3x^3 - 4x^2 + 5x - 6.$
   k) $\frac{dy}{dx} = x^{-1} + 3x^2.$
   l) $\frac{dy}{dx} = e^{5x}.$
   m) $\frac{dy}{dx} = 2e^{5x}.$
   n) $\frac{dy}{dx} = (20x + 2)e^{5x^2+x}.$
   o) $\frac{dy}{dx} = \ln(3x).$
   p) $\frac{dy}{dx} = \ln(x^2).$

4. Take the derivative of the answer to each of the problems in the previous question to check your work.

5. Which of the options below best describes $\int_a^b \frac{dy}{dx} dx$?

   a) It is the indefinite integral of the derivative of y.
   b) It is the area from y to x for the function defined in a and b.

c) It is the integral of the derivative of y with respect to x over the range a to b.

6. Compute the following integrals:

a) $\int (a_n x^n + a_{n-1}x^{n-1} \ldots + a_0)dx$. You get a bonus for expressing the integral as a sum.

b) $\int (3x^{3/2} - 2x^{-5/4} + 4^x)dx$.

c) $\int_1^{16} (5x^{3/2} - 2x^{-5/4})dx$.

d) $\int (-\frac{1}{x}\ln(\frac{1}{x}))dx$.

e) $\int (xe^{3x^2+1})dx$.

f) $\int (x^2(x^3+15)^{3/2})dx$.

g) $\int (\frac{12x^2-16x+20}{x^3-2x^2+5x})dx$.

h) $\int (xe^x)dx$.

i) $\int_2^4 (3x^2 + x + 5)dx$.

j) $\int_2^2 (3x^2 + x + 5)dx$.

k) $\int_2^4 (3x^4 + 2x^3 + x^2 + x + 1)dx + \int_4^6 (3x^4 + 2x^3 + x^2 + x + 1)dx$.

l) $\int_2^6 (3x^4 + 2x^3 + x^2 + x + 1)dx$.

Chapter Eight

Extrema in One Dimension

We have noted repeatedly over the past few chapters that a major motivation for computing the derivative is to find the maximum or minimum of a function. Maxima and minima are both types of *extrema*, and this chapter is devoted to finding them. Thus, in some ways this chapter is the payoff to this part of the book.

Finding extrema is useful in optimization theory, a topic that comes up fairly often in political science, and one to which we return in Chapter 16. Theorists often assume that an actor wants to maximize or minimize something (e.g., power, utility, time in office, etc.). If one can write out an explicit function (i.e., identify the variables that would influence the thing the actor wants to minimize or maximize and their interrelationships), then one can use calculus to further evaluate those relationships and deduce hypotheses. That is, maximizing a utility function (or minimizing a loss)[1] is the mathematical tool that is used to produce the deductions of many game theoretic models. To illustrate, in the chapter's conclusion we offer a simplified utility maximization example.

Similarly, statistical analysis often hinges on locating the maximum or minimum value of a function. Ordinary least squares (OLS) is a technique that *minimizes* the squared error between the observed data and the predicted values of a regression model. Maximum likelihood estimation *maximizes* a likelihood function. Though many textbooks present these models without reference to calculus, presentation is much easier and more straightforward if one is familiar with calculus and specifically with the use of first- and second-order derivatives to find extrema. In the conclusion we briefly walk through the OLS regression model.

Before providing a method for finding extrema, we need to introduce a few concepts. In Section 1 we define the notion of an extremum in general, along with the maximum and minimum, supremum and infimum. In Section 2 we discuss higher-order derivatives, which describe the shape of a function beyond its instantaneous slope. In Section 3 we talk through the method for finding extrema informally before providing a formal guide. Finally, in Section 4 we offer two worked examples to provide practice and to illustrate the method's utility in political science.

---

[1] A utility function mathematically describes the goals or preferences of an actor in much formal theory work, as discussed in Chapter 3.

8.1 EXTREMA

We discussed many properties of functions in Chapter 3 but put off a discussion of several important properties: a function's maximum and minimum, as well as what it means for a function to be concave or convex. We did so because these concepts relate naturally to the derivative, and are essential to understanding optimization. We discuss **maxima** and **minima** here, and concave and convex functions in the next section.

Given the conventional use of the words maximum and minimum, you might wonder why we need the derivative at all. Or if you're of a different mindset, you might wonder whether the formal definitions of these words are somehow more complex than you had thought. We explain the first in Section 3, but you need not worry about the second. Maximum and minimum are concepts just as straightforward as you might expect. They are the high and low points of a function, or the "peaks" and "valleys" in the graph of a function. A high point is called a **maximum** and a low point is called a **minimum**. Together these two points are referred to as the **extrema** of a function.

For example, consider the function $f(x) = x^2$ in Figure 8.1, which has a minimum at $x = 0$. To the left of this point, the function is decreasing, to the right it is increasing. If we flip the parabola, i.e., $f(x) = -x^2$ in Figure 8.2, we get an example of a maximum, again at $x = 0$, and conversely the function is increasing to the left and decreasing to the right of the maximum. Furthermore, the first function has no maximum, while the second has no minimum, because both keep increasing to infinity and decreasing to negative infinity, respectively.

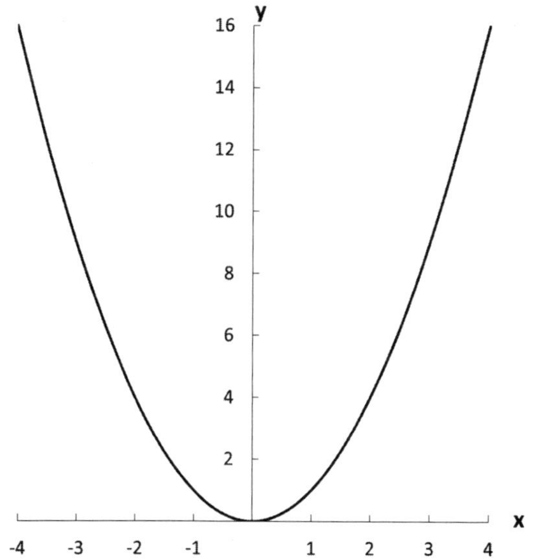

Figure 8.1: Graph of $f(x) = x^2$

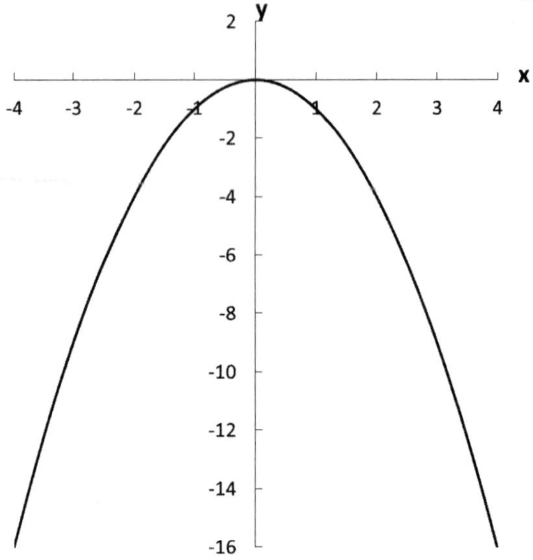

Figure 8.2: Graph of $f(x) = -x^2$

As you saw in Chapter 5, the first derivative of a function tells us the slope of lines that are tangent to the points on a graph of the function. For $f(x) = x^2$ ($f(x) = -x^2$), the slope of the tangents to the left of the maximum (minimum) would be positive (negative), and the slope of the tangents to the right would be negative (positive). This definition is just another way of restating the definition of extrema given in the example above, but it also has two additional and useful implications. First, the slope of the line tangent to an extremum itself will always be zero, and hence the first derivative of the function at a point that is an extremum will always equal zero as well. Second, if we were to graph the first derivative, an extremum will always be a point on that graph that crosses the x-axis, i.e., a point where the slope of the tangents to points in the original function changes signs. Again, this is just another way of saying that the original function changes from increasing to decreasing or vice versa. To illustrate these statements, we may consider Figures 8.3 and 8.4, which respectively graph the first derivative of $f(x) = x^2$, which is $f'(x) = 2x$, and $f(x) = -x^2$, which is $f'(x) = -2x$. As we can see, the derivative of $f(x) = x^2$ is negative for negative x, zero at $x = 0$, and positive for positive x, implying that the function is decreasing for negative x, increasing for positive x, and unchanging at $x = 0$. The same is true in reverse for $f(x) = -x^2$.

Note that an extremum does not have to be, but may be, the absolutely highest or lowest value in a function. We say an extremum is **local** whenever it is the largest (or smallest) value of the function over some interval of values in

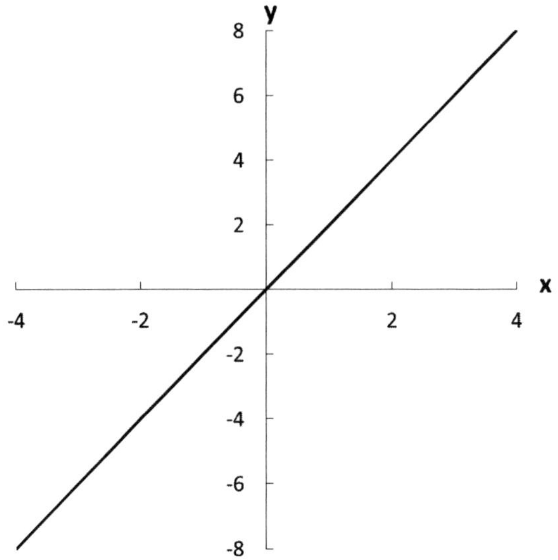

Figure 8.3: Graph of First Derivative of $f(x) = x^2$

the domain of the function, e.g., over some interval on the x-axis.[2] So we can have local maxima, and local minima. A **global** extremum, in comparison, is the highest (or lowest) point on the function.[3]

What is the difference? In our two examples above there was none: the minimum and maximum were indeed global extrema because no other point in the respective function had a corresponding y-value lower or higher than the extrema, respectively. However, many functions have local extrema, points that are the biggest or smallest values in their neighborhoods but that are less or greater, respectively, than other values the function takes. For example, the function $f(x) = (x-1)(x-3)(x-4)$ in Figure 8.5 has two local extrema between 1 and 4, one maximum and one minimum. But the function shoots off to ∞ as x gets bigger than 4 and to $-\infty$ as x gets less than 1, so neither of these two local extrema is a global extremum if we expand the range over which we're looking. One way to avoid confusion is to think of these sorts of local extrema as "valleys" and "peaks" in a function, rather than necessarily as the highest or lowest values in that whole function.

We discuss ways to find local *and* global extrema in Section 3, but a key concept involved is the domain of the function over which we are looking for extrema. Let's consider the preceding example. If we constrain the domain

[2]Formally, a local maximum exists at x^* if there exists some $\epsilon > 0$ such that $f(x^*) \geq f(x)$ whenever $|x - x^*| < \epsilon$. Similarly, a local minimum exists at x^* if there exists some $\epsilon > 0$ such that $f(x^*) \leq f(x)$ whenever $|x - x^*| < \epsilon$.

[3]Formally, a global maximum exists at x^* if $f(x^*) \geq f(x)$ for all x, and a global minimum exists at x^* if $f(x^*) \leq f(x)$ for all x.

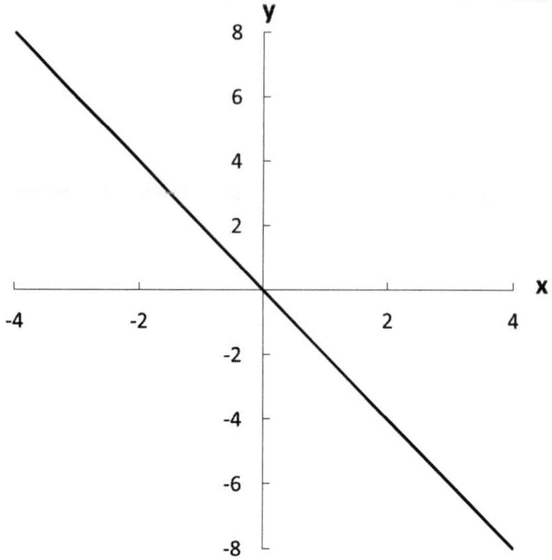

Figure 8.4: Graph of First Derivative of $f(x) = -x^2$

to $[1, 4]$, then our two local extrema are global extrema on this domain—there are no larger or smaller values of $f(x)$ with $x \in [1, 4]$. On the other hand, if $x \in [0, 6]$, then neither extremum is global. As we show, distinguishing global from merely local extrema will involve consideration of the bounds we place on the maximization or minimization problem. Extrema at the boundary of the domain need not correspond to points at which the first derivative is zero. We call extrema that occur within the domain—i.e., not at its boundary—**interior** extrema.

This brings up a related point to address before moving on. What happens when we have a function as basic as $f(x) = x$ on the domain $(0, 1)$? This function has no maximum on this domain of any sort since one can always increase x toward 1, which increases the value of the function, without ever reaching 1 (recall our discussion of open sets in Chapter 4). Despite the strict accuracy of this statement, most observers glancing at the plot of the function would say that the function clearly has a "maximum" at 1 or so, and a "minimum" at 0 or so, on this domain.

While such observers would be wrong in the technical sense, they are correctly identifying the general concept, and two other mathematical definitions exist to encode this concept formally. These are the **supremum** and the **infimum**, which are similar to the maximum and the minimum, respectively. The supremum can be thought of as the least upper bound of any set, and the infimum can be thought of as the greatest lower bound of any set. How does this help us? Let's consider again our example. On the domain $(0, 1)$, the values of $f(x)$ are also in the set $(0, 1)$. This set has a least upper bound of 1 and

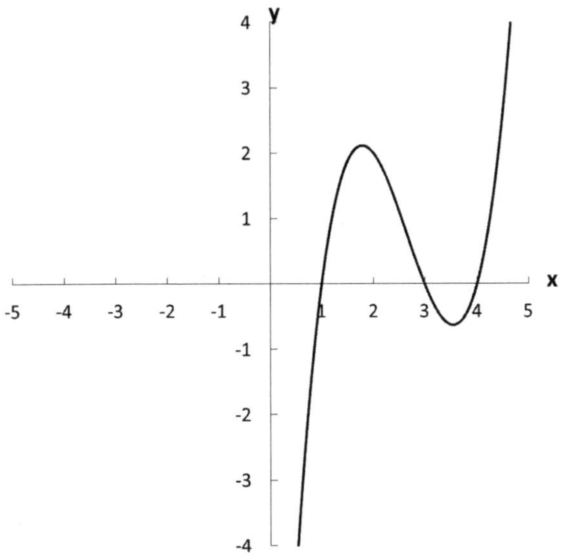

Figure 8.5: Graph of $f(x) = (x-1)(x-3)(x-4)$

a greatest lower bound of 0; hence, the supremum is 1 and the infimum is 0. Even though the function has no maximum or minimum on this domain, it has a well-defined supremum and infimum, which allow us to speak intelligibly about extreme values of the function. Consequently, you will sometimes see these words used when one is searching for the greatest (or least) value when no maximum or minimum exists, particularly in formal theory.

8.2 HIGHER-ORDER DERIVATIVES, CONCAVITY, AND CONVEXITY

To this point we have concentrated on the first-order derivative. But we can also take higher-order derivatives. The first-order derivative is the instantaneous rate of change at a specific point. The second-order derivative is the rate of change of the instantaneous rate of change. The third-order derivative is the rate of change of the second-order derivative, and so on. We refer to these as the first, second, and third derivatives.

The procedure for computing these is exactly the same as that for computing the first derivative. That is, you differentiate the function at hand, whether it is the original function, its derivative, its derivative's derivative, and so on. One can in theory continue to compute these indefinitely, as long as each new derivative remains differentiable. Since a function must be continuous to be differentiable, we sometimes use a notation to specify how many times a derivative may be taken. You will see this in some formal theory work in particular. A continuous function is a member of the set C^0, a function that has a continuous first derivative is a member of C^1, one that has a continuous second derivative is a member of C^2, and so on.

As with the first derivative, there are several notations for higher-order derivatives you might see. If $\frac{df(x)}{dx}$ is a first derivative, then $\frac{d^2f(x)}{dx^2}$ is a second derivative, $\frac{d^3f(x)}{dx^3}$ is a third derivative, and so on. Note the placement of the exponents. In general, an nth-order derivative is written $\frac{d^nf(x)}{dx^n}$. Similarly, if $f'(x)$ is a first derivative, then $f''(x)$ is a second derivative and $f'''(x)$ is a third derivative. As this notation would get messy if we just kept adding primes, higher-order derivatives are usually specified as $f^{(n)}(x)$ for an nth-order derivative.[4] Finally, if D_x is a first derivative in x (or if ∂_x is a first partial derivative in x), then D_x^2 is a second derivative (or ∂_x^2), D_x^3 is a third derivative (or ∂_x^3), and so on. This notation is more common in multivariate calculus, where you can have what is known as a cross-partial: $\frac{\partial f(x,y)}{\partial x \partial y}$. We discuss this more in Part V.

To illustrate the computation of higher-order derivatives, let's consider the first derivative of $y = f(x) = 2x^3$: $\frac{dy}{dx} = 6x^2$. The second-order derivative is $f''(x) = \frac{d^2y}{dx^2} = (f'(x))' = (6x^2)' = 12x$. The third-order derivative is $f'''(x) = \frac{d^3y}{dx^3} = (f''(x))' = (12x)' = 12$. And the fourth-order derivative is $f^{(4)}(x) = \frac{d^4y}{dx^4} = (f'''(x))' = (12)' = 0$. Since the derivative of 0 is 0, further derivatives would just be 0.

For another example, let $f(x) = 6x^4 + x^2 - 3$ and compute $\frac{d^3f(x)}{dx^3}$. We compute this third-order derivative in stages. First, $f'(x) = 24x^3 + 2x$. Second, $f''(x) = 72x^2 + 2$. Third, $f'''(x) = 144x$, which is our answer. Observe that any power of x less than the order of the derivative will not be relevant in the answer.

What do higher-order derivatives represent conceptually? In physics, where one is concerned with the motion of physical bodies in space, the first derivative is the speed or velocity of an object, its rate of change in position with respect to time. The second derivative is the rate at which speed changes, which is known more commonly as the *acceleration* of an object. You will at times see this word used in reference to the second derivative of variables in political science as well.

More generally, though, derivatives of any order tell us something about the *shape* of the function. The first derivative tells us in which direction it is trending: is it increasing or decreasing? The second derivative tells us the most basic curvature of the function: is the rate at which it is increasing or decreasing getting faster or slower? Higher-order derivatives provide more nuanced information about the shape of functions; however, we will typically find first and second derivatives sufficient for our purposes.

---

[4] You might sometimes see roman numerals instead, e.g., f^{iv} for a fourth derivative.

8.2.1 Concavity and Convexity

To understand what the second derivative tells us, let's start by considering an increasing function. One with a rate of increase that slows as the value of the function gets bigger is an example of a **concave function.**[5] One with a rate of increase that speeds up as the value of the function gets bigger is an example of a **convex function.**[6] Distinguishing convex from concave functions proves particularly useful in formal theory when defining utility functions because they capture risk preferences, as we discuss in Part III. Concave functions are also used fairly often in statistical models when expressing the belief that the marginal effect of a variable is decreasing, as we noted in Chapter 3 in discussing the logarithm.

Here we define concave and convex functions more generally. In general, a function is concave if for any points x_1, x_2 in its domain and any weight $\lambda \in [0, 1]$:

$$f(\lambda x_1 + (1 - \lambda)x_2) \geq \lambda f(x_1) + (1 - \lambda)f(x_2).$$

A function is **strictly concave** if the inequality is strict, so that the LHS is strictly greater than the RHS, rather than greater than or equal to it.

When this is so, a secant drawn on a graph of the function between any two points must be below the function itself. Figure 8.6, of $\ln(x)$ with a dashed line for the secant between the values of the function evaluated at $x = 2$ and $x = 4$, illustrates this definition. Why is this so? The expression $\lambda x_1 + (1 - \lambda)x_2$ is known as a convex combination of x_1 and x_2; for a function to be concave, the value of the function at a convex combination of any two points must exceed the weighted sum of the values of the function at those two points. And all such weighted sums lie along the secant, which must therefore be below the function itself.

You may be asking why, given this definition, we have included concavity (the property of being concave) in this chapter. To see the reason, note that this function is increasing, but at a decreasing rate. The first derivative is $\frac{1}{x}$, which is positive, supporting this. The second derivative is $-\frac{1}{x^2}$, which is negative, implying that the rate of increase in the first derivative is decreasing. That is, the function is "decelerating." Concave curves are characterized by their shape, and a negative second derivative is another way of expressing this shape. Since this manner of determining concavity is often much more convenient than the previous definition we provided, the second derivative is very helpful for being able to characterize the concavity of a function.

As another example, consider the function $f(x) = -x^2$ we discussed above. We noted there that the function increases to the left and decreases to the right of the maximum. That implies that the rate of increase in the function is decreasing for $x < 0$, which implies that its second derivative is negative. We can check this: $f' = -2x > 0$ for $x < 0$, so the function is increasing left of

---

[5]Sometimes this is confusingly called *concave down*. We avoid this phrase.

[6]Sometimes this is confusingly called *concave up*. We avoid this phrase.

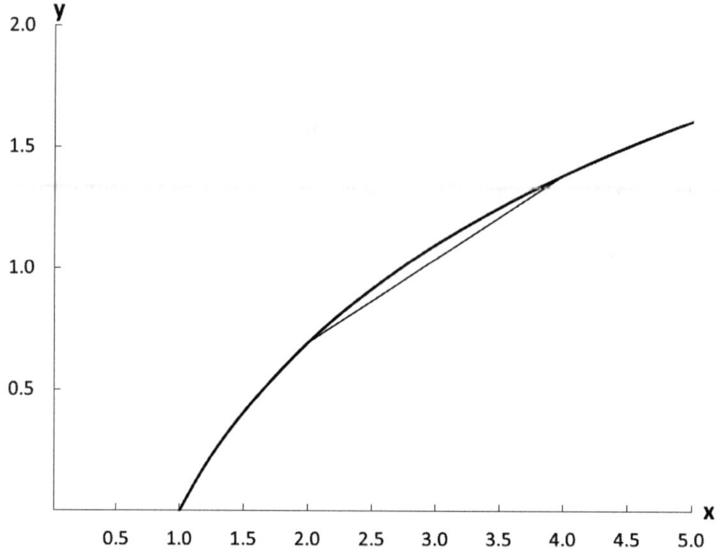

Figure 8.6: Graph of $f(x) = \ln(x)$ with Secant

its maximum, but $f'' = -2 < 0$, and thus the rate of increase is decreasing. Therefore this function is concave as well.

A convex function, in contrast, has exactly the opposite properties. A function is convex if for any points x_1, x_2 in its domain and any weight $\lambda \in [0, 1]$:

$$f(\lambda x_1 + (1 - \lambda)x_2) \leq \lambda f(x_1) + (1 - \lambda)f(x_2)).$$

A **strictly convex** function uses a $<$ rather than a $\leq$. In words, a secant drawn between any two points must be above the function itself. Figure 8.7, of x^2 with a dashed line for the secant between the values of the function evaluated at $x = 1$ and $x = 2$, illustrates this definition.

We discussed this function above, noting that the function decreases to the left and increases to the right of the minimum. That implies that the rate of increase in the function is increasing, which implies that its second derivative is positive. We can check this: $f' = 2x > 0$ for $x > 0$ and $f'' = 2 > 0$. So positive second derivatives correspond to convex functions.

8.2.2 Taylor Series

At the beginning of this section we noted that higher-order derivatives encode information about the shape of the function. The first derivative tells us whether the function is increasing or decreasing, the second what the curvature is, and so on. This suggests that one could build up a function by incorporating all the information encoded in these derivatives. It turns out that one can do this for a large class of functions known as analytic functions. The series that expresses a

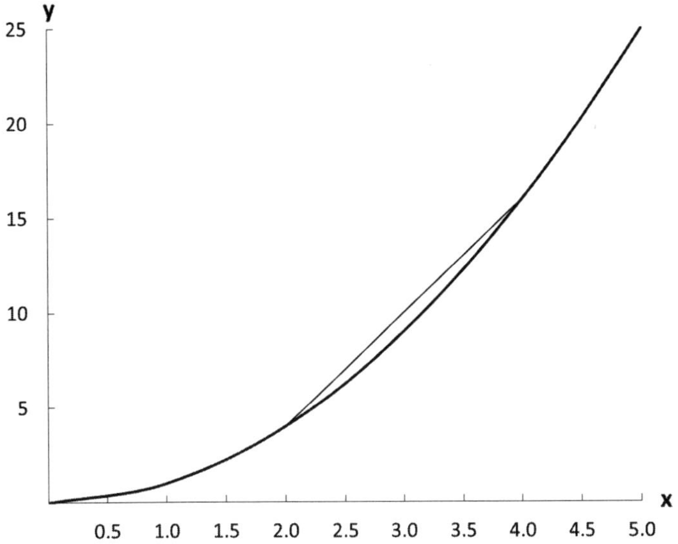

Figure 8.7: Graph of $f(x) = x^2$ with Secant

function in terms of its derivatives is known as a **Taylor series**, named after the English mathematician Brook Taylor. If an analytic function $f(x)$ is infinitely differentiable close to some number a, then the Taylor series is given by the infinite sum[7]

$$\begin{aligned} f(x) &= f(a) + \frac{f'(a)}{1!}(x-a) + \frac{f''(a)}{2!}(x-a)^2 + \frac{f'''(a)}{3!}(x-a)^3 + \ldots \\ &= \sum_{n=0}^{\infty} \frac{f^{(n)}(a)}{n!}(x-a)^n. \end{aligned}$$

The Taylor series is useful for many reasons, but the primary one for our purposes is that it allows one to replace a complex function with a bunch of powers of x. We already used this to calculate the derivative of e^x; recall our second definition of $e^x = \sum_{n=0}^{\infty} \frac{x^n}{n!} = 1 + x + \frac{x^2}{2} + \frac{x^3}{6} + \ldots$. It is no accident that this is equal to the Taylor series of e^x with $a = 0$ (sometimes called the Maclaurin series), since all derivatives of e^x are e^x, and $e^0 = 1$.

[7]One can derive the Taylor series via repeated use of integration by parts and the fundamental theorem of calculus. By the latter, $\int_a^x f'(t)dt = f(x) - f(a)$. Rearrange to get $f(x) = f(a) + \int_a^x f'(t)dt$. This gives the $f(a)$. Now we integrate $\int_a^x f'(t)dt$ by parts (letting $u = f'(t)$ and $dv = dt$, so $du = f''(t)dt$ and $v = t - x$, where $C = -x$ is the constant of integration) to get $f'(a)(x-a) + \int_a^x f''(t)(x-t)dt$. This yields the first derivative term (to get it, we evaluate uv at the bounds of the integral). Now we integrate $\int_a^x f''(t)(x-t)dt$ by parts, this time letting the constant of integration be 0, to get $\frac{1}{2}f''(a)(x-a)^2 + \frac{1}{2}\int_a^x f'''(t)(x-t)^2 dt$. This produces the second derivative term. Repeating this process gives the Taylor expansion to as many derivatives as one would like.

More usefully, the Taylor series provides a good approximation of a function near any point. Note that the expansion is in powers of $(x-a)$. If we only care about values of x really near a, and so consider only these, then $(x-a)$ is small, $(x-a)^2$ is smaller, and so on. At some point, adding an additional term doesn't change the sum enough to be worth it, particularly with the $n!$ in the denominator. So we can cut off the approximation there.

For example, consider the function $\ln(1+x)$ expanded around $a=0$, and let's say we are only concerned with values of x very close to zero. The full Taylor series for this function is $\ln(1+x)=\sum_{n=1}^{\infty}(-1)^{n+1}\frac{x^n}{n}$ for $-1<x\leq 1$ (left as an exercise), but it might be well approximated by the first two terms: $\ln(1+x)\approx x-\frac{x^2}{2}$, which can be much easier to manipulate than the log.

8.3 FINDING EXTREMA

We are now almost ready to offer a guide to extrema. Before doing so, however, we must define two more terms.

8.3.1 Critical Points and Inflection Points

We saw in Section 1 that the first derivatives of functions at interior extrema are equal to zero. Does this have some larger meaning? The second derivative can be zero as well. Does this mean something? As you might have guessed by the existence of this subsection, the answer to these questions is a qualified yes. Points at which the first and second derivatives change from positive to negative or vice versa, implying a stopover at zero, have substantive meaning.

Let's begin with the first derivative. A **critical point** is any point x^* such that either $f'(x^*)=0$ or $f'(x^*)$ doesn't exist. Loosely, critical points are points in the function's domain at which things *happen*. Either the function blows up, or it jumps, or it is stationary. The last occurs when $f'(x^*)=0$, and such points are known as **stationary points** because the value of the function is, however briefly, not changing at that point. We are interested in such points because Fermat's theorem (not his last one!) tells us that local extrema occur at critical points, and we are interested in finding extrema.

However, just because local extrema occur at critical points does not mean all critical points are extrema. Some are instead **inflection points**, which are points at which the graph of the function changes from concave to convex or vice versa. For example, up to a certain point a function may be increasing at a slower and slower rate, but after that point it might increase at a faster and faster rate. Such a function is $f(x)=x^3$, which has an inflection point at $x=0$, even though $f'(0)=0$. First, note that the function is increasing on both sides of $x=0$, as we can see in Figure 8.8. (We can also check that $f'(x)=3x^2\geq 0$ for all x, implying an increasing function.)

Now consider the curvature of the function. To the left of zero the function is increasing more and more slowly as we get closer to $x=0$, whereas to the right of zero the function is increasing more and more quickly as we move to larger

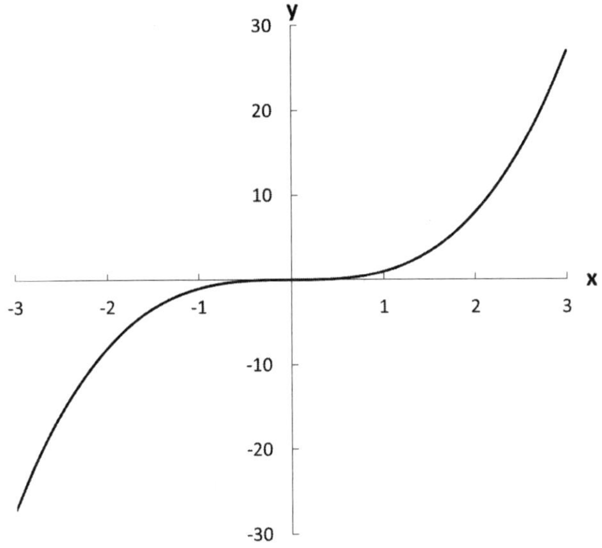

Figure 8.8: Graph of $f(x) = x^3$

x-values away from zero. Thus, to the left of zero the function is concave, while to the right the function is convex. For this reason $x = 0$ is an inflection point of the function, and not an extremum. (We can also check that the second derivative $f''(x) = 6x$ changes sign at $x = 0$: it is negative for $x < 0$ and positive for $x > 0$.) We can verify this conclusion by simply looking at the graph: the value of $f(0)$ is not the biggest or the smallest the function takes in a neighborhood around $x = 0$, regardless of how small the neighborhood is.

When the slope of the tangent line to the function is not zero at the inflection point, the existence of inflection points gives us no trouble in finding extrema. Given that our interest is in finding extrema and not inflection points, we needn't concern ourselves with these (called nonstationary points of inflection). However, as in our example, in some instances the slope of a tangent to the inflection point will equal zero. That is, both $f'(x^*) = 0$ and $f''(x^*) = 0$, and x^* is further an inflection point. Such points are also known as **saddle points** (owing to their appearance in two dimensions) and are not extrema. Thus, a stationary point for which $f'(x^*) = 0$ can be a local minimum, a local maximum, or an inflection point.[8] The key to identifying extrema is to figure out in which case one finds oneself.

[8]It is also possible that the function may oscillate near the critical point; however, this is not typical for functions we commonly use in political science, and we do not discuss it further.

8.3.2 Stationary Points and the First Derivative Test

In political science we are often interested in the largest achievable likelihood function or utility function, or the smallest achievable mean-squared error. One way we could find these would be to manually examine the values of $f(x)$ for different values of x and slowly figure out which value of x leads to the highest or lowest $f(x)$. This brute force approach, of course, is likely to be time-consuming and imprecise.[9] A much more elegant solution to this problem involves using the derivatives of a function to determine the interesting points in that function.

We have already seen several times now how this can work. For example, ignoring any domain boundary, as we will do until further notice, the highest value of a function can only be a point at which the function is increasing before the point and decreasing after the point (think of the peak of the graph of a function). This implies that the function is increasing to one side of a maximum or minimum and decreasing on the other. In turn, this implies that the derivative must be positive to one side and negative to the other side of the extremum, which means that at the extremum, the derivative is stationary. In other words, the extremum occurs at the stationary point. Thus, our first step in finding extrema is to find all stationary points.

To find the stationary points of a function $f(x)$ we use the first derivative test: we set the first derivative, $f'(x)$, equal to zero and solve for x. This yields the equation $f'(x^*) = 0$, which is often called the first-order condition, or FOC for short. We can find the y-values for these stationary points by plugging the values of x^* we obtained back into the original function, $f(x)$. The points we thus obtain are candidates that could be either extrema, i.e., either minima or maxima, or inflection points.

For example, consider the function $f(x) = x^3 - 3x^2 + 7$. The first derivative of this function is $f'(x) = 3x^2 - 6x$. Now we set the first derivative equal to zero:

$$3(x^*)^2 - 6x^* = 0$$
$$x^*(3x^* - 6) = 0$$
$$x^* = 0 \text{ or } x^* = 2.$$

For this example, there are two stationary points, at zero and two. Substituting these values of x back into $f(x)$ yields the y-values seven and three. We now know that these two points are stationary points for $f(x)$; however, while this is a *necessary* condition for local extrema, it is not a *sufficient* one. We still must determine whether they are minima, maxima, or inflection points.

[9]That said, an organized, fine-grained version of such a search performed by a computer instead of by you—known as a grid search—can be a good first pass for finding an extremum computationally.

8.3.3 Second Derivative Test

We use the second derivative test to determine whether the stationary points we obtained in the first derivative test are extrema or inflection points. First we have to determine the second derivative, $f''(x)$, of the original function $f(x)$. Then we substitute the stationary points x^* we determined from the FOC into $f''(x)$. If the answer is negative, i.e., if $f''(x^*) < 0$, the stationary point is a maximum since the function is concave near x^*. If, in contrast, $f''(x^*) > 0$, the stationary point is a minimum since the function is convex near x^*. Finally, if $f''(x) = 0$, the stationary point may be an inflection point. These three conditions are collectively referred to as the second-order condition, or SOC. The third one is the more complex case, and bears some elaboration.

An inflection point occurs whenever the sign of the second derivative changes, since this implies a shift from convex to concave, or vice versa. For the same reason that we can't simply check when $f'(x) = 0$ to find extrema, however, we can't simply check when $f''(x) = 0$ to find inflection points. We also need to make sure that the second derivative actually changes sign. One way to do this mimics our procedure for finding extrema: we can check whether the potential inflection point x_i is an extremum of the function $f'(x)$. If it is, then the second derivative must change sign just as the first derivative must change sign at an extremum of the function $f(x)$. This entails computing the third derivative at x_i. If it is non-zero, x_i is an inflection point, but if it is zero, it is not. For example, if $f(x) = x^3$, $f'''(x) = 6 > 0$, so zero is an inflection point, as we have found. However, if $f(x) = x^4$, $f'''(x) = 24x = 0$ at $x = 0$, so this is not necessarily an inflection point. In fact, one can check by simply graphing the function that it is convex and $x = 0$ is a minimum, not an inflection point.[10]

There is an easier method than graphing, however, for more complicated functions than x^4. Whenever $f''(x) = 0$, the second derivative test is inconclusive, and we must check to see whether it is an inflection point. We can do this by continuing to compute derivatives at the stationary point x^*. Eventually one will be non-zero, if we assume the possibility that an inflection point exists. If this derivative is of odd order—e.g., $f'''(x), f^{(5)}(x), f^{(7)}(x)$, etc.—then the stationary point is an inflection point (and so a saddle point). If it is of even order, it is not a point of inflection and is thus an extremum. If the value of this non-zero derivative (of even order) evaluated at x^* is positive, then the point is a local minimum. If it is negative, it is a local maximum. This is known as a higher-order derivative test.

This covers the necessary tools needed to find local extrema; we provide a briefer summary at the end of the section.[11] First let's return to the example from the previous subsection, in which we found two stationary points. The

[10] We could also just compute the second derivative to both sides of the possible inflection point to see if it changes. Just plug $x_i - \epsilon$ and $x_i + \epsilon$ into $f''(x)$ if x_i is the possible inflection point and ϵ is some small number. If they produce different signs, then it is an inflection point. You might want to verify that this works to distinguish x^3 from x^4.

[11] This will be particularly useful if you found a bit confusing our discussion of even and odd order derivatives and points of inflection.

original function was $f(x) = x^3 - 3x^2 + 7$, and the first derivative was $f'(x) = 3x^2 - 6x$. Taking the derivative again, we get $f''(x) = 6x - 6$. The first stationary point we obtained previously was 0, which when we plug it into the second derivative yields a value of $-6 < 0$. Since it's negative, $x^* = 0$ is a local maximum of the function. The second stationary point was 2, which gives us $6 > 0$, and since this is positive we know that $x^* = 2$ is a local minimum.[12] This gives us all the local extrema of the function.

In general, functions may have any combination and number of local minima, local maxima, and inflection points, and some functions may not have any at all. For example, a linear function of the type $f(x) = a + bx$ will never have local extrema or inflection points. However, this doesn't mean they have no extrema within the bounds of a limited domain.

8.3.4 Boundaries and Global Extrema

Consider the linear function $f(x) = 3x + 2$, and define it on the domain $[-1, 3]$. We've just said it has no local extrema, as one could verify by noting that there is no solution to the FOC: $f'(x) = 3 \neq 0$ for any x. But it must have some maximum and minimum on this domain, right?

Of course it does. This function increases monotonically from -1 to 3, so we can guess that its global maximum occurs at 3 and its minimum at -1 in this domain. Let's try a slightly harder one, the example we used earlier of $f(x) = x^3 - 3x^2 + 7$. We graph it below on the domain $[-4, 4]$.

We've marked the local maximum and local minimum on the graph. The local maximum is at $x = 0$, and at this point the function has a value of $f(0) = 7$. Similarly, the local maximum is at $x = 2$, and at this point the function has a value of $f(0) = 3$. A glance at the graph supports that these points are extrema relative to nearby points—that is, *local* extrema—but are they *global* extrema on this range? Another glance says no to both. The function shoots off upward past 2, and has a value clearly greater than 7 well before it hits 4. It also shoots off downward past 0, and has a value less than 3 well before -4. Consequently, on this domain the global maximum occurs at $x = 4$, and the global minimum at $x = -4$.

This analysis raises two questions. How did we know there were global extrema, and how do we find them in general? A couple of theorems will be of great use to us. There is the **extreme value theorem**, which states that a real-valued function that is continuous on a closed and bounded interval $[a, b]$ (i.e., a function that spits out real numbers that is continuous on a closed, finite interval of the real line) must attain both its global minimum and its global maximum on that interval, at least once each.[13] Karl Weierstrass, a German

[12] Solving $f''(x_i) = 0$ gives a potential inflection point at $x_i = 1$, and we can verify it by checking that $f'''(1) = 6 \neq 0$.

[13] Formally, $\exists x_{min}, x_{max} \in [a, b]$ such that $f(x_{min}) \leq f(x) \leq f(x_{max}) \forall x \in [a, b]$.

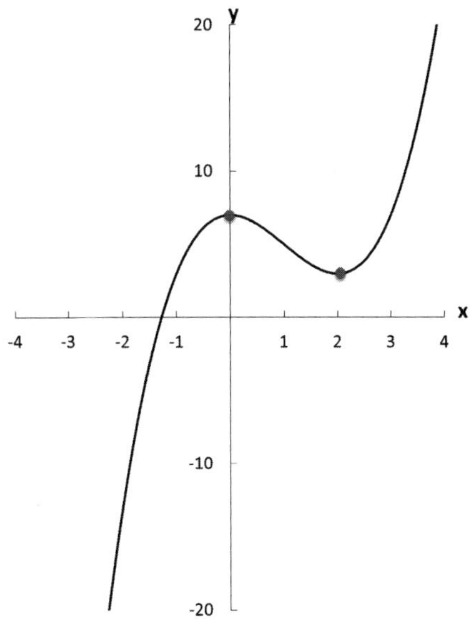

Figure 8.9: Local Extrema for Cubic Equation

mathematician, extended this result to continuous (real-valued) functions on any compact set, rather than just $[a, b]$.[14]

These theorems have an important consequence: there are only so many places a global extremum can be, as long as the function is continuous and its domain is bounded. Either it is at a critical point or it is at a point at which the procedure we used to find critical points would fail because we were stopped from moving further along the function—namely, the boundary of the domain. Thus, not only do the global minimum and global maximum exist, as long as the domain is bounded, but either they are at the points of local extrema of the function within that domain or they are at the boundaries. If the function does not have any critical points, as the linear function does not, then both minimum and maximum must be at the boundaries of the domain, as we found in our example. Similarly, if the function has only one critical point in any domain, as for example $x^2 - 2x$ does at $x = 1$, then the other extremum (minimum or maximum) must be found at the boundary of the domain.

So, to find the global minimum and maximum, one need compare the value of the function at only a few points: (1) at all the local minima and maxima, and (2) at the boundaries of the domain. For our example of $f(x) = x^3 - 3x^2 + 7$,

[14]A closely related theorem that you may encounter is the mean value theorem, which states that for a differentiable function f on an interval (a, b), there exists a c such that $f'(c) = \frac{f(b)-f(a)}{b-a}$. In words, this means that there is a point c at which the instantaneous rate of change equals the mean, or average, rate of change across the interval. In other words, there exists a point between two other points at which the tangent has the same slope as the secant between the two endpoints.

the value of the local minimum is $f(2) = 3$ and the local maximum is $f(0) = 7$. If we assume the boundaries match the figure, so $x \in [-4, 4]$, then the values of the function at the boundaries are $f(-4) = -105$ and $f(4) = 23$. Since $-105 < 3$ and $23 > 7$, the global minimum and maximum are at $x = -4$ and $x = 4$, respectively, within this domain.

8.3.5 Summary of Method

Having worked out a procedure to find the global minimum and maximum of a function within any domain, we now summarize the method. To find the extrema and inflection points of any given (at least) twice-differentiable function $f(x)$, follow these steps:

1. Find $f'(x)$.
2. Set $f'(x^*) = 0$ and solve for all x^*. These are stationary points of the function.
3. Find $f''(x)$.
4. For each stationary point x^*, substitute x^* into $f''(x)$.
   - If $f''(x^*) < 0$, $f(x)$ has a local maximum at x^*.
   - If $f''(x^*) > 0$, $f(x)$ has a local minimum at x^*.
   - If $f''(x^*) = 0$, x^* may be an inflection point. To check this:
     a) Calculate higher-order derivatives ($f'''(x), f^{(4)}(x)$, etc.) until you find the first one that is non-zero at x^*. Call the order of this derivative n.
     b) If n is odd, then this x^* is an inflection point and not an extremum. Do not include it in further steps.
     c) If n is even and $f^{(n)}(x^*) < 0$, $f(x)$ has a local maximum at x^*.
     d) If n is even and $f^{(n)}(x^*) > 0$, $f(x)$ has a local minimum at x^*.
5. Substitute each local extremum into $f(x)$ to find the function's value at that point.
6. Substitute the lower and upper bounds of the domain over which you are attempting to find the extrema into $f(x)$ to find the function's values at those points.
7. Find the smallest value of the function from those computed in the previous two steps. This is the global minimum, and the function attains this at the corresponding x^* or boundary point. Find the largest value of the function from those computed in the previous two steps. This is the global maximum, and the function attains this at the corresponding x^* or boundary point.

8.4 TWO EXAMPLES

8.4.1 Maximizing Utility

Game theoretic models assume that individuals' preferences are represented by utility functions. One can use the same framework to make decisions for a single person; this is a form of decision theory. Let's assume that you have \$1 million that you could donate to a campaign. Call the amount of money you choose to donate, in millions of dollars, x. Let's also assume that you get some nice benefit for donating, either via feeling good about yourself or, more cynically, via future favors expected, and that this benefit can be represented by the function x. That is, you get the same benefit in terms of utility as the amount of money you donate, x. Finally, let's assume that you could have done other things with your money, so donating occasions an opportunity cost that can be represented by x^2. Finally, let's assume x is measured in millions of dollars, so you can donate any $x \in [0, 1]$.

Putting these assumptions together gives a utility function: $u(x) = x - x^2$. The optimal choice for a donation maximizes this function in the domain $[0, 1]$. Again, we follow the steps. First, we find $u'(x) = 1 - 2x$. Second, we solve for x^*: $1 - 2x^* = 0 \Rightarrow x^* = \frac{1}{2}$. Third, we find $u''(x) = -2 < 0$. This is negative for all x, so step four tells us we have a local maximum. We do have bounds to the domain this time, so we check all the values of the function: $u(\frac{1}{2}) = \frac{1}{4}, u(0) = u(1) = 0$. So the global maximum occurs at a donation of $x = \frac{1}{2}$. Note that when maximizing utility we are often far less concerned with what the maximum value of the function is than with what value of x maximizes the function. If the general maximization problem is written $\max_x u(x)$, which means "maximize $u(x)$ with respect to x" but we are interested in the *argument* that maximizes the utility, we can write $\arg\max_x u(x)$, which means "find the value of x that maximizes $u(x)$."

This maximum would not change if we were to change the bounds to, say, $[0, 2]$, or anything else for that matter. What if we changed the benefit to a concave function such as $\ln(1+x)$, and the cost to a linear function, x, however? This would give a utility function of $u(x) = \ln(1+x) - x$. (Let's use a domain of $[0, 1]$.) Repeating our procedure, first, we find $u'(x) = \frac{1}{1+x} - 1$. Second, we solve for x^*: $\frac{1}{1+x^*} - 1 = 0 \Rightarrow x^* = 0$. Third, we find $u''(x) = -\frac{1}{(1+x)^2} < 0$. This is negative on our domain for all x, so step four tells us we have a local maximum. This time we need to check only one of the bounds, since our local maximum is actually coincident with the other bound. So $u(0) = 0, u(1) = \ln(2) - 1 \approx -0.3$. The global maximum thus occurs at a donation of $x = 0$. You can check that this too doesn't change by increasing the larger bound on the domain.

8.4.2 Linear Models and OLS Regression

Consider first a linear model of some political phenomenon in which an independent variable, x, is thought to directly affect some dependent variable, y. The variable x might be, for example, an individual's education or income or a nation's GDP or number of military allies, and y might be the likelihood one turns out to vote or votes for a particular party, or the expected number of trading partners or militarized disputes a nation experiences. To say this relationship is linear means a change in x causes a proportional change in y. We can write this as $y = \beta x$, where β is the constant of proportionality.

To account for the uncertainty in this relation, we can add an error term, ϵ, to the RHS of our linear equation: $y = \beta x + \epsilon$. The goal of OLS is to find a β that minimizes the squared error, ϵ^2, given some set of data in y and x. We have to wait until the introduction of vectors and matrices to deal with this problem when we have many data points, but let's say for the sake of argument that we have one data point for x and y. What is the value of β that minimizes the squared error (ϵ^2)?[15]

We start by writing out ϵ^2: $\epsilon^2 = (y - \beta x)^2 = f(\beta)$. We want to find the minimum of the RHS with respect to β, so we follow the steps from the previous section. First, we find $f'(\beta)$, which equals $-2x(y - \beta x)$.[16] Next, we solve for β^*, yielding $-2x(y - \beta^* x) = 0 \Rightarrow 2xy = 2\beta^* x^2 \Rightarrow \beta^* = \frac{xy}{x^2}$. Third, we find $f''(\beta)$. Doing so generates the inequality $f''(\beta) = 2x^2 > 0$, which is true for all β, so the fourth step tells us that $f(\beta)$ has a minimum at β^*, as we wanted. We have been given no bounds for the domain, and there's only one minimum, so this is our global minimum. In other words, the optimal choice of β to minimize the squared error is $\beta^* = \frac{xy}{x^2}$. We can see the similarities between this and the OLS estimator for the vector $\vec{\beta}$ in a multiple regression framework, where $\mathbf{y}$ is a vector and X is a matrix: $\vec{\beta^*} = (X^T x)^{-1} X^T \mathbf{y}$. We cover matrices and vectors in Part IV of this book, and return to this example at the end of Chapter 12.

8.5 EXERCISES

1. Find all extrema (local and global) of the following functions on the specified domains, and state whether each extremum is a minimum or maximum and whether each is only local or global on that domain.

   a) $f(x) = x^3 - x + 1, x \in [0, 1]$.
   b) $f(x) = x^2 - 2x + 17, x \in [0, 5]$.
   c) $f(x) = \frac{x^2}{e^x}, x \in [0, \infty)$.
   d) $f(x) = 2 - 3x, x \in [-3, 10]$.
   e) $f(x) = 6x - x^2 + 12, x \in [0, 10]$.
   f) $f(x) = -4 + 3x^3, x \in [0, 10]$.

[15] The following derivation when all variables are vectors is not appreciably different from the scalar case presented here, at least as far as the calculus goes.

[16] To get this, one could either expand the square and then differentiate the sum, or treat $f(\beta)$ as a composite function of $g(u) = u^2$ and $h(\beta) = (y - \beta x)$, with $u = h(\beta)$, and use the chain rule.

g) $f(x) = 4x^3 + x^2 - 2x + 3, x \in [-1, 1]$.

h) $f(x) = 4x^3 + x^2 - 2x + 3, x \in [-3, 3]$.

i) $f(x) = \ln(1 + x) - x^2, x \in [-2, 2]$.

j) $f(x) = \frac{1}{4}x^4 - \frac{4}{3}x^3 + \frac{1}{2}x^2 + 6x + 2, x \in [-2, 3]$.

2. The utility that a legislator extracts from a policy in a one-dimensional policy space varies as a quadratic loss function, $U(x)$, of the form $U(x) = -a(x - x_0)^2$, where x_0 represents the ideal location of the legislator and x represents the location of the current policy in the one-dimensional policy space. Prove that the legislator's utility is maximized when the policy is located at $x = x_0$.

3. Show that the Taylor series for $\ln(1+x) = \sum_{n=1}^{\infty}(-1)^{n+1}\frac{x^n}{n}$ for $-1 < x \leq 1$.

Part III

Probability

Chapter Nine

An Introduction to Probability

Probability is central to the work of political scientists, as uncertainty is prevalent in every aspect of the social world and probability is the formal language of uncertainty. You need to care about probability because it can help you develop explicit statements about uncertainty. This will be necessary when you are trying to develop a theory to explain situations where people are acting under uncertainty. For example, individuals could be uncertain of each other's benefits or costs for taking some action, which of a number of potential strategies they might use in any instance, or even just what outcome is likely to occur given a particular behavior, as in a lottery. Formal representations of these and other forms of uncertainty underlie games of incomplete information in game theory, as well as many models of bounded rationality. Bayes' rule, which we discuss in this chapter, plays a particularly key role in game theoretic solution concepts.

Explicit statements about uncertainty are also necessary when challenging theories empirically. In statistics you want to be able to make a claim about the confidence of your findings; this requires the formal use of uncertainty. Further, probability is the cornerstone of both frequentist and Bayesian statistics, so you cannot take proper advantage of the powerful tool of statistical inference unless you have a firm grasp of probability theory.

In this chapter we introduce the basics of probability theory, including how to calculate what are known as *simple event* probabilities by way of combinations. The basics are useful to all political scientists, while calculating event probabilities is more typically useful in game theory. Accordingly, Section 1 focuses on the basics, and we thread the use of probability in statistics through this section as motivation. Section 2 covers the computation of probabilities using uncertainty in game theory as motivation; this sets up our discussion of expectations and expected utilities in Chapter 10. Section 2 also introduces Bayes' rule, which is central to several solution concepts in game theory. Finally, Section 3 offers a few specific measures of probability that may turn up in statistics classes.

9.1 BASIC PROBABILITY THEORY

9.1.1 Objective versus Subjective Probability

One can distinguish two different accounts of probability, objective and subjective. The objective account is based on theory or observations from data, whereas the subjective account concerns the beliefs of human beings. There are two types of objective probability, classical and empirical. The classical account focuses on the ratio of an outcome to all possible outcomes. The empirical account focuses on the ratio of the frequency with which a given outcome actually occurs relative to all other outcomes that occur. Subjective probability is based on people's expertise (or lack thereof) and judgments "and is basically an educated guess as to the chances of an event occurring" (Bluman, 2005, p. 16).

The classical and empirical accounts played fundamental roles in the development of frequentist (aka Pearson-Neyman) statistics. The classical and subjective accounts played fundamental roles in the development of rational choice (aka expected utility) theory and Bayesian statistics. Frequentist statistical theory was introduced to political scientists by the work of the Chicago school in the 1920s and '30s (Almond, 2004) and was the centerpiece of the behavioral revolution in the 1950s and 1960s (Easton, 1969). Today one or more courses in frequentist statistical theory are required in almost every PhD program in political science. The Rochester school was largely responsible for introducing rational choice theory to political science in the 1960s and 1970s (Amadae and Bueno de Mesquita, 1999), but Bayesian statistics has made inroads only in the past decade or so. Required literacy courses in rational choice theory are still unusual in political science, though the number of programs requiring such a course is growing. Courses in Bayesian statistics are taught only in a few political science departments at present, but most people familiar with political methodology expect that number to grow substantially over the next two decades.

While both views of probability, objective and subjective, have influenced political science, they have had an even greater impact on fields such as statistics, economics, philosophy, and others. The key point is that they have proved fruitful and helped launch major research programs across multiple disciplines, and that is why it is important to gain familiarity with both views.

The good news is that the differences are philosophical and have little practical effect *with respect to the basics.* That is, rather than having to learn two sets of basics of probability theory, you need learn only one, as both the objective and subjective accounts draw from the same basics.

With that as background, it is useful to provide an intuitive introduction to probability. Probability can be thought of as a formal means to help us model uncertainty. All human beings make judgments and decisions under uncertainty. If we replace the word "judgments" with "inferences," the connection to statistical theory becomes apparent.[1] In daily life, human beings make both descriptive inferences (or judgments) about available facts and causal inferences (or judgments) about the way the world works. Statistics is little more than the formal study of how one should best make such judgments in the face of

[1] Inference is a "best guess" about an unknown (often an unknowable), given known information (e.g., King, Keohane, and Verba, 1994, pp. 46–49). In statistics, inference involves claims about an unknown value of X, or relationship between X and Y, in a population, given the known values of x, or a relationship between x and y, from a sample.

uncertainty. Further, human beings make choices under uncertainty, from trivial decisions about whether to commute to work via one route versus another based on one's belief about prevailing traffic to potentially more consequential decisions, such as whether to cross through an intersection given a stale yellow light. Many rational choice theories of politics are nothing more than formal theories of choice under uncertainty.

9.1.2 Classical Probability

Classical probability is the theoretical analysis of events in the absence of data. The three key concepts are outcome, event, and sample space. Outcomes are anything that might happen in the world. Events are composed of one or more outcomes. The canonical example is a die. There are six possible outcomes when one rolls a die. Each of those outcomes can be defined as an event, but we also want to be able to define the event that one rolls *an even number* or *a number greater than three*, etc. That is, events can be composed of multiple outcomes.

Events can be divided into two groups with respect to their cause, those that will happen with some probability given certain conditions and those that will happen (or not happen) with certainty given certain conditions. The first group of events is known as random events, and the study of probability applies to them. The second group of events is known as deterministic events; we do not need probability here because events either always happen or never do, based on the situation.

The phenomena that political scientists study are random events. This may strike you as an odd claim. In common usage the term "random" is often used to indicate something that cannot be anticipated: we cannot describe a causal process that produces it. When social scientists refer to events produced by random processes they have a more precise meaning in mind: such events are probabilistic (aka stochastic) as opposed to deterministic events. In other words, for a random (or probabilistic or stochastic) event one might be able to specify causal processes that alter the chance that it occurs, but one cannot specify causal processes that *guarantee* the event will occur, as would be the case for a deterministic event.

Consider an election as an event. Imagine that you could know all of the causes that determine which candidate wins that election. Could you predict the event with certainty? Most of us think not. For starters, we find it difficult to believe that anyone would know *all* of the causal factors that produce the winner of an election. Further, while we may know what factors are important in winning elections, we cannot collect all of the information we would need to know the precise value of all of those factors. The best we can do is develop a theory that specifies the factors that are important to determining election winners and then conjecture that those factors increase (or decrease) the odds of a given candidate's prospects. In other words, most political scientists believe that the best we can do is make probabilistic statements about the "events" we study. As such, classical probability is relevant to political scientists.[2]

A **sample space** is the set of all possible outcomes: it is a list of each event we *might* observe. In June 2004 three parties faced off in an election in Uruguay: Colorado, Encounter-Wide Front (EP-FA), and National. Ignoring ties, we can define six different possible outcomes: Colorado wins a majority of seats; EP-FA wins a majority; National wins a majority; no party wins a majority but Colorado wins the largest block of seats; no party wins a majority but EP-FA wins the largest block; and no party wins a majority but National wins the largest block. The sample space is the list of possible outcomes, so the sample space consists of six outcomes in this example.

By convention, the sum of the probabilities that each outcome in the sample space, $o_1, o_2, o_3 \ldots o_n$, occurs is set equal to 1. That is, if we define the event of interest as the sample space, S, then the probability of that event is the sum of the probabilities of each outcome, which is 1: $Pr(S) = Pr(o_1) + Pr(o_2) + Pr(o_3) + \ldots + Pr(o_n) = 1.0$. Why? Because *one* of the outcomes in the sample space has to occur, and a probability of 1 indicates that the event will happen with certainty. The probability that something in the sample space occurs is thus always 1, since something must happen. Similarly, a probability of 0 indicates that the event cannot happen. The values of 0 and 1 therefore represent deterministic outcomes, and values between 0 and 1 represent uncertain outcomes.

With outcome, event, and sample space defined, we can define the classical probability of an event:

$$Pr(e) = \frac{\text{No. of outcomes in event e}}{\text{No. of outcomes in the sample space}}.$$

In our electoral example there are six outcomes, so the denominator is six. If we are interested in the probability that the election must produce a minority or coalition government, then there are three outcomes that produce that event: no party wins a majority but Colorado wins the largest block of seats; no party wins a majority but EP-FA wins the largest block; and no party wins a majority but National wins the largest block. We assign the numerator the value 3 to determine the probability: $\frac{3}{6} = \frac{1}{2} = 0.5$.

Consider another event, EF-PA governing as a majority party or as the largest party in a coalition that includes the plurality winner of the election. This event is composed of two outcomes: EP-FA wins a majority *and* no party wins a majority but EP-FA wins the largest block. The classical probability of this event is $\frac{2}{6} = \frac{1}{3} \approx 0.33$.

[2]We are eliding an important theoretical distinction in this example between stochastic and deterministic events: the election might in theory be deterministic but in practice stochastic, for the reasons presented in this paragraph. We do this for a simple reason: quantitative political scientists live "in practice." It is simply impossible to specify every last element that enters into each voter's decision, not to mention the myriad factors that affect who are the candidates. Consequently, whether or not a full accounting would in theory predict the election with certainty, in practice the election will always be uncertain. (And we would argue that it is uncertain for all finite populations in theory as well, but this is a different matter.)

Note that we have determined these probabilities without reference to any voting data: classical probability concerns the probability of events absent reference to data. It is objective: we do not reference any people's beliefs but rather reference the sample space and the number of outcomes in the event. We could have referenced past electoral outcomes and used those to develop an expected probability about an upcoming election. Doing so would have appealed to the frequency distribution of past elections and would have produced an empirical probability (which is also objective). We discuss empirical probabilities in detail in the following two chapters and offer a brief introduction here.

The empirical probability of an event is the relative frequency, or proportion, of a given value of a variable in a population (or sample). Because they are proportions (described below), empirical probabilities also range between a minimum value of 0 and a maximum value of 1. We discuss this in more detail in the following chapters. For the present, we need only know that the probability that a given value turns up for a given variable is the proportion of the number of times that value turned up in a sample relative to all other values that might have turned up for that variable.

As an example, we would determine the empirical probability that a randomly selected Uruguayan voter cast his ballot for the Colorado party in the June 2004 primary by looking at the vote share for each party: the EP-FA had 454,011 votes, the NP had 440,774 votes, and Colorado had 159,280 votes (United Press International, 2004). With 1,054,065 votes cast, the probability that a Uruguayan voter selected at random voted for Colorado is

$$\frac{159,280}{159,280 + 454,011 + 440,774} = \frac{159,280}{1,054,065} \approx 0.15.$$

9.1.2.1 Simple and Compound Events

A **simple event** is a single outcome that we represent as having occurred to an individual or group. That is, we cannot break down a simple event into constituent parts (i.e., multiple outcomes). A **compound event**, on the other hand, is composed of two or more simple events; we can break it down into constituent parts (i.e., outcomes).

As an example, the vote choice of a given voter is a simple event, as is that voter's gender, her partisan identification, and whether she voted in the previous election. The voter's vote choice *and* her gender together form a compound event, as do her vote choice *or* gender. Put differently, then, a compound event is a set of simple events joined by the word *and* or the word *or* (in the language of set theory, the intersection or union of two sets).

9.1.3 Independence, Mutual Exclusivity and Collective Exhaustivity

Compound events are either independent of one another or conditional on one another. We cover conditional compound events in the following subsection.

Here we introduce independence, mutual exclusivity, and collective exhaustivity and briefly explain why they are important distinctions.

Two events are **independent** if the probability that one occurs does not change as a consequence of the other event's occurring. Put in terms of random variables, about which we say more in the next chapter, two variables are independent if the probability of variable a taking value i is not changed as a result of variable b taking a particular value, j. To take an example, vote choice is not affected by the temperature: if we ran an experiment where people cast a ballot multiple times and the only difference was the temperature of the room where they cast their ballot, we would expect them to vote the same way each time.[3]

Two events are **mutually exclusive** when one cannot occur if the other has occurred. The values of a variable are a good example of mutual exclusivity: the values of the *highest* level of education achieved on the scale "(1) some high school, (2) high school graduate, (3) some college" are mutually exclusive because each individual can be placed in only one category.

Collective exhaustivity is the characteristic that each and every event fits into at least one of the categories. Again, the values of a variable provide a convenient example: neither a person who dropped out of school in seventh grade nor a person who completed a JD can be assigned a proper value on the scheme in the preceding paragraph. To make that scheme collectively exhaustive we would need to add (at least) two categories: "no high school" and "postgraduate degree."

Note that events that are mutually exclusive are not independent. Further, events that are collectively exhaustive are not independent. If one happens, the others cannot happen.

9.1.4 Joint and Conditional Probabilities

A **joint probability** is the probability of a compound event. If the simple events of a compound event are independent, then their joint probability is the product of the probabilities of each simple event. Thus, if the probability that a US voter drawn at random voted Democratic is 0.47 *and* the probability that the temperature is 83 degrees is 0.15, then the joint probability is $0.47 \times 0.15 \approx 0.07$. The joint probability of two independent events is the probability that both events occur: that the voter casts a ballot for the Democratic candidate *and* the temperature is 83 degrees. The joint probabilities of independent events can be uninteresting (as in this example), but people using statistical inference to conduct hypothesis tests are often interested in such joint probabilities. The key point to take away is that the joint probability of independent events is the probability of both events' occurring, and we calculate it as the product of the probabilities of each individual event.

---

[3]The candidate who wins the election may very well be affected by weather. For example, one party's supporters may be less likely to turn out in rainy weather than another's (Gomez, Hansford, and Krause, 2007). To the extent that this is so, we would observe that *turnout* is not independent of weather. But *vote choice* can be distinguished from *turnout*.

Let us now consider the joint probability of mutually exclusive compound events, for example the joint probability that a US voter voted for the Democratic Party *or* that she voted for the Republican Party. The joint probability of two mutually exclusive events is *not* the product of the simple probabilities. Instead, it is the sum of the simple probabilities: given that $p(y_D) = .4$ and $p(y_R) = .5$, $p(y_D \text{ or } y_R) = 0.4 + 0.5 = 0.9$.[4] It should be apparent to you that the joint probability of a set of mutually exclusive, collectively exhaustive events is 1.0, since just events make up the entire sample space.[5]

Like mutually exclusive and collectively exhaustive events, conditional events are not independent: the probability of one event occurring is affected by whether another event occurs and is referred to as a **conditional probability**. We need some notation to indicate that the probability of an event is conditional on other events: $p(y|x, z)$, which is read "the probability of y given x and z."

9.1.5 Why Should I Care?

The independence, mutually exclusive, and collectively exhaustive properties are useful because one cannot construct probabilities without understanding them. To see this, consider what a sample space is: the set of all possible outcomes. If a set of events is not collectively exhaustive, then that set does not form the sample space: some of the outcomes that could occur are not included. The other two properties are critical with respect to determining the probabilities of compound events, which is important because not all events of interest in political science are simple. If two simple events are not independent, then the rules for computing the probability of compound events in which both simple events occur requires conditional probabilities; the same is often true when simple events are not mutually exclusive and one is interested in compound events in which at least one of the simple events occurs. Rules for computing the probability of such compound events are given in the next section.

Because few events are truly independent, conditional probabilities are central to understanding political phenomena. Even in an act such as voting we see conditionality: if the weather is good, I will vote; otherwise I will stay home. The act of turning out to vote is conditioned on the weather.

This sort of conditionality is central to formal theory, and we discuss that below in this chapter as well as in the next one. But it is also central to statistics. Consider the linear model $y = \beta x + \epsilon$. This means, in words, that the value

[4]The complete equation is a bit more complicated than indicated in the text: $p(y_D \text{ or } y_R) = p(y_D) + p(y_R) - p(y_D \text{ and } y_R)$. However, insofar as y_D and y_R are mutually exclusive, $p(y_D \text{ and } y_R) = 0$, we drop it above. We discuss this more in the next section.

[5]If this is not apparent, work out the following by assigning your own subjective probabilities to these five mutually exclusive, collectively exhaustive events: the next president of the United States will (1) place a trade embargo on North Korea, (2) order air strikes against North Korea, but not use ground troops, (3) launch a military invasion of North Korea, (4) limit interactions with North Korea to diplomacy, (5) cut off diplomatic relations with North Korea but take no other actions. Having specified the individual probabilities, determine the joint probability of all five events.

of the dependent variable, y, depends on the values of both the independent variable x and some random variable ϵ. Here ϵ represents all the stuff we don't know about how y varies, stuff that's not captured in the variance of x. This is close to the simplest model one might specify for a linear regression.

Let's say one had measured x and had a value for β, perhaps via ordinary least squares (OLS) regression, and was interested in the likelihood that $y > 0.5$. Well, since $y = \beta x + \epsilon$, $Pr(y > 0.5|\beta, x) = Pr((\epsilon + \beta x > 0.5)|\beta, x)$. If one knows how ϵ is distributed, a concept we address at length in the next two chapters, one knows the RHS of this equation, and so knows the *conditional* probability that y exceeds one-half. Note that this probability is conditional because it depends fundamentally on the values of x and β.

9.2 COMPUTING PROBABILITIES

9.2.1 Notation and Some Rules

As we've introduced concepts, we've implicitly defined some notation. Here we gather what we've defined, add a few more things, and use them all to represent our concepts a bit more formally.

First, consider the probability that an event A occurs. We denote this $Pr(A)$. This is an unconditional or marginal probability; it assumes either that the probability that A will occur is completely independent of everything else in the sample space or that we have already "averaged over" everything else on which it is conditional (more on this in the next two chapters).

All probabilities lie between zero and one, so $Pr(A) \in [0, 1]$. We say A is a deterministic event if $Pr(A) \in \{0, 1\}$ and a random, probabilistic, or stochastic event otherwise. If S is the sample space containing all events that might happen, then $Pr(S) = 1$. If $Pr(A) = 0$, then A cannot happen.

$Pr(A|B)$ is the conditional probability of A on B. In other words, it is the probability that A occurs given that B has already occurred. If A and B are independent events, then the fact that B has already occurred doesn't influence the probability that A will occur. So, for independent events, $Pr(A|B) = Pr(A)$.

The symbols for set union ($\cup$) and intersection ($\cap$) also apply to events. $A \cup B$ is the compound event where *either* A or B happens, or both. Thus, we read $A \cup B$ as "A or B." We can also use $A \vee B$ to symbolize the notion of **or**. The idea is roughly that $A \cup B$ contains all the events in A and B, so the compound event happens if *any* of the events in A or B happen. $A \cap B$ is the compound event where *both* A and B happen. Thus, we read $A \cap B$ as "A and B." We can also use $A \wedge B$ to symbolize the notion of **and**. The idea is roughly that $A \cap B$ contains only those events that are common to both A and B, so the compound event happens only if both constituent events A and B happen.

Both *and* and *or* have rules for their computation. We've already discussed simplified versions of them. The rule for *and* looks like this:

$$Pr(A \cap B) = Pr(B|A)Pr(A) = Pr(A|B)Pr(B).$$

To see why the last two expressions are equal, consider what they say in words. The first is the probability that B happens conditional on A's happening, times the chance that A happens. So both A and B happen, with A happening first. The second is the probability that A happens conditional on B's happening, times the chance that B happens. So again, both A and B happen, with B happening first this time. Since we have no order on A and B happening, these are equivalent ways of saying that A and B both happen, which is what we mean when we write $A \cap B$. So both are equivalent mathematically.[6]

When A and B are independent, $Pr(A|B) = Pr(A)$ and $Pr(B|A) = Pr(B)$, and this rule reduces to $Pr(A \cap B) = Pr(A)Pr(B)$. When they are dependent, the joint probability includes the conditional probability. This is so because the sample space shrinks by the number of outcomes represented by event A when we calculate the probability of event B (or vice versa). That is, the probability that A occurs is the number of outcomes event A represents divided by the sample space, but the probability that B occurs given that A has also occurred is the number of events B represents divided by the number of outcomes left in the sample space given that event A has occurred. When A and B are mutually exclusive there is no overlap in the chances that they occur, so $Pr(A \cap B)$=0. In other words, for mutually exclusive events, $Pr(B|A) = Pr(A|B) = 0$.

The rule for *or* looks like this:

$$Pr(A \cup B) = Pr(A) + Pr(B) - Pr(A \cap B).$$

The first two terms just tell us that the chance that either A or B happens is closely related to the chance that each happens on its own. The third term is there to avoid double counting. That is, if A and B have some overlap—i.e., if they are not mutually exclusive—then when you add the probabilities that each occurs, you are including twice in the sum the probability that the overlapping part, $A \cap B$, occurs, so you need to subtract one of those probabilities out. When events are mutually exclusive that overlap is zero, though, so you just get $Pr(A \cup B) = Pr(A) + Pr(B)$. If we have a set of collective exhaustive events, then together they make up the whole sample space S, i.e., $S = \bigcup_i A_i$ for some collectively exhaustive group of events indexed by i. Since $Pr(S) = 1$, it must be true that for a set of collectively exhaustive events, $Pr(\bigcup_i A_i) = 1$ as well.

All these rules apply for more than two events as well, though with multiple conditional events one must be careful to keep the conditionalities straight. So, for example, $Pr(\bigcup_{i=1}^n A_i) = \sum_{i=1}^n Pr(A_i)$ for mutually exclusive events and $Pr(\bigcap_{i=1}^n A_i) = \prod_{i=1}^n Pr(A_i)$ for independent events. With an understanding of the properties discussed in the previous section and these rules you can determine the probability of any compound event.

---

[6]Of course, they may not be equal substantively. We might conceptualize B as dependent on A, as in turning out to vote based on weather, but A as independent of B, since the weather isn't affected by one's choice. So we'd use only $Pr(A \cap B) = Pr(B|A)Pr(A)$ in this case.

9.2.2 Combinations and Permutations

The rules in the previous subsection provide in theory the tools you need to compute relevant probabilities, but they can be difficult to apply to complex situations and events. In particular, though classical probability tells you to divide the number of ways an event can happen by the number of total ways any event can happen (i.e., the size of the sample space), it does not tell you how to compute these two numbers. Two computational aids will make things easier: the combination and the permutation. Each has a straightforward meaning. A **combination** is a way of choosing k objects from n objects when one does not care about the order in which one chooses the objects. A **permutation**, in comparison, is a way of choosing k objects from n objects when one *does* care about the order in which one chooses the objects.

Let's consider the set $\{a, b, c, d, e, f\}$. This set has $n = 6$ elements in it. How many ways are there to draw $k = 2$ elements from this set (without replacing any letters) if we don't care about the order in which we draw them? This question is equivalent to asking how many different combinations of two letters are in the set, so we can just count them: $ab, ac, ad, ae, af, bc, bd, be, bf, cd, ce, cf, de, df, ef$ are all the two-letter combinations, and there are fifteen in total. This tells us the size of the sample space consisting of two-letter combinations. If we then wanted to know what the chance of having an a in one of these two-letter combinations was, we could add the number of times this happens, which is five, and divide by the sample space, fifteen, to get one-third.

This was somewhat tedious, but there's a faster way. The expression $\binom{n}{k}$ is to be read "n choose k." (You will sometimes also see the notation ${}_nC_k$ or $C(n, k)$.) In other words, take n things and choose k of them, ignoring the order in which you choose them. It is the case that $\binom{n}{k} = \frac{n!}{k!(n-k)!}$. So, for our example, we have $n = 6, k = 2$, and $\binom{6}{2} = \frac{6!}{2!4!} = \frac{6\cdot5\cdot4\cdot3\cdot2\cdot1}{(2\cdot1)(4\cdot3\cdot2\cdot1)} = \frac{30}{2} = 15$. We hope you can see that this will be much faster than writing out all combinations when n gets large!

Permutations arise much less frequently in political science, but behave similarly. Returning to our example, what if we now cared about order? Then in addition to all fifteen of the two-letter combinations we found, we'd have another fifteen with the order of the letters reversed, for thirty total permutations. (There will always be more permutations than combinations, as ignoring order always decreases the number of options that matter.) We can compute this more easily as well: $P(n, k) = \frac{n!}{(n-k)!}$. This is the same as our equation for the combination, save a missing $k!$ in the denominator. This term is what allows us to ignore order in the equation for combinations: the number of ways one can arrange k objects is $k!$.

9.2.2.1 *Why Should I Care?*

Combinations and permutations, along with the concepts of classical probability, allow us to compute a wide array of probabilities, and there are several examples at the end of the chapter. But why would we want to do this?

One answer to this is that those who go on to study game theory will need a working facility with uncertainty, expected utility, and Bayes' rule. Probability is the formalization of uncertainty, and in order to compare probabilities in a quantitative fashion, one must compute them. These computed probabilities are necessary components of expected utility, which is the amount of utility an actor expects to get. Utility may be uncertain for many reasons. For example, in a lottery the payoffs are uncertain: one could win a million dollars or, far more likely, receive nothing. This is a true lottery, but game theorists call any random variable that describes stochastic payoffs a **lottery**, and lotteries are pretty common in game theory.

Lotteries can occur in decision theoretic contexts: i.e., situations in which one must decide on one's own what one's best action is. However, when other players are added, new forms of uncertainty arise. One form arises because of the possibility that another actor may play what is known as a **mixed strategy**. In a mixed strategy, what one does is stochastically determined. The canonical example is a game called matching pennies. In this game, each of two players can pick the heads side of a coin (H) or the tails side (T), and both players make their pick simultaneously. One player wants to match both choices, and so get either both H or both T, while the other wants one H and one T in any order. Because the players have opposite preferences, neither can settle on a single pick with certainty. If the first player always picks H, the second player will know this and always pick T, which is bad for the first player. Similarly, if the second player always picks H, the first one will always pick H too, which is bad for the second player. Thus, each player must act in a random fashion, choosing H sometimes and T sometimes according the flip of a coin.[7] Acting randomly in this fashion is part of playing a mixed strategy. Because of this behavior, any particular action yields uncertain utility.

A second form arises when one is not sure what *type* of player one is playing, as in what are known as incomplete or imperfect information games. In these games one may, for example, not know if one is facing a nice type of person who would prefer cooperation to any sort of defection, or a mean type of person who will always defect. To deal with this, one can assign likelihoods to each type, quantifying the uncertainty, and then compute expected utility to discern the best action.[8]

Because the calculation of expected utility is more naturally discussed in

[7]Another common example is the penalty kick in soccer (football). If one always kicks or leaps to block a kick to the left or the right, then the other player can react accordingly, so each kicks or leaps to block a kick randomly.

[8]One can find equilibrium behavior in this case by making use of what is known as a Bayesian Nash equilibrium solution concept (McCarty and Meirowitz, 2007).

the context of expectations and distributions of random variables, we put it off until the next chapter. However, there is one more aspect of uncertainty that is useful in game theory: Bayes' rule, which addresses how one updates subjective belief. Bayes' rule is necessary to understand how a rational actor should optimally incorporate new information during the course of a game and is central to solving dynamic games of incomplete information. The perfect Bayesian equilibrium solution concept is central to this analysis, and its name is taken from this rule. Moreover, Bayes' rule forms the basis of Bayesian statistics. As it can be derived from the rules above, we turn to it now.

9.2.3 Bayes' Rule

Thomas Bayes, an English reverend who lived during the eighteenth century, was responsible for a theorem that explains how one ought to change one's mind in response to evidence about the world. Bayes was concerned with subjective probability, and especially with how people ought to develop reasonable beliefs about the probability of events. A conventional story concerns gambling, such as with dice or the card game blackjack, and observes that a person with unreasonable beliefs about the appropriate circumstances in which to place bets in these games will soon be relieved of her money. Bayes wanted to develop a rule for how a naïve gambler would develop reasonable beliefs by observing repeated throws of a dice or hands of blackjack. His theorem provides that rule.

To introduce Bayes' rule we must distinguish between prior and posterior beliefs. A prior belief is the belief one has before new information (evidence) is introduced. A posterior belief is the belief one holds subsequent to the consideration of new information (evidence). Thus, if we use standard time notation, where t represents one moment in time and $t + 1$ represents the subsequent moment in time, and use B to represent a belief, then B_t is the prior belief and B_{t+1} is the posterior belief if an event that provided new information occurred at time t.

The *Economist* (2004) offers the following illustration of Bayes' rule:

> The canonical example is to imagine that a precarious newborn observes his first sunrise, and wonders whether the sun will rise again or not. He assigns equal prior probability to both possible outcomes, and represents this by placing one white and one black marble into a bag. The following day, when the sun rises, the child places another white marble in the bag. The probability that a marble plucked randomly from the bag will be white (i.e., the child's degree of belief in future sunrises) has thus gone from a half to two-thirds. After sunrise the next day, the child adds another white marble, and the probability (and thus the degree of belief) goes from two-thirds to three-quarters. And so on. Gradually, the initial belief that the sun is just as likely to rise as not to rise each morning is modified to become a near-certainty that sun will always rise.

To summarize, the child relies on evidence to change his beliefs. Bayes' rule describes how one should change one's beliefs about the probabilities of events occurring. Some readers may find Corey Chivers's visualization of a Bayesian updating process useful (`http://tinyurl.com/BayesUpdate`).

To consider a more formal description, let B and A be two events of interest and, to make it simple, we will make B and A the positive values of binary variables such that $\sim B$ (read "not B") and $\sim A$ represent the absence of the events. Bayes argued that one should update one's prior beliefs about B by comparing the probability that one would observe A given one is correct about B with the joint probability of observing A given one is correct *and* given one is mistaken. That is, he suggested that one normalize the probability that one is correct by the probabilities associated with all other possible states of the world (i.e., one is correct *and* one is not correct). More explicitly, Bayes suggested that we examine the ratio of the joint probability of B **and** A (i.e., $Pr(A|B)Pr(B)$) with the joint probability of B *and* A **or** $\sim B$ *and* A (i.e., $Pr(A|B)Pr(B) + Pr(A|\sim B)Pr(\sim B)$). We can write Bayes' theorem in this simple case as follows:

$$Pr(B|A) = \frac{Pr(A|B)Pr(B)}{Pr(A|B)Pr(B) + Pr(A|\sim B)Pr(\sim B)}. \tag{9.1}$$

> One can read equation (9.1) as follows: *the posterior probability of B given A is the product of the prior probability of B and the probability of A given B divided by the product of the prior probability of B and the probability of A given B plus the product of the prior probability of not B and the probability of A given not B.*

In more intuitive terms it is (1) the joint probability that one is right *and* the probability that one will observe something given that one is right ($Pr(A|B)Pr(B)$) weighted by (2) that same probability ($Pr(A|B)Pr(B)$) plus the joint probability that one is mistaken and the probability that one will observe that same something given that one is instead mistaken ($Pr(A|\sim B)Pr(\sim B)$).

This might seem to be coming from left field, but in actuality we've already seen most of its derivation above. Recall the rule for computing the joint probability in which two events must both happen (i.e., *and*): $Pr(A \cap B) = Pr(B|A)Pr(A) = Pr(A|B)Pr(B)$. Let's divide the $Pr(A)$ from the second expression into both that expression and the third one. This yields $Pr(B|A) = \frac{Pr(A|B)Pr(B)}{Pr(A)}$. This expression looks a lot like Bayes' rule already, except for the denominator. But remember that $Pr(A)$ is the unconditional or marginal probability. In other words, it is the probability that A happens, taking into account all the other things that might also happen and their probabilities of happening. In the case of this example, only two things are possible, B and $\sim B$. So we can write $Pr(A)$ as $Pr(A|B)Pr(B) + Pr(A|\sim B)Pr(\sim B)$. Either B happens, in which case we need to know the probability of A given B, or $\sim B$ happens, in which case we need to know the probability of A given $\sim B$.

That is the origin of this sum. Replacing the denominator with this sum yields Bayes' theorem.

This same idea can be extended. Start by breaking up the sample space into n disjoint, mutually exclusive pieces. Label each of these B_i, so that $\bigcup_{i=1}^{n} B_i = S$. Since one and only one of the B_i can happen at any given time, $Pr(A) = \sum_{i=1}^{n} Pr(A|B_i)Pr(B_i)$. Then Bayes' rule looks like $Pr(B_k|A) = \frac{Pr(A|B_k)Pr(B_k)}{\sum_{i=1}^{n} Pr(A|B_i)Pr(B_i)}$ for any particular B_k.

The theorem tells us how to calculate a posterior belief about B conditional on having observed A. Students often want to know where these beliefs (probabilities) come from. They can come from the past, or they can simply be beliefs that the actor (or scholar) makes up. We consider two examples, one of which finds the values for beliefs in the historical record and the other of which uses values we make up. We begin with Stokes (2001).

In her book on elections and mandates in Latin America, Stokes (2001) explores the implications of politicians who run for office promising to pursue one set of policies, then implement a different policy set once in office. She asks whether the presence of such "switchers" implies that campaigns are uninformative (because people expect politicians to lie) and therefore that electoral mandates are meaningless, thus undermining democracy.

In her model voters believe that there are two types of politicians: *ideologues*, who will pursue the policies they campaign on, and *power seekers*, who will lie during the campaign when they know their preferred policy is unpopular and then switch once in office. Stokes asks whether campaigns can provide information to voters to help them revise their beliefs about whether a given candidate is an *ideologue* or a *power seeker*. More specifically, Stokes asks whether a campaign can inform a voter's belief about the probability that a given candidate will implement one or another of two economic policies. She turns to the historical record and assumes that a Latin American voter will be reasonably informed about the number of elected leaders who implemented one policy or the other, as well as about the number of elected leaders who implemented the policies they promised to implement during the campaign (and those who switched).

Focusing on economic policies, Stokes provides a table of Latin American leaders between 1982 and 1995 and determines whether the politician who won the election promised to pursue a security-oriented policy, s, where wages were indexed to inflation and parastatal corporations remained intact, or an efficiency-oriented policy, e, where wage indexing was relaxed and parastatal corporations were privatized. She also records whether, once in office, the politician adopted a security-oriented or an efficiency-oriented policy. The data reveal that 33 of 43 Latin American leaders elected between 1982 and 1995 adopted efficiency policies: $Pr(e) = \frac{33}{43} = 0.77$.[9] Given $Pr(e)$, we can calculate the empirical probability that a politician adopted a security-oriented policy: $Pr(s) = 1 - Pr(e) = 1 - 0.77 = 0.23$. These prior beliefs suggest that

[9]Stokes reports that 29 of 39 (0.74) leaders adopted efficiency policies (p. 14), but her table reveals that it is actually 33 of 43.

the typical voter in Latin America will expect 77% of candidates to implement efficiency-oriented policies and 23% of candidates to implement security-oriented policies.

To use Bayes' rule to update those beliefs in response to a campaign in which candidates make promises, we need to define the situation. During the campaign the voter will monitor the promises made by the candidates. Consider a contest with two candidates, one of whom campaigns on an efficiency-oriented platform, the other of whom campaigns on a security-oriented platform. If Bayes' rule leads to the conclusion that the voter ought to revise his beliefs about the candidates' probability of implementing the policy on which they campaign then we will conclude that Latin American campaigns are informative and the conventional wisdom is poor.

In other words, we want to know, for example, whether campaigning on an efficiency-oriented platform increases voters' beliefs that the candidate will adopt an efficiency policy in office. If we let ϵ indicate a campaign promise of efficiency-oriented economic policy and e indicate the adoption of an efficiency policy in office, then this belief is $Pr(e|\epsilon)$. Bayes' rule lets us calculate this, if we know $Pr(\epsilon|e)$, $Pr(\epsilon)$, and $Pr(e)$.[10]

We already know that $Pr(e) = 0.77$. Next we need the conditional probability that a candidate campaigned on an efficiency-oriented platform given that he adopted an efficiency policy in office: $Pr(\epsilon|e)$. Stokes's table reveals that 16 of the 33 candidates who adopted efficiency policies also campaigned on an efficiency platform: $Pr(\epsilon|e) = \frac{16}{33} = 0.48$. Finally, consider $Pr(\epsilon)$. We expand this as above to get $Pr(\epsilon) = Pr(\epsilon|e)Pr(e) + Pr(\epsilon| \sim e)Pr(\sim e)$. We know the first term in the sum already, and also that $Pr(\sim e) = 0.23$. This leaves $Pr(\epsilon| \sim e)$: the conditional probability that a candidate campaigned on an efficiency-oriented platform given that he adopted a security-oriented (i.e., "not efficiency") policy in office. It turns out that none of the ten Latin American politicians who enacted security-oriented policies once in office campaigned on efficiency: $Pr(\epsilon| \sim e) = \frac{0}{10} = 0$.

Plugging these into Bayes' rule yields

$$\begin{aligned} Pr(e|\epsilon) &= \frac{Pr(\epsilon|e)Pr(e)}{Pr(\epsilon|e)Pr(e) + Pr(\epsilon| \sim e)Pr(\sim e)} \\ &= \frac{0.48(0.77)}{0.48(0.77) + 0(0.23)} = \frac{0.37}{0.37 + 0} = 1. \end{aligned}$$

The voter's posterior belief is 1.0, which is a considerable increase from 0.77, the voter's prior belief. Thus, the campaign has a substantial impact: on knowing that the candidate is promising to implement efficiency-oriented policies, the voter shifts from being confident that the candidate will do so (0.77 probability) to being certain that the candidate will do so (1.0 probability). Why? The zero

[10]This suggests a good way of approaching problems involving Bayes' rule: identify all the information you are given, convert it to conditional and marginal probabilities, as appropriate, and then note which information you are missing.

in the denominator above (i.e., $Pr(\epsilon| \sim e)$) is responsible. If there is no contrary evidence, the voter changes his belief to certainty.

Next we turn to the issue of whether a candidate will implement a security-oriented platform, given that he campaigned on a security-oriented policy. Letting σ be a security-oriented campaign and s be a security-oriented implementation in office, this conditional probability is $Pr(s|\sigma)$. To compute this we'll need to know the conditional probability that a candidate campaigned on a security-oriented policy given that he adopted a security-oriented policy in office, which is $Pr(\sigma|s)$, as well as $Pr(s) = 0.23$ and $Pr(\sigma) = Pr(\sigma|s)Pr(s) + Pr(\sigma| \sim s)Pr(\sim s)$. Stokes's data indicate that all ten of the leaders who implemented a security-oriented policy also campaigned on it: $Pr(\sigma|s) = \frac{10}{10} = 1.0$, where σ (sigma) represents a candidate who campaigns on a security-oriented policy. The final piece of information we need is the number of leaders who campaigned on a security platform but adopted an efficiency-oriented policy. The data reveal that 12 of the 33 leaders who adopted an efficiency-oriented policy campaigned on a security platform: $p(\sigma| \sim s) = \frac{12}{33} = 0.36$.

Plugging these into Bayes' rule yields

$$\begin{aligned} Pr(s|\sigma) &= \frac{Pr(\sigma|s)Pr(s)}{Pr(\sigma|s)Pr(s) + Pr(\sigma| \sim s)Pr(\sim s)} \\ &= \frac{1.0(0.23)}{1.0(0.23) + 0.36(0.77)} = \frac{0.23}{0.23 + 0.28} = \frac{0.23}{0.51} = 0.45. \end{aligned}$$

The voter's posterior belief after observing the campaign is 0.45. Again, the campaign is informative, leading the voter to change his beliefs from a 0.23 probability that the candidate will implement a security-oriented policy to a 0.45 probability that he will do so.

This demonstration of using Bayes' rule to update beliefs relies on observable data, but we can also use it when we have much less concrete beliefs (or information). This is typically the case in game theory, wherein one assumes prior beliefs and the informativeness of subsequent signals. To illustrate, let's try a second example, but this time we will make up the values of the probabilities. Consider a state legislator deciding whether to enter a race for the US Congress. To make it interesting, assume that he does not face a term limit. The decision to enter is rather complex, but, given our purpose, we will consider a simple (and incomplete) model. Assume that the seat is presently held by an incumbent from the other party, but that the district has changed sufficiently over the years the incumbent has held office, and the state legislator believes his party can carry the district. Imagine that the key issue for the legislator is whether the incumbent runs for a new term. If the incumbent is going to run, then the legislator will bide his time in the Statehouse. But if the incumbent is not going to run, then he wants to begin quietly fund raising and building his campaign team so as to get a head start on his competitors. In effect, we expect him to "read the tea leaves" that the incumbent's behavior provides.

Let's set the legislator's prior belief about the probability that the incumbent runs at 50:50: $Pr(I_r) = 0.5$. We will further assume that the legislator believes that the probability that the incumbent will hold a fundraising dinner in her district in December of the year prior to the election is 0.6 if she is running and 0.4 if she is not: $Pr(f|I_r) = 0.6$, $Pr(f| \sim I_r) = 0.4$.

Let's consider how the legislator would update his beliefs about the incumbent's running using Bayes' rule, assuming the incumbent hosted fundraising dinners in December. We do so by plugging the relevant values into equation 9.1 as follows:

$$\begin{aligned} Pr(I_r|f) &= \frac{Pr(f|I_r)Pr(I_r)}{Pr(f|I_r)Pr(I_r) + Pr(f| \sim I_r)Pr(\sim I_r)} \\ &= \frac{0.6 \cdot 0.5}{0.6 \cdot 0.5 + 0.4 \cdot 0.5} = \frac{0.3}{0.3 + 0.2} = \frac{0.3}{0.5} = 0.6. \end{aligned}$$

Thus, having observed a dinner in December, the legislator updates his belief that the incumbent will run from 0.5 to 0.6. Bayes' rule provides us with an explicit formula for describing such a process.

9.2.3.1 Why Should I Care?

As mentioned above, there is a long-standing debate about the best foundation for statistics. The frequentist (classical) account of statistical inference is the venerable school and by far the most common in political science and other disciplines. The Bayesian account is the new kid on the block, and though it is presently used infrequently in applied work, it is becoming increasingly more common, and we'd recommend supplementing training in frequentist statistics with at minimum an introductory course in Bayesian statistics. In brief, the frequentist approach was developed in the early twentieth century by Ronald Fisher, Jerzy Neyman, and Karl Pearson and relies to a strong degree on the central limit theorem, which concerns the distribution of distributions and will be taught in your statistics courses (but is briefly mentioned in Chapter 11). The theory of Bayesian statistics is due to Pierre-Simon Laplace (1749–1827), and was developed further in the mid-twentieth century by Harold Jeffreys and Edwin Jaynes. Bayesian estimation as practiced today is not feasible to implement without powerful computers, whereas frequentist statistics can be implemented "by hand," especially when what is known as a closed form solution for a given estimation problem can be found. Though the Bayesian–frequentist debate has produced considerable philosophical discussion (e.g., Fisher believed Laplace's work on probability was rubbish), modern computers have produced a Bayesian revolution in statistics, and Gill (1999) presents a critique of the frequentist approach from a Bayesian perspective. Regardless of one's views on the Bayesian–frequentist debate, the central point is that since Bayes' rule is the foundation of the Bayesian approach, it is important.

In addition, students who plan to study game theory will also need to be intimately familiar with Bayes' rule as a number of equilibrium concepts, which are solutions to game theory models, rely on it. We mentioned the most prominent one earlier in the chapter: perfect Bayesian equilibrium (McCarty and Meirowitz, 2007). This equilibrium concept has two components. First, everyone must respond optimally conditional on all prior actions and everyone's beliefs about all uncertain factors. Second, everyone must update beliefs by using Bayes' rule to incorporate new information, whenever new information is available. Here new information usually consists of the actions of other actors, which can provide *signals* as to what types of player they are. Signaling games, which focus on the strategic use of such signals both to inform and to misinform, are fairly common in political science. For example, in Chapter 4 we discussed the sequence of events in Gilligan and Krehbiel (1989) as an example of why sequences are important in political science. The last two actions in that sequence are (1) a congressional committee proposes a bill and (2) the floor of Congress votes on that proposal. In their game theoretic model, the committee can have information on the worth of each bill that the floor does not. This is known as private information. If the committee provided this information to the floor, the floor would vote to optimize its own utility given this accurate information. But this might lead to outcomes the committee doesn't like if the committee has different preferences from the floor. So it is in the committee's interest to signal false information to the floor, in the hope that the floor will pass a bill closer to the committee's preferences. The floor's challenge is to incorporate the signal the committee sends optimally, taking into account the committee's incentive to provide false information. Bayes' rule is central to this analysis.

9.3 SOME SPECIFIC MEASURES OF PROBABILITIES

When reading the substantive literature in political science you will encounter works that interpret statistical findings in terms of probabilities. You will learn how to do this in your statistics courses. In this section we briefly introduce two indicators that you are likely to come across. The goal is to aid you in understanding the meaning of those findings, not to show you how to compute them.

9.3.1 Odds and the Odds Ratio

One way in which probabilities frequently enter the reporting of results is the odds ratio. The odds of an event is defined as the ratio of the probability of the event's occurring and the probability that it does not occur: $\frac{Pr(y)}{Pr(\sim y)}$. The odds ratio of two events, x_1 and x_2, then, is the ratio of the individual odds:

$$\frac{Pr(x_1)/Pr(\sim x_1)}{Pr(x_2)/Pr(\sim x_2)}. \tag{9.2}$$

The odds ratio provides a useful way to summarize the relationship between the odds of two independent events and is foundational to maximum likelihood statistics.

9.3.2 The Relative Risk Ratio

The relative risk ratio is the ratio of two probabilities and can be quite useful for comparing relationships. It is often useful to be able to summarize the risk of two groups to some outcome. Comparing the ratio of those risks gives us a good measure of the relative risks. Relative risk ratios have a range from zero to infinity. A value of 1 indicates that the risks are equal across the pair. A value below 1 indicates that the probability in the first part of the ratio is smaller than the probability in the second part, while values greater than 1 represent the opposite.

To illustrate, consider the following relative risk ratios reported in Raknerud and Hegre's (1997) study of pairs of countries going to war. They report that the relative risk ratio for war is 1.95 for a pair of countries with one major power relative to a pair in which neither is a major power. That means that the country pairs with at least one major power are nearly twice as likely to go to war as country pairs without a major power (i.e., almost 2.0). Raknerud and Hegre further report that relative to a pair of countries made up of one democracy and one autocracy (aka a mixed dyad), the relative risk ratio for a pair of autocracies (aka an autocratic dyad) going to war is 0.67. This value is less than 1, and indicates that a pair of autocracies are less likely to go to war than a mixed dyad, and we might claim that they are almost half as likely (i.e., near 0.5).

Rather than describe raw ratios some people like to convert relative risks to percentages and report the percentage increase (decrease) in the likelihood of outcomes. Recall that to convert a proportion into a percentage we multiply it by 100%. It might be helpful to observe that if the relative risk ratio is 1, then $1 \times 100\% = 100\%$, so that the risk of war for one dyad is 100%, i.e., the same as the risk of war for the other dyad in the comparison. Continuing with the Raknerud and Hegre study, if the relative risk of going to war for major power dyads relative to other pairs of countries is 1.95, then $1.95 \times 100\% = 195\%$, or a 95% increase for major power dyads relative to other pairs of countries. This suggests that a pair of countries with one major power is 95% *more* likely to go to war than a pair of countries in which neither is a major power.

They report that the relative risk ratio for a pair of autocracies (aka an autocratic dyad) going to war relative to a pair of countries with one democracy and one autocracy is 0.67, whereas the ratio for a pair of democracies (aka a democratic dyad) relative to the same base is 0.43 (p. 394). Again, to convert a proportion to a percentage change we multiply by 100. As such, two autocracies are 33% *less* likely (since $0.67 \times 100\% = 67\%$) to go to war than a pair of countries with one democracy and one autocracy. Two democracies, on the other hand, are (since $0.43 \times 100\% = 43\%$) 57% *less* likely to go to war with one another than a pair of countries with one democracy and one autocracy.

9.4 EXERCISES

1. Identify the following as objective or subjective probability claims:

   a) Using her old gradebooks, Professor Lindström determines that the likelihood that one of her students earns an A is 0.18.

   b) Professor Turan suspects that few of his students have arrived to class well prepared.

   c) Professor Long learns that over the past five years, 79% of his students have given him "Very Good" or "Excellent" ratings.

   d) Professor Lee tells her students that nuclear war between Pakistan and India is unlikely.

   e) Responding to a reporter doing background work on a story, Professor Tures explains that the incumbent's popularity has risen over the past several months.

2. Identify each of the following as a *classical* (objective), *empirical* (objective), or *subjective* probability claim:

   a) Gelman, Silver, and Edlin (2012) examined registered voter rolls and the Electoral College rules and determined that "on average, a voter in America had a 1 in 60 million chance of being decisive in the presidential election."

   b) Ghanain politician Ursula Owusu stated that there was a greater than than 70% chance that the economy would improve next year.

   c) New Mexico governor Gary Johnson vetoed more than 200 bills during his first term.

3. Identify the following as simple or compound events:

   a) Country A invades Country B.

   b) The president vetoes a bill.

   c) Dissidents drive the government from power and convene a constitutional convention.

   d) The Supreme Court issues a unanimous or split decision.

   e) Changes to overtime regulations reduce payrolls and lawsuits.

4. Characterize the following as independent, mutually exclusive, and/or collectively exhaustive:

   a) 33 year-old, middle income, Asian American, male.

   b) Strongly disagree, neutral, agree.

   c) Vote share, size of the economy, education level.

d) War, not war.

e) Less, same, more.

5. If a and b are independent events, are the following true or false?

a) $Pr(a \cap b) = Pr(a)Pr(b)$

b) $Pr(a|b) = Pr(a) + Pr(a)Pr(b)$

c) $Pr(b|a) = Pr(b)$

6. If a and b are mutually exclusive and collectively exhaustive, what is the joint probability of (a or b)?

7. If $a, b, c,$ and d are mutually exclusive and collectively exhaustive, and $Pr(a = 0.23)$, $Pr(b = 0.15)$, and $Pr(c = 0.46)$, then what is the joint probability of (a or d)?

8. If $a, b, c,$ and d form a set of mutually exclusive, collectively exhaustive events, what is the joint probability of (a and b and c and d)?

9. If a and b are independent, and $Pr(a = 0.13)$ and $Pr(b = 0.36)$, what is the joint probability of (a and b)?

10. Let $P(A) = 0.3$ and $P(A \cup B) = 0.5$. Find $P(B)$, assuming both events are independent.

11. Let $P(A) = 0.4$ and $P(A \cup B) = 0.7$. Find $P(B)$, assuming both events are independent.

12. Let $P(A) = 0.4$ and $P(A \cup B) = 0.6$. Find $P(B)$, assuming both events are independent.

13. Compute each of the following:

a) $\frac{12!}{7!}$.

b) $\frac{5!}{6!}$.

c) $\binom{12}{5}$.

d) $\binom{7}{2}$.

e) Same as (c), but now order matters.

f) Same as (d), but now order matters.

14. A committee of five members is to be formed consisting of two representatives from labor, two from management, and one from the public. If there are six representatives from labor, five from management, and four from the public, how many different committees can be formed?

15. In how many different ways can 6 differently colored marbles be arranged in a row?

16. A committee contains fifteen legislators with ten men and five women. Find the number of ways that a delegation of six:

    a) Can be chosen.

    b) With an equal number of men and women can be chosen.

    c) With a proportional number of men and women can be chosen.

17. Assume that four cards are independently drawn from a (fifty-two-card) deck with replacement. What is the probability that the jack of hearts is drawn exactly once assuming that all four selected cards were jacks?

18. Assume five independent draws with replacement from a deck of cards. Assuming all cards selected were queens, what is the probability of the queen of spades being selected exactly two times?

19. Assume a fair pair of dice are rolled. What is the probability of rolling a 7? How about rolling a 3?

20. Return to the problem described in section 9.2.3 and assume that he further believes that the conditional probability of fundraising dinners in the district changes over time as follows: January, $Pr(f|I_r) = 0.7$; February, $Pr(f|I_r) = 0.8$; March, $Pr(f|I_r) = 0.9$; April: $Pr(f|I_r) = 0.99$. How should the legislator update his beliefs about whether the incumbent will run as a consequence of observing fundraising dinners in the district in each of the successive months? Calculate the legislator's updated beliefs given that he observes a fundraising dinner in February but in neither January nor March.

21. Work through the tutorial at `http://bayes.bgsu.edu/nsf_web/tutorial/a_brief_tutorial.htm`.

22. Solve what is known as the Monte Hall problem. There are three doors. Behind two of these are goats, while behind the third is a new car. You choose one door. Monte Hall opens one of the other two doors, revealing a goat, and asks if you'd like to stick with the door you have, or switch to the other door he did not open. You get whatever is behind the door you choose. Should you switch doors? Why or why not?

23. In a certain city, 30% of the citizens are conservatives, 30% are liberals, and 40% are independents. In a recent election, 50% of conservatives voted, 40% of liberals voted, and 30% of independents voted.

    a) What is the probability that a person voted?

    b) If the person voted, what is the probability that the voter is conservative?

c) Liberal?

d) Independent?

24. If $Pr(y) = 0.62$, what are the odds that y occurs?

25. If the odds of x_1 are 3:1 and the odds of x_2 are 1:2, what is the odds ratio of $x_1 : x_2$?

26. A study reports that the relative risk ratio of voting for the National Front in a French election is 2.42 for an unemployed person relative to an employed person, and 0.38 for a person with a college degree relative to someone who did not complete high school. Write a sentence that describes the impact of the value of those variables on the probability of voting for the National Front.

9.5 APPENDIX

It is possible to use set theory to develop the main points we have presented here. For a remarkably sparse presentation see Morrow (1994, pp. 320–21). Gill (2006, chap. 6), provides a considerably more detailed presentation.

Those interested in a companion introductory guide with problems will find Bluman (2005) useful. Students interested in a comprehensive introduction to probability will find Ross (2009) helpful. DeGroot and Schervish (2001) provide a classic and comprehensive introduction to probability and statistics.

The following Wikipedia entries (and associated links) provide an overview of the Bayesian (subjective) and frequentist (objective) theories of probability: `http://en.wikipedia.org/wiki/Bayesian_probability` and `http://en.wikipedia.org/wiki/Frequency_probability`.

James Jones's "Stats Lecture Notes" pages on "Probability" and "Conditional Probability" are useful: `http://people.richland.edu/james/lecture/m170/ch05-rul.html` and `http://people.richland.edu/james/lecture/m170/ch05-cnd.html`.

You can find a somewhat cheeky but very detailed explication of Bayes' rule with lots of (medical) examples at: `http://yudkowsky.net/rational/bayes`. The Wikipedia entry on Bayes' Theorem is pretty good: `http://en.wikipedia.org/wiki/Bayes'_theorem`. Finally, you might be interested to know that Bayes' rule is considered a useful alternative to classical inductive reasoning. You can read more about that at the Wikipedia entry on the Raven paradox: `http://en.wikipedia.org/wiki/Raven_paradox`.

Chapter Ten

An Introduction to (Discrete) Distributions

As discussed in Chapter 1, variables are indicators of concepts, and they take several values. If we look at a population (or sample), we often want to know how many people or states or other variables of interest hold each value. Put differently, we want to know the *distribution* of cases across the values of the variable. One can use mathematics to develop (1) an understanding of how the cases are distributed in a given population (sample) or (2) an expectation of how cases should be distributed given one's beliefs about the process that produces the variation in that concept (or variable). The first task is an empirically based exercise while the latter is conceptual, but we generally undertake the exercise as a precursor to conducting statistical analyses that compare our actual data against our theoretical expectations. The second task is also fundamental to game theory, as it relates to the utility an actor expects to receive in an uncertain scenario. In this chapter we introduce you to the mathematics involved in understanding (1) frequency distributions (which are empirical) and (2) probability distributions (which are theoretically constructed). The discussion of probability distributions is limited to concepts (variables) that take discrete values. In Chapter 11 we discuss the probability distributions of concepts (variables) that can take continuous values. We stress that the difference between discrete and continuous probability distributions is largely technical, and we separate them into different chapters because calculus is needed for continuous distributions but not for discrete ones.

The first section of this chapter covers distributions of one variable generally, and defines random variables, i.e., those variables that can take on multiple values with some likelihood. The second section details empirical sample distributions. The third discusses empirical joint and marginal distributions. The fourth details the theoretical probability mass function. The fifth presents the cumulative distribution function for discrete random variables. The sixth section presents examples of probability mass functions. Finally, the seventh section describes the concept of an expectation of a random variable, its relation to the moments of a distribution, and the notions of expected value and expected utility, which are fundamental to game theory.

10.1 THE DISTRIBUTION OF A SINGLE CONCEPT (VARIABLE)

There are several ways to characterize the distribution of a concept (variable). You are already familiar with the frequency and relative frequency distributions (though you may not know they have names) as they are widely used in media reports and textbooks. Before introducing frequency distributions, we would like to make a point about why you should study distributions. This is the headline: science involves generalization, and generalization involves the distributions of concepts (variables).

10.1.1 From Specifics to Generalizations

Science involves the identification of general causal processes. By causal we mean that one concept influences another such that change in the first consistently produces change in the latter.[1] By general we mean processes that apply to all or most members of a class or population: the causal process is not unique to individual members of the class or population. Most causal theory in political science is probabilistic: the hypothetical relationship between a causal variable X and a caused variable Y is not expected to hold in every case. In fields such as physics, theorists often posit laws that are expected to hold in all cases (e.g., the laws of thermodynamics). Political scientists rarely (if ever) posit laws. Instead, we posit probabilistic hypotheses about the impact of X on Y. That is, theories in political science posit causal relations that apply to most members of a class or population at most moments in time. For example, theories of voting that posit the hypothesis that party identification (e.g., Republican) is positively associated with vote choice (e.g., George W. Bush in 2000) are probabilistic: finding a registered Republican who voted for a Democratic candidate does not falsify the hypothesis.

Whether probabilistic or nomologic,[2] theories about political behavior and institutions are generalizations: they do not apply to a single specific individual. To be clearer, an article that discusses Charles de Gaulle's impact on French politics is descriptive; an article that posits a theory about the impact of the office of the president on French politics is general. An even more general theory would develop hypotheses about the impact of presidents on politics in democracies (and would require one to define the terms used in this sentence, as there is a great deal of variation among presidencies in democracies). Once we shift our attention away from specific individual members of a class or population toward most members of a class or population, we have already begun to think

[1] This is similar to the "inductive regularity" definition of causation found in Little (1991). As Little demonstrates, the "causal mechanism" definition of causation is superior to this one, but that distinction is not of immediate concern here. See Little (1991, pp. 13–29) for a detailed discussion.

[2] A nomologic theory proposition applies to all members of a class or population at all times.

about distributions. And the mathematical study of distributions provides some powerful tools to aid one in both developing theory and testing the hypotheses implied by theories. Put differently, if one is interested in theory (which, by definition, involves generalization) rather than description, then probability distributions are a useful tool at one's disposal. As such, it is important to develop a working familiarity with probability distributions.

Before moving on we would like to advance an important claim. While political scientists widely recognize the centrality of probability distributions to statistical analysis (an important tool for testing hypotheses), a considerably smaller group recognizes the usefulness of understanding them for developing theory. As such, instruction in distributions is almost exclusively restricted to the first statistics course and generally consists of a single lecture.[3] While recognizing that probability distributions are critical to the study of statistics, and thereby hypothesis testing, we argue that thinking about distributions is equally important for creating theories (and developing hypotheses). By showing how an understanding of probability distributions sharpens one's theorizing about politics, this chapter is intended to help correct an imbalance we have in our field. We discuss this specifically in the final section covering expected utility.

However, much as in the previous chapter, the formalism of distributions is the same whether or not we are referring to statistical or measurement error, uncertainty, errors in decision making, or any other source for probabilistic hypothesizing or theorizing. Along these lines, the same formalism is used whether we're discussing, for example, the distribution of individual characteristics or behavior across a population or the distribution of individual behavior across opportunities to act (e.g., vote sometimes and not others). In short, we reiterate that distributions are a remarkably powerful and useful tool in political science.

10.1.2 Random Variables

In a footnote we noted that a distribution defines the set of values that a random variable may take, but this notion is too important to leave to a footnote, as this connection implies that random variables are as fundamental to political science as distributions. For the same reason we noted that while hypotheses in political science (and many theories as well) are probabilistic, the components of hypotheses (and many theories) are random variables. So what are random variables?

Random variables are those that have their value determined, in part, by chance: the value they take in any given circumstance can be described probabilistically. More precisely, a **random variable** is a variable that can take some array of values, with the probability that it takes any particular value defined according to some random process. The set of values the variable might take is the **distribution** of the random variable and, as we show below, the function that defines the probability that each value occurs is known as the probability

[3] What we are calling distributions is also often discussed under the label "random variables," as distributions describe the set of values that random variables may take.

mass function or the probability distribution function, depending on whether or not the distribution of values is discrete or continuous.

Saying that the variables in our hypotheses (and many theories) are random variables amounts to saying that we expect the value of each variable in our hypotheses (and many theories) to be a draw from some associated distribution. A simple example of this would be the roll of a fair die. The random variable corresponding to the die's value would have an equal probability (one-sixth) of having each of the integer values from 1 to 6. This is called a uniform distribution. However, more complicated distributions are allowed, particularly when we're discussing dependent random variables. For example, we could say that one's decision to vote is a random variable Y that can take the values of 1 (for yes) and 0 (for no). The probability that it takes the value 1 will depend on many other factors, such as past voting behavior, education, income, etc. In other words, the probability distribution of values for Y is a conditional probability distribution: one for which the likely outcomes vary as one or more other variables vary in value (more on this below). Increasing the value of an independent variable such as education or income might shift the distribution of 1s and 0s toward more 1s, implying that a more educated or higher-earning person will be more likely to vote, and more likely to vote more often, given multiple opportunities to do so.[4] This is obviously weaker than a deterministic statement like "If you earn more than $100k a year you will vote," but political scientists generally believe such probabilistic statements are relevant to the real world.[5]

This is most of what one needs to know about random variables, but a little bit of terminology is helpful before moving on. As we've stated, a random variable can take many values.[6] The **realization** of a random variable is a particular value that it takes. When it's not confusing, we often use capital letters, such as Y, to correspond to random variables and lowercase letters, such as y, to correspond to particular realizations of a random variable. The statement $Pr(Y < y)$ then reads "the probability that Y is less than a particular value y." This will prove important later.

The probability that a random variable has *any* value is given by its probability distribution; the **support** of this distribution is the set of all values for which the probability that the random variable takes on that value is greater than zero. This terminology is commonly used for continuous probability distributions. Discrete random variables have discrete associated probability dis-

[4] The connection between these two statements relates to the expectation of a random variable, which we discuss in Section 7 of this chapter.

[5] Even theories that make deterministic claims typically highlight the way in which the claim varies with parameters of the model (i.e., independent variables). These *comparative statics* of theories, which we'll discuss in Chapter 17 of this book, are weaker statements than the theory's point predictions and more directly testable. Consequently, these are generally more believable.

[6] A slightly more complex way of thinking about this is that a random variable stochastically determines that some event happens, out of all the events that might have happened, and assigns a value (typically a real number) to that event.

Table 10.1: Top Problem Facing the United States, September 2001

Problem	% Listing it No. 1
Terrorism	60
Economy	8
No opinion	7
Noneconomic, other	6
Moral decline	6

tributions; continuous random variables have continuous associated probability distributions.

10.2 SAMPLE DISTRIBUTIONS

Sample distributions are empirical constructs. We distinguish them from probability distributions, which are constructed theoretically using the rules of classical probability. We discuss probability distributions below.[7]

Sample distributions are representations of the number of cases that take each of the values in a sample space for a given portion of the population of cases. Put differently, a sample distribution is the distribution of values of a variable resulting from the collection of actual data for a finite number of cases. It turns out that you have encountered many sample distributions in various textbooks, news articles, and other places.

10.2.1 The Frequency Distribution

The first sample distribution to consider is the frequency distribution. It is a count of the number of members that have a specific value on a variable. Public opinion scholars often ask a question about the issues that are most important to the country, thereby producing a rank order of several issues (e.g., crime, the environment, poverty, terrorism). The frequency distribution for that variable would be the number of respondents to the survey who listed the first issue as most important, the second issue as most important, etc. (Table 10.1 reports such a survey conducted in Georgia).[8]

Students of coalition governments are often interested in a variable that records the number of seats a party won in a given election. The frequency distribution for that variable simply lists the number of seats each party obtained (an example from Lithuania is given in Table 10.2).[9]

Studies in international relations often include a variable that records whether a given country initiated a militarized dispute with another country during a

[7]Many statistics textbooks refer to probability distributions as *population distributions*.

[8]Source: Peach State Poll, `http://www.cviog.uga.edu/peach-state-poll/2001-12-07.pdf`.

[9]Source: `http://www.electionworld.org/lithuania.htm`.

Table 10.2: Lithuanian Parliamentary Seats, 2000

Party Abbreviation	Seats Won
ABSK	51
LLS	33
NS	28
TS-LK	9
LVP	4
LKDP	2
LCS	2
LLRA	2
KDS	1
NKS	1
LLS	1
JL/PKS	1

Table 10.3: Militarized Interstate Dispute Initiators, 1816–2002

MID Initiator	No. of Countries	% of Countries
No	67	31
Yes	147	69

given period of time. In Table 10.3 the frequency distribution for this variable is the number of countries that initiated a dispute at some time between 1816 and 2002 and the number that did not.[10]

One can create a frequency distribution for any variable. The level of measurement is irrelevant: whether discrete (i.e., nominal [aka categorical], ordinal, or integer) or continuous, we can produce a frequency distribution if we collect data on a sample. That said, a frequency distribution is only of interest if all members of the population do **not** have unique values on the variable of interest. If each case has a unique value, then all of the values in the sample frequency distribution will be 1.

10.2.1.1 Why Should I Care?

The frequency distribution is widely used; you have encountered it countless times in textbooks, news articles, and elsewhere. Given that it is so simple and widely used, why dedicate space to it in a graduate-level text? The reason is that it is critical to think in terms of distributions over concepts when developing theory, and the frequency distribution is very convenient and useful for that purpose. When thinking about a political process one often thinks about a specific example of the process in question (e.g., one's own vote in an election, the formation of a specific coalition government, or a particular militarized dispute).

[10]Source: Militarized Interstate Dispute Data as available in the EUGene software, `http://www.eugenesoftware.org`.

This is natural and especially common for new graduate students. However, it is not a very good practice for developing general theories of politics, and failure to recognize this not only reduces the likelihood that one will develop a useful theory but can also lead one to errors in research design when it comes time to test hypotheses (e.g., selection on the dependent variable; see King, Keohane, and Verba, 1994, pp. 129–37). Thinking about specific examples is definitely a good place to *begin* theory development, but having established a potential causal relation from the example in question, one must move to thinking about all members of the class or population in question and ask whether the relationship might hold for most members most of the time. Simple thought experiments combined with rough knowledge of the distribution of the measure of a concept can lead one to reject ideas without having to formally test them, and also illuminate new puzzles that warrant explanation.

To be a political scientist (rather than a scribe of politics) one must shift one's thinking away from specific examples toward general patterns. The frequency distribution is the most intuitive, simple distribution and thus one with which you will want to become very comfortable not only with respect to actual sample data (i.e., the values of variables) but also in the abstract, when you are theorizing (i.e., concepts).

10.2.2 The Relative Frequency Distribution

To this point, we have discussed distributions without concerning ourselves with probability. Let us remedy that. The probability that a specific case drawn at random[11] from a sample has a specific value, i, is the relative frequency of the value i.[12] The relative frequency of value i is the frequency (i.e., the number of cases with the value i) divided by the total number of cases. Put differently, the relative frequency of value i is the proportion of cases that have that value. As such, the relative frequency of a given value lies between 0 and 1, and the sum of all relative frequencies equals 1. Note that a probability of 0 indicates that there is no chance that a given value can be drawn from a sample, and a probability of 1 indicates that the value in question will be drawn with certainty. More generally, larger probability values indicate a greater likelihood that the value in question will be drawn.

The relative frequency distribution is a transformation of the frequency distribution (to reiterate and be specific, we divide the frequency by the total number of cases). It can also be represented in tabular or graphical form, is defined for all variables regardless of their level of measurement, and is uninteresting in samples where all cases have unique values. Finally, because most people are more familiar with percentages than with proportions, relative frequency distributions

[11] Note that this definition holds only when the case is *drawn at random*. This issue is discussed in more detail in most statistics texts or in a good research design text.

[12] We can define $i \in I$, where I is defined over the range from the minimum to maximum value in the sample.

are sometimes transformed to percentages (this transformation is conducted by multiplying the proportion by 100%).

10.2.2.1 Histograms

A histogram is a specific representation of the relative frequency distribution: it is a bar chart of the distribution of the relative frequencies in which the area under each bar is equal to the relative frequency for that value. In other words, the sum of the areas of each bar equals 1, and the area of each bar equals the probability that the value represented by the bar would be chosen at random from the sample depicted. The formula for the bars in a histogram can be represented by equation (10.1):

$$Pr(Y = y) = f_y \Delta_i, \tag{10.1}$$

where $Pr(Y = y)$ is the area covered by the bar (i.e., the probability that value y would be drawn at random), f_y is the height of the bar for value y, and Δ_i is the width of the bar.[13] By convention, one holds the width of each bar constant at 1 when dealing with a discrete distribution, thus making Δ_i known. Statistics packages that produce histograms (most of them do) need only solve for the unknown height, f_y, to produce the bar chart. Because the variation is only in the height of the bars (the widths are constant at 1), the histogram often provides an appealing graphical form to use to display a relative frequency distribution.

10.2.2.2 Why Should I Care?

The relative frequency distribution is relevant to political scientists for the same reason that the frequency distribution is of interest. As an example, Fearon and Laitin (1996) is a response to a spate of books and articles that sought to explain ethnic conflict.[14] These studies observed that an ethnic war occurred in one or more locations and offered explanations for the outbreak of such conflict. Fearon and Laitin begin by thinking about the distribution of ethnic conflict and observe that interethnic cooperation is far more common than interethnic conflict. They do not produce an actual relative frequency distribution to establish this point. Instead, they simply observe that socio-econo-political interaction across ethnic groups is extremely common: few human beings live their lives within ethnically homogeneous societies. Next they observe that violent conflict is rare. Again, they do not need to produce a relative frequency distribution over a specific spatial temporal domain to establish their point: most people most of the time are not engaged in violent conflict with other people. When one puts both

[13]Recall from geometry that the area of a rectangle equals the product of its height and its width.

[14]This interest was spurred by the collapse of the Soviet bloc, which (1) eliminated the Cold War as a topic of interest and (2) contributed to the collapse of Yugoslavia, which produced wars that demanded explanation.

of these points together, one observes that ethnic cooperation is common and ethnic conflict is rare.[15]

This claim causes problems for many of the theories that others have put forth: they focus their attention on explaining the presence of ethnic conflict without attending to the (relative) frequency distribution of ethnic conflict. Fearon and Laitin's attention to that distribution led them to recognize that a useful theory of ethnic conflict needed not only to account for the outbreak of such events but also to explain ethnic cooperation and the relative distribution of the two. That is, by thinking abstractly about the distribution of ethnic conflict and cooperation (note that they did not even have to collect any data or do any statistical tests—this was theoretical thinking linked to common knowledge about the rough empirical distribution) Fearon and Laitin cast dispersion on existing theories and provided themselves with a useful starting point for developing a new, better theory.

Fearon and Laitin (1996) show that by focusing on specific cases and theorizing only about the rare outbreak of such conflict, the entire field of ethnic conflict studies had erred.[16] Once one identifies this failure to think about the distribution of ethnic cooperation and conflict the weakness in such an approach seems obvious. Yet if that weakness was so obvious, then dozens of bright, talented political scientists would not have made the error, and the point raised by Fearon and Laitin would have been made long before 1996.

To summarize, the relative frequency distribution can be useful both as an abstract theoretical tool and as a concrete empirical tool. Becoming comfortable with it will prove useful to both the development and the testing of hypotheses.

10.3 EMPIRICAL JOINT AND MARGINAL DISTRIBUTIONS

The distribution of a single concept or variable can be of considerable theoretical and statistical interest, but causal theories of politics necessarily involve expected relationships among concepts or variables. As such, we want to study joint distributions; marginal distributions are a natural extension. For readability, we will focus on the case of two variables in this section.

10.3.1 The Contingency Table

A contingency table is the joint frequency distribution for two variables. While it can be created for both discrete and continuous variables, it is of considerably more value for discrete than for continuous variables.

With respect to construction, the contingency table is a matrix with one variable's values represented in the rows (typically the dependent or caused variable in empirical work) and the other variable's values represented in the

[15] This is so even if we attribute all violent conflict to ethnic cleavages.

[16] The study of war similarly suffered from this problem for decades (Most and Starr, 1989, pp. 57–58).

columns (typically the independent or explanatory variable). The cell entries record the number of cases that have the row value for the row variable and the column value for the column variable. The resulting matrix provides a quick summary of the joint distribution of the two variables.

Lest one think that contingency tables are only of value for empirical work, reconsider the discussion of Fearon and Laitin's (1996) article of interethnic cooperation and conflict. Though they did not produce a contingency table to illustrate their thought experiment, they could have done so. We have produced such a contingency table here.

Table 10.4: The Fearon and Laitin (1996) Contingency Table

	Ethnic Homogeneity	Ethnic Heterogeneity
Cooperation	Rare	Common
Conflict	Rare	Rare

One can readily see that ethnic heterogeneity is a poor explanation for the outbreak of ethnic conflict.[17] Of more importance, one can see that cooperation in ethnically heterogeneous communities is the modal[18] outcome, and thus it is important that theories of ethnic conflict be able to explain interethnic cooperation (i.e., theories need to account for the full range of phenomena, in this case both common and rare outcomes).

That said, contingency tables are particularly useful for the analysis of empirical relationships between discrete variables. You will learn more about contingency tables in your introductory statistics course.

10.3.2 Marginal Probabilities

Marginal probability is a label that often confuses students, likely because it arises from a practice that is rarely used apart from in contingency tables. Simply put, as noted in the previous chapter, the marginal probability of an event A is the probability that A will occur unconditional on all the other events on which A may depend. To make this work, one must in general sum the conditional probabilities of A on all the other mutually exclusive, collectively exhaustive events B_i on which it may depend, each weighted by the chance that the particular B_i will occur. From the discussion of Bayes' rule in the previous chapter, we know this amounts to writing the marginal probability $Pr(A) = \sum_{i=1}^{n} Pr(A|B_i)Pr(B_i)$. In words, this means that one averages over other events and focuses on the one event, A, of interest.

For example, let's say we had computed the joint probability of rolling a 7 on two dice and drawing a king from a deck of cards. If all we cared about

[17] Some of the scholars Fearon and Laitin (1996) criticize recognize this point. See, for example, Posen (1993).

[18] The modal outcome is the most common outcome.

Table 10.5: Militarized Disputes, 1946–92

	Nonterritorial	Territorial	
State A or B Democracy < 10	43	31	74
State A or B Democracy $= 10$	21	2	23
	64	33	97

was the chance of rolling a 7—the marginal probability of rolling a 7—then we'd ignore the deck of cards entirely and just stick to the chance of rolling a 7, which is one-sixth. This example is artificially easy because the two parts of the joint event, rolling dice and drawing a card, are independent. But the argument holds for dependent events. For example, let's say we are interested in the probability of voting, but voting is conditional on whether or not it is raining. The conditional probability of voting given rain might be $Pr(V|R) = 0.4$. But what about the marginal probability of voting? To get this we need the conditional probability of voting given that it is not raining, as well as the probability of rain. Let's say the former is $Pr(V| \sim R) = 0.6$ and $Pr(R) = 0.3$. Then $Pr(\sim R) = 0.7$ and so the unconditional, marginal probability of voting is $Pr(V) = Pr(V|R)Pr(R) + Pr(V| \sim R)Pr(\sim R) = (0.4)(0.3) + (0.6)(0.7) = 0.54$.

However, when we are looking at empirical probability only, particularly when there are only two variables, the concept becomes much easier and the word "marginal" makes much more sense. To find the marginal probabilities, we sum the simple conditional probabilities of one variable across all values of the other variable.[19] The label "marginal" comes from the fact that the marginal probabilities are written in the margins of $n \times n$ (read "n by n") tables.

For example, Mitchell and Prins (1999) are interested in the frequency with which countries with strong democratic institutions get in militarized disputes over territory relative to countries without strong democratic institutions. Table 10.5 reproduces a portion of the evidence they report. With that information we can determine the empirical probabilities that (or the relative frequency with which) the compound events occurred. Tables 10.6 and 10.7 contain cell entries that represent the empirical probabilities. The marginal entries are the sum of the probabilities in the row or column, respectively, relative to all events. Each of these marginals sums to 1.

Let's walk through it. To calculate the cell entries in the center of Table 10.6 we need to determine the empirical probabilities across the rows. The upper left cell entry is the number of dyads (i.e., country pairs) without a strong democracy that became involved in nonterritorial disputes (43) relative to the total number of dyads without a strong democracy (74): $\frac{43}{74} \simeq 0.58$. The lower right cell is the number of dyads with a strong democracy that became involved in territorial disputes relative to the number of dyads with a strong democracy: $\frac{2}{23} \simeq 0.09$.

[19] A more accurate statement is that the above holds for discrete variables. For continuous variables one integrates over, rather than sums across, the values of the other variable. We discuss this issue in more detail in the next chapter.

Table 10.6: Row Probabilities

	Nonterritorial	Territorial	Row Marginals
State A or B democracy < 10	0.58	0.42	0.76
State A or B democracy $= 10$	0.91	0.09	0.24

Table 10.7: Column Probabilities

	Nonterritorial	Territorial
State A or B democracy < 10	0.67	0.94
State A or B democracy $= 10$	0.33	0.06
Column marginals	0.66	0.34

The same logic holds for the other two entries in the center of the table. The values of the row marginals at the right are the ratio of all the cases in each row to the total number of cases. The upper value is the number of dyads without a strong democracy (74) relative to all dyads (97): $\frac{74}{97} \simeq 0.76$. Note that the two row marginals in Table 10.6 sum to 1: $0.76 + 0.24 = 1.0$. The same is true of the values in each row in the center of the table.

We calculate the column probabilities in the same way except that we calculate the column totals, not the row totals. For example, the lower left cell in the center of Table 10.7 is the frequency of dyads with a strong democracy that became involved in a nonterritorial dispute (21) relative to all dyads that became involved in a nonterritorial dispute (64): $\frac{21}{64} \simeq 0.33$. Similarly, the upper right cell is the number of dyads without a strong democracy that engaged in territorial disputes (31) relative to the total number of dyads involved in territorial disputes (33): $\frac{31}{33} \simeq 0.94$. Finally, the cells listing the column marginals represent the relative frequency of each column to the total number of cases. Thus, there are 33 territorial disputes out of 97 total disputes: $\frac{33}{97} \simeq 0.34$. Observe that the marginals sum to 1: $0.66 + 0.34 = 1.0$.

Marginal probabilities are used in statistical inference. For example, the χ^2 (chi squared) statistic is used to evaluate the statistical significance of an association between two ordinal level variables.

10.4 THE PROBABILITY MASS FUNCTION

Above we discussed frequency distributions, which one can use to describe how the values of a variable for which we have collected data are distributed within a population or sample. It is also possible to use the laws of probability to develop expectations about how one *expects* the values of a concept (variable) to be distributed *given* one's beliefs about the process that generates the values that concept will take. Those of you who have had a course in statistics have already studied a number of these distributions, but even those of you who have never studied statistics have probably heard of the "bell curve." The bell, normal, or Gaussian curve is a probability distribution, though since it describes concepts (variables) that can take continuous values, we do not discuss it in this chapter.

All functions are a mapping of the values in one set to another, and a probability function describes the likelihood of each of the values a concept (variable) might take given a description of the process that generates it. It turns out that it is possible to develop a number of different probability functions to describe the distribution of a concept (variable), but we limit the discussion to two of them: the probability mass (or distribution) function and the cumulative density function. You should know, however, that there are other functions: the hazard, cumulative hazard, survival, inverse survival, and percent point functions can also be used to characterize different distributions. While the probability mass (or distribution) function and the cumulative density function are the most widely used in statistics texts popular in political science, a full understanding requires additional coursework or self-study. But this is just an introduction, and a sound understanding of the material contained here will nicely pave the route for additional study for those of you who become interested in pursuing it.

In the chapter on probability we looked only at the classical probability of a single value being drawn randomly from a sample. Political scientists are often more interested in being able to say something about a range of values. The probability mass function (PMF) allows us to do this. As important, it also allows the chance of drawing any particular value to vary by value. The PMF is a function that specifies the probabilities of drawing discrete values (we discuss the density function of continuous variables in the following chapter).[20] As we show below, the PMF of a discrete variable is related to the relative frequency distribution.[21] More specifically, the PMF is a function that connects the various classical probabilities of specific values for a sample.

Another way to think about this is that the PMF is a function that allows one to sum a series of weights. More specifically, the weights are the probabilities that each value will be randomly drawn from the sample. The PMF makes it possible to identify different probability distributions, and being able to do so turns out to be very important for developing statistical models that can produce valid hypothesis tests (more on why you care below).

The PMF of a discrete (i.e., nominal, ordinal, or integer) variable assigns probabilities to each value being drawn randomly from a population. The relative frequency distribution of a discrete variable is a tabular or graphical representation of the empirical probabilities that each value is drawn at random from the sample. In other words, the PMF is the function that describes the expected relative frequency distribution.

More formally, the PMF of a discrete variable, Y, may be written as $p(y_i) = Pr(Y = y_i)$ where $0 \leq p(y_i) \leq 1$ and $\sum p(y_i) = 1$, and Y is the variable and

[20] More formally, let X be a discrete random variable that maps elements in a sample space to the real numbers. Then a probability mass function is a map from the set of real numbers to probabilities in $[0, 1]$: $f_X(x) : \mathbb{R} \rightarrow [0, 1]$.

[21] These functions are sometimes referred to more generically as the probability distribution.

y_i is a specific value of Y. This function describes the height of the bars of a histogram, as depicted in Figure 10.1.

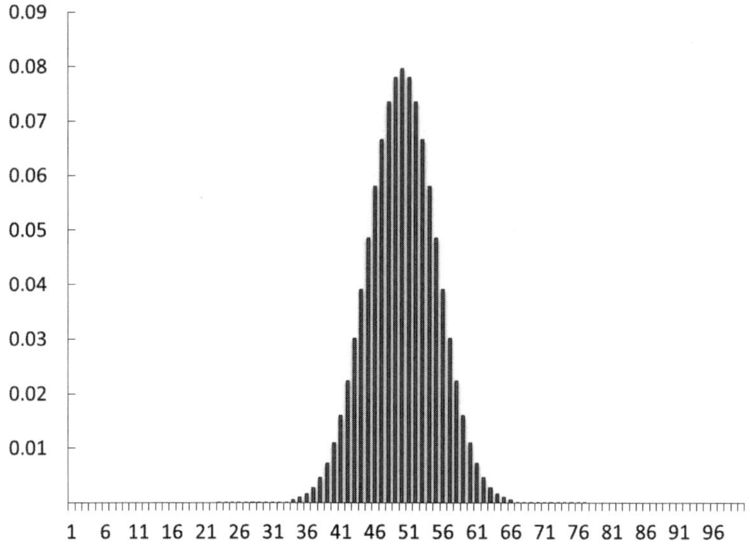

Figure 10.1: PMF of a Binomial Distribution, $n = 100, p = 0.5$

You might be wondering why we discuss discrete variables here but leave a discussion of continuous variables for the next chapter. The reason is that continuous variables have an infinite number of possible values, and the probability that we will randomly select any given value from a set with infinite support is zero.[22] As such, the probability function of a continuous variable is a bit more complex.

10.4.1 Where Does a Specific PMF Come From?

The representation of a PMF above is general—we did not identify a specific functional form. Many people who look at a given PMF find it far from obvious where it came from or what it actually means. Here we briefly describe the mathematical process used to create a PMF. Below we introduce a number of specific PMFs and verbally describe the data-generating process that the PMF describes.

In Chapter 3 we observed that one does not want to memorize a set of functions and then build theories of politics by applying each until one finds one that fits both one's conjectures and the relevant data. Rather, one wants to become familiar with functions so that one can discipline one's speculations and

[22] We see in the next chapter that the way around this is to ask instead what is the probability of selecting any value within a small region. This will be non-zero, but answering this question requires calculus. Hence we discuss it in the next chapter.

conjectures and make them more explicit. King (1989, p. 38) makes the same point with respect to the distributions that underlie our statistical models: "it is not necessary to fit one's data to an existing, and perhaps inappropriate, stochastic model. Instead, we can state first principles at the level of political science theory and derive a stochastic model deductively."[23] King is arguing that rather than simply employ a statistical model off the shelf because someone put it there, we want to think about the political process that produced the concept or variable that interests us and then use the rules of probability theory to write down a function that describes that process. A PMF is precisely that: a function that uses probability theory to describe the process that generated the expected frequency of a discrete concept (variable) across cases. King (1989, p. 41) explains it this way:

> each distribution [presented in statistics textbooks] was originally derived from a very specific set of theoretical assumptions. These assumptions may be stated in abstract mathematical form [i.e., as a function], but they may also be interpreted as political assumptions about the underlying process generating the data. The ultimate mathematical form for most distributions is usually not very intuitive, but the first principles from which they were derived represent models of interesting political science situations and are much closer to both data and theory. When a list of these principles is known, understanding them is critical to the correct application of a particular probability distribution.

We reviewed the principles of probability in the previous chapter. While few political scientists actually write down their own probability functions in empirical work, it is important to gain a familiarity with how one can go about it so that one can understand where the different probability functions that are commonly used come from and make informed choices about which is most appropriate.[24] Further, those who study game theory will come to use several probability functions more directly, particularly when computing expected values and expected utilities, the topic of the last section of this chapter.

10.4.1.1 Why Should I Care?

It is possible to assign qualitative (i.e., nominal and ordinal) and integer values to a wide array of political phenomena. We can measure political attitudes, record vote choice, count the number of seats a party holds in the legislature or

[23] A stochastic model is just a model that employs random variables. The most common way we see this in political science is via statistical models, and statistics courses tend to talk most about stochastic models. However, many formal models are also stochastic, including game theoretic models that involve lotteries over payoffs or opponents' strategies or uncertainty, and many models of bounded rationality.

[24] Those who wish to respond to King's call and develop their own probability functions will need to pursue additional studies in probability and statistics, most likely in their statistics department.

the number of militarized disputes in which a country has participated, etc. It should not surprise you that if one graphed the relative frequency distributions of the many and varied variables that political scientists have created, these graphs would not all look the same. It might surprise you, however, to learn that each and every one of them (roughly) approximates a PMF (for discrete variables) that statisticians have studied and named. Further, statisticians have gone to the trouble of developing and naming distributions for the express purpose of developing statistical models that can be used to test causal hypotheses. So if you are interested in applying the considerable power of inferential statistics to your own research, it is critical that you gain a working familiarity with PMFs. Doing so will put you in a position to choose statistical models appropriately and thus ensure that your statistical hypothesis testing is valid. Further, as King urges, some of you may want to focus on political methodology in which case you may end up writing down new distributions that are more appropriate for given theories in political science than any of the probability distributions developed to date. And, as we noted, if you read or employ game theory in your research, you will use probability distributions in computing expectations.

10.4.2 Parameters of a PMF

We introduced the concept of a sample space in the previous chapter. Here we need to define the terms **parameter** and **parameter space**. In discussions of probability functions, the term "parameter" refers to a term of known or unknown value in the function that specifies the precise mathematical relationship among the variables. Further, parameters are independent of the values of the sample space. Parameters can take multiple values, and the parameter space is the set of all values the parameters can take.[25] More specifically, "the functional form of [a probability distribution] and the value of the parameters ... together determine the shape, location, and spread of the distribution" (Hendry, 1995, p. 34).

That description is jargon-laden, and we offer some examples momentarily. But first, note that the general representation above did not contain any parameters: it is not specific about the mathematical relationships among the variables. The specific PMFs that we introduce below contain parameters.

To illustrate, let's consider the case of voter turnout where we ask, "Which registered voters cast ballots?" There are two outcomes for each voter: (0) *did not cast a ballot* and (1) *cast a ballot*. We can write the following PMF:

$$\begin{aligned} p(y_i = 0) &= \pi, \\ p(y_i = 1) &= 1 - \pi. \end{aligned}$$

[25]We referenced the parameter space way back in Chapter 1, as an example of a space of interest in game theory. As you can see, it is of interest in statistics as well. It means the same thing in both cases: parameters help to dictate the dependence of functions on variables, each parameter can take a range of values, and the space of all values all parameters can take is known as the parameter space.

Here, π is the parameter of this PMF.[26] Because probabilities range between 0 and 1, $\pi \in \Theta = [0, 1]$. And Θ is the parameter space, as it is the set of all values the parameter π can take. We do not know what value to assign π, but we could turn to voter turnout data from previous elections to develop an expectation about π.

To further illustrate, let's consider the example of tossing a fair coin. It turns out that the PMF is precisely the same as above (e.g., [0] *heads* and [1] *tails*), except that in this example we know the value of π: it equals 0.5. If the coin is biased and turns up "heads" twice as often as "tails," then $\pi \simeq 0.67$ and $1 - \pi \simeq 0.33$. If the coin is biased and produces "tails" three-fourths of the time, then $\pi = 0.25$ and $1 - \pi = 0.75$.

Note that in these examples, when we change the value of the parameter the distribution of outcomes change. That is precisely what a PMF should tell us: the distribution of outcomes across the values of the variable given parameter values. The parameter space is the set of values the parameter might possibly take, and it thus dictates the set of possible distributions of outcomes.

Two important parameters are the location and scale (dispersion) parameters, and we introduce those below. It is important to note that not all probability distributions have parameters. When we discuss a specific distribution below we note whether it has any parameters, and if so, which ones.

10.4.2.1 Location and Scale (Dispersion) Parameters

Many distributions have a location and a scale (dispersion) parameter. Some have only a location parameter, and still others do not have any parameters.

The **location parameter** specifies the location of the center of the distribution. Thus, as one changes the value of the location parameter, one shifts the graph of the distribution's PMF to the left or right along the horizontal axis. For some distributions (especially the normal distribution, a continuous distribution we discuss in the following chapter), the location parameter has an empirical referent known as the **mean**. The location parameter (mean) is often represented in classical (empirical) probability by the Greek letter μ (mu).

For example, Figure 10.2 displays graphs of the PMF of a Poisson distribution (we introduce the PMF of the Poisson and other distributions below). It has only one parameter, a location parameter, μ. We have set μ equal to 1, 3, and 5. As the location parameter gets larger (i.e., changes from 1 to 3 to 5), the center of the distribution (i.e., its highest point) moves to the right. For all distributions with a location parameter, larger values will move the center of the PMF to the right and lower values will move the center to the left.

Second, the **scale parameter** provides information about the spread (or scale) of the distribution around its central location. As such, changing the scale parameter stretches or squeezes the graph of the PMF. Compared with a scale parameter equal to one, values greater than one increase the width of the graph

[26] It turns out that this is the PMF for the Bernoulli distribution. We discuss it in more detail below.

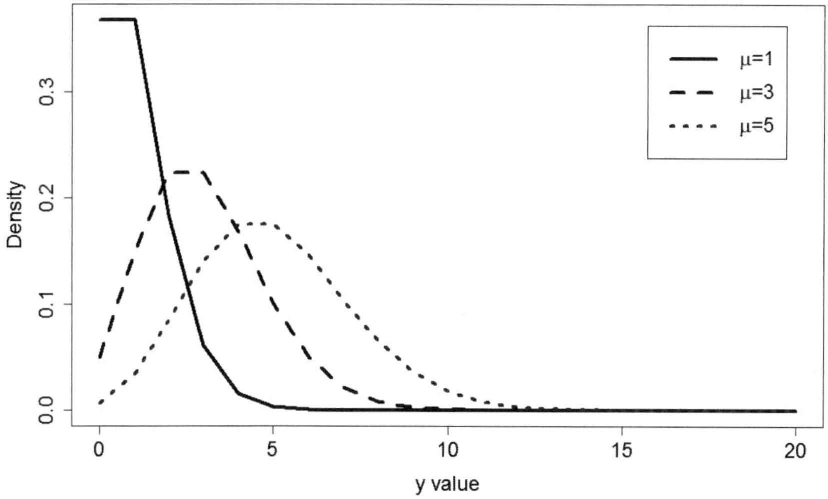

Figure 10.2: PMF of a Poisson Distribution, $\mu = 1, 3, 5$

(i.e., indicate both a lower minimum and higher maximum value than a PMF with a scale value of one). Similarly, scale parameters between zero[27] and one squeeze the PMF relative to one with a scale value of one: the minimum value is greater and the maximum value lower than they are for a PMF with a scale value of one. The scale parameter has an empirical referent known as the **standard deviation**, which is a measure of the distance of the distribution's values from its mean (or average) value. Both the scale parameter (classical probability) and the standard deviation (empirical probability) are usually represented with the Greek letter σ (sigma).

The **dispersion parameter** is the square of the scale parameter. As such, it also describes the spread about the central location of the distribution, except it emphasizes extreme values. Most treatments consider the decision to identify the scale or the dispersion parameter when defining the parameters of a classical probability distribution an arbitrary choice. In statistics, the dispersion parameter corresponds to the **variance** of an empirical distribution, and both are typically identified as σ^2 (sigma squared).

10.4.2.2 The Standard Form

The standard form of a distribution (or standard form PMF) is one in which the location parameter is set to zero and the scale parameter is set to one.

[27]Scale parameters cannot take negative values.

10.4.2.3 Why Should I Care?

The parameters of a classical probability distribution are important because they help us define the probability function (or PMF, for discrete distributions). We can also use the parameters of a distribution to help us determine whether data we have collected closely match the expected distribution given our beliefs about the process that produced the data. For example, we might think that the number of appointments a US president makes to the Supreme Court is produced randomly by what is known as a Poisson process (described below). The Poisson distribution has one parameter (location). Ulmer (1982) proffers just that hypothesis. He collected data on the number of Supreme Court appointments over the period from 1790 to 1980 and used it to estimate the location parameter to see whether it is likely that those data were produced by a Poisson process (you will learn about parameter estimation in your statistics courses).[28] He finds that the data were likely produced by a random Poisson process. Similarly, Midlarsky, Crenshaw, and Yoshida (1980) use the Poisson distribution to study contagion among transnational terror events data.

10.5 THE CUMULATIVE DISTRIBUTION FUNCTION

When conducting hypothesis tests it is often useful to determine the probability that a value drawn at random from a sample is above or below a specific value. The cumulative distribution function (CDF) describes the function that covers a range of values below a specific value and is defined for both discrete and continuous random variables.[29]

The CDF for a discrete random variable is, hopefully, intuitive: if we want to know the cumulative (or total) probability that a random draw from a population produces a value less than some quantity, then we need to add together the individual probabilities of each of the values below that number. We sum the individual probabilities because the values are mutually exclusive, and the joint probability of mutually exclusive events is the sum of the probabilities of the individual events. We can write the CDF for discrete variables as

$$Pr(Y \leq y) = \sum_{i \leq y} p(i). \tag{10.2}$$

Equation (10.2) states that we sum the probabilities of each value for all values less than or equal to y.[30] Sometimes you will see the notation $f(x)$ for a probability distribution function (PDF or PMF) and $F(x)$ for a CDF; using that notation makes clearer the connection between the two functions. We discuss this

[28] The term *parameter* can be used somewhat differently in probability theory and statistics, though at times they mean the same thing. See the wikipedia entry at `http://en.wikipedia.org/wiki/Parameter`.

[29] The CDF is also sometimes called the distribution function (see, e.g., the MathWorld entry at `http://mathworld.wolfram.com/DistributionFunction.html`).

[30] To add some precision, we can note that $0 \leq Pr(Y \leq y) \leq 1$, and that $Pr(Y \leq y)$ is increasing in y (i.e., $Pr(Y \leq y)$ gets larger as y gets larger).

more in the next chapter. Note that, since the values are mutually exclusive and all the values together are collectively exhaustive, $Pr(Y \leq y) + Pr(Y > y) = 1$, which implies that $Pr(Y > y) = 1 - Pr(Y \leq y)$. In words, the probability that a random draw exceeds some quantity is equal to one minus the probability that it does not exceed that quantity. Further, if y is the highest value that Y can take, then $Pr(Y \leq y) = 1$, since in this case we are adding the probability of all outcomes in the sample space. So all CDFs plateau at one.

Let's consider a concrete example. We might be interested in knowing the probability that a potential voter in the United States is partisan (e.g., self-identifies with either the Democratic or Republican Party). Imagine that we have a randomly drawn sample of survey data that records whether the respondents identify with the Democratic Party (value = 1), the Republican Party (value = 2), or a third party/no party at all (value = 3). Our party identification variable is discrete (in this case it is nominal), and we want to know the probability that a respondent drawn at random from the sample is partisan (i.e., has a value less than 3). Assume that the frequencies for the sample are

1. Democratic: 330,
2. Republican: 240,
3. Other: 180.

Given that there are 750 people in our fictitious survey, the relative frequencies are

1. Democratic: 0.44,
2. Republican: 0.32,
3. Other: 0.24.

To determine the probability that a respondent drawn at random from our sample is partisan, we add the probability that she self-identifies with the Democratic Party to the probability that she self-identifies with the Republican Party: $0.44 + 0.32 = 0.76$.

That is simple enough, but it is only one value, and the CDF is a function, so we need to specify the value of the CDF at all values the variable might take. For discrete variables this is straightforward (if tedious for variables with a large number of values). In the present example we need to know the probability that a randomly drawn respondent has a value less than or equal to 1 (0.44), the probability that the value is less than or equal to 2 (0.76), and the probability that it is less than or equal to 3 (1.0). The function that traces the graph of these values is the CDF for our example, as depicted in Figure 10.3.

10.5.1 Why Should I Care?

The CDF is widely used in the construction of hypothesis tests in inferential statistics. For example, we often want to know whether a given value is likely to

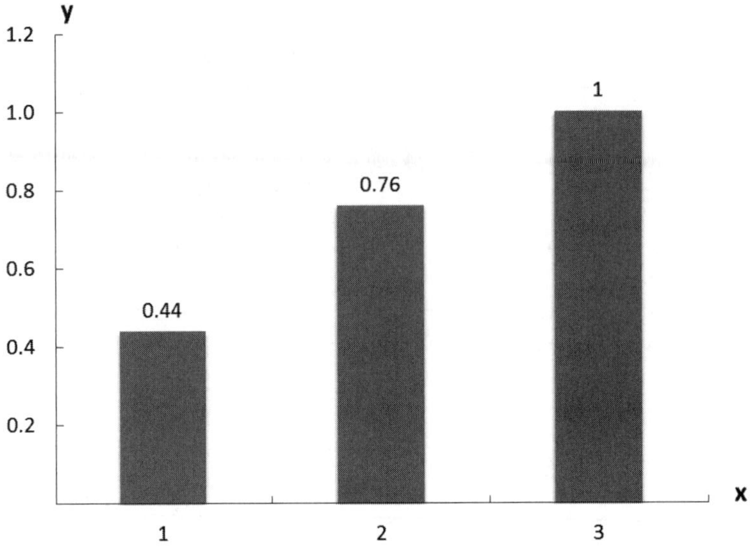

Figure 10.3: CDF of Party ID

have been drawn from a specific portion of a distribution (i.e., the chance that we would observe an outcome greater or less than some specific value). The CDF helps us answer that question. Pragmatically, you will find a nontrivial portion of the material in your introductory statistics course considerably easier to master if you have a working familiarity with the CDF.

In addition, the CDF is also used in some areas of game theory. Often one is interested in whether or not one thing that exhibits an element of randomness exceeds another—for example, whether the outcome from a policy is greater than a cutoff, or, in a Bayesian game, whether the type of opposing player has preferences sufficiently aligned so as to make a bargain possible. In cases like these, we need to know the probability that the random variable corresponding to outcome or type exceeds some value, which is just one minus the CDF of the distribution function at that value, as noted above.

10.6 PROBABILITY DISTRIBUTIONS AND STATISTICAL MODELING

Hendry (1995, p. 21) explains that to begin statistical modeling, "we first conjecture a potentially relevant probability process as the mechanism that generated the data" we are studying. Put differently, one begins by specifying a probability function that one suspects generated the data one is trying to explain. Thus, one reason we are interested in probability distributions is that they are formal statements of our conjecture about the **data-generating process** (DGP).

As an example, we might be interested in studying the probability that an American voter voted for the Democratic or Republican Party, the duration of a government in office in a parliamentary democracy, or the number of wars in which a country participated. Following King (1989, p. 38), the DGP is a formal statement of our beliefs about the probability process that produced the party the voter voted for, the length of time the government held office, or the number of wars in which a country was involved. We can use the rules of probability to specify a PMF or CDF of a given DGP.

Probability distributions are of interest to political scientists in large part because an adequate understanding of them makes it possible to use statistical inference to test the hypotheses implied by theories of politics. Let us be more specific. A person with no understanding whatsoever of distributions can read about (or develop) a theory of politics, identify one or more hypotheses implied by the theory, assemble a relevant dataset, read the data into a statistical software package (or even spreadsheet software), click some buttons, evaluate the output, and thereby invoke statistical inference as a test of the hypothesis that some X has an association with some Y. However, there is a wide (and growing) variety of statistical models available to test a given hypothesis, and the vast majority of them are *not* appropriate for any given hypothesis. The appropriateness of a given statistical model for a given hypothesis depends in large part (though not exclusively) on the distributional assumptions of the statistical model and the distribution of the dependent variable that measures the concept that is hypothesized to be caused by various factors. In other words, if one wants to draw valid inferences (and there is little reason to be interested in drawing an invalid inference), then one must match the distributional assumptions of the statistical model to the distribution of one's dependent variable. To do that, one must have a working knowledge of probability distributions. As noted above, PMFs identify different probability distributions of discrete variables (we discuss the distributions of continuous variables in the following chapter).

We present in this section the distributions most commonly used by political scientists by first writing an equation that identifies the PMF. We then describe the types of processes or events that most often produce variables with that distribution, and provide examples. It is important to understand that these are theoretical distributions or—if you will—the distributions of populations. Few samples of data will fit these distributions perfectly: a sample is a specific realization of a population, and any given sample will likely differ from its population. In your statistics courses you will learn more about the difference between population and sample distributions, including some formal tests one can invoke to determine the probability that a given sample was drawn from a specific distribution.

10.6.1 The Bernoulli Distribution

The first PMF we will consider applies to binary variables only and can be written as

$$Pr(Y = y|p) = \begin{cases} 1 - p & \text{for } y = 0, \\ p & \text{for } y = 1. \end{cases} \tag{10.3}$$

Equation (10.3) states that the probability that $Y = 0$ is $1-p$ and the probability that $Y = 1$ is p, where $0 \leq p \leq 1$ (or $p \in [0, 1]$). Put differently, this says that if the probability that $Y = 1$ is 0.4, then the probability that $Y = 0$ is $1 - 0.4$, or 0.6.

We can also write the PMF for the Bernoulli distribution as

$$Pr(Y = y|p) = p^y(1 - p)^{1-y}, \tag{10.4}$$

where $y = 0$ or $y = 1$. If we solve equation (10.4) for $y = 0$ and $y = 1$, we get the information provided in equation (10.3): $Pr(Y = 0) = p^0(1-p)^{1-0} = 1-p$, and $Pr(Y = 1) = p^1(1 - p)^{1-1} = p$.[31]

The Bernoulli distribution is the building block for other discrete distributions (e.g., the binomial and negative binomial), and we will use the representation in equation (10.4) when we introduce other distributions.

The Bernoulli distribution describes randomly produced binary variables and is generally introduced using the example of flipping coins. But we can also think of political science events. For instance, we might use the Bernoulli distribution to model the expected frequency of valid versus spoilt ballots in an Australian national election (e.g., Mackerras and McAllister, 1999). Voting is compulsory in Australia, so those who wish to protest often mangle or otherwise spoil their ballot rather than cast a valid one.[32] Why might the Bernoulli distribution be useful for describing this process? The Bernoulli distribution describes the frequency of two outcomes over repeated observations. Each voter is an observation in this example. If the process that determines whether a given voter casts a valid versus a spoilt ballot is random (i.e., is not deterministic), then we can use the Bernoulli distribution as long as we are willing to assume that one voter's decision does not influence another's.

That is because the Bernoulli distribution is built on an assumption that the individual events are *independent* of one another (e.g., the outcome of one flip of a fair coin does not influence the outcome of the subsequent flip of a fair coin). So we need to assume that the probability that a given eligible voter casts a ballot in an election is independent of other eligible voters' decisions to cast a ballot. The assumption of independence often (but not always!) underlies a given distribution, and political scientists have to make judgments about whether they can assume independence. While it is certainly likely that a given

[31] Recall that any number to the zero power is equal to one: $p^0 = 1$.

[32] To be sure, some invalid ballots are due to error rather than intention, but the percentage of invalid ballots in compulsory voting systems considerably exceeds that in voluntary systems, and it is widely understood among the Australian electorate that a spoilt ballot is a protest vote for "none of the above."

eligible voter's decision to cast a ballot is influenced by one (or a handful of) other voter(s), we might think it less likely that it is influenced by many of the other voters. If we draw a random sample (i.e., choose the survey respondents at random), then the probability that any given respondent's decision to vote was influenced by another respondent's decision is effectively zero.[33]

We can use the Bernoulli distribution to describe the relative frequency distribution of the outcomes over "not vote, vote" in an Australian election. If, for example, 96% of the electorate submitted valid ballots, then the relative frequency distribution would look like Figure 10.4.

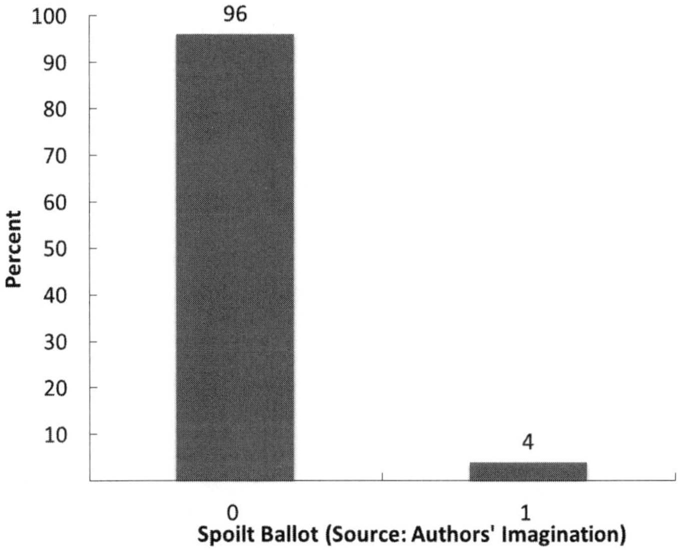

Figure 10.4: Bernoulli Distribution, $p = 0.04$

The Bernoulli distribution provides an important foundation for building more complex distributions, as we show below. It is useful for both statistical and theoretical models where one is interested in sequences of independent binary choices.

A more detailed overview of the Bernoulli distribution can be found online at `http://mathworld.wolfram.com/BernoulliDistribution.html`.

10.6.2 The Binomial Distribution

The PMF for the binomial distribution is defined by the equation:

$$Pr(Y = y|n, p) = \binom{n}{y} p^y (1 - p)^{n-y} \tag{10.5}$$

[33] If this point does not make sense, please review a discussion of random sampling in a good research design text.

Table 10.8: Unanimous Court Decisions

Case 1	Case 2	Case 3	No. of Unanimous Cases
D	D	D	0
U	D	D	1
D	U	D	1
D	D	U	1
U	U	D	2
U	D	U	2
D	U	U	2
U	U	U	3

where $n \geq y$, n, and y are positive integers and $0 \leq p \leq 1$.[34] The variables n and y in equation (10.5) represent the number of cases (or observations) and the number of positive outcomes, respectively. We recognize that this is the sort of equation that makes many political scientists blanch and ask why they are messing with such exotica! So let's work it out via a concrete example, after first describing the assumptions that underlie it. You can also flip back to Figure 10.1 to recall what an example of the binomial distribution looks like.

The binomial distribution can describe any discrete distribution with three or more observations where (1) each observation is composed of a binary outcome, (2) the observations are independent, and (3) we have a record of the number of times one value was obtained (e.g., the sum of positive outcomes). As an example, a data source might record the number of unanimous votes by a court (e.g., Epstein, Segal, and Spaeth, 2001) but not provide us with the individual vote breakdown for each case. If we assume that the justices' votes are independent across cases, then the binomial distribution should be useful for describing the DGP.

To keep the example simple, we will assume that the court rules on only three cases per term. This is not terribly realistic, but one could extend it to twenty-five, thirty, or however many cases there actually are per term. Limiting the example to three keeps things tractable.

The first thing to do is to list the possible outcomes as ordered sets and count them. Since there are two possibilities (divided decision, unanimous decision) and three cases, there are $2^3 = 8$ possibilities, as listed in Table 10.8.

To develop the binomial distribution, we start with the Bernoulli distribution, which says that $Pr(Y = 1) = p$ and $Pr(Y = 0) = 1 - p$ (see equation (10.3)). We will assign a unanimous case (U) the value 1 and a divided case (D) the value 0. Since we have assumed that the three cases are independent, the probability that there are zero unanimous (i.e., three divided) cases is the product of the

[34] Recall from the previous chapter that $\binom{n}{y} = \left(\frac{n!}{y!(n-y)!}\right)$ is shorthand for choosing y from n, i.e., the number of combinations that involve choosing y elements of some type from n total elements.

marginal probabilities that each case is divided, or $Pr(Y = 0, 0, 0)$: $(1-p) \times (1-p) \times (1-p) = (1-p)^3$. This matches equation (10.5) when $n = 3$ and $y = 0$.

The probability that there is only one unanimous case is the sum of the products of the marginal probabilities over the three ordered sets that might produce that outcome. That's a mouthful, so let's break it down. In Table 10.8, D represents a divided decision and U represents a unanimous decision. The table indicates that there are three different ways we might end up with one unanimous decision. So we will need to sum the probabilities over those three ways. What is the probability that we will observe only one unanimous case in each manner it can be achieved? Again, we take the product of the marginal probabilities. In the first row in Table 10.8 with only one unanimous case, that product is $p \times (1-p) \times (1-p)$. In the second row with only one unanimous case, the joint probability is $(1-p) \times p \times (1-p)$. Finally, the third row produces the joint probability $(1-p) \times (1-p) \times p$. When we add those three together we get $3p(1-p)^2$. This matches equation (10.5) when $n = 3$ and $y = 1$.

We determine the other outcomes the same way: we take the sum of the joint probabilities, each of which is the product of marginal probabilities. Table 10.8 indicates that there are again three ordered sets that yield two unanimous decisions. Thus, the probability that we observe two unanimous decisions is the sum of the joint probabilities of each of those combinations: $p \times p \times (1-p)$ plus $p \times (1-p) \times p$ plus $(1-p) \times p \times p$, or $3p^2(1-p)$. This matches equation (10.5) when $n = 3$ and $y = 2$.

Finally, there is only one ordered set that produces the outcome of three unanimous decisions. So the probability that there are three unanimous decisions is the joint probability $p \times p \times p = p^3$. This matches equation (10.5) when $n = 3$ and $y = 3$.

Equation (10.5) is simply a general representation of the sum of the joint probabilities that we discussed in the preceding paragraphs as individual equations. To get a graphical sense of what the binomial distribution looks like, please point your browser to Balasubramanian Narasimhan's "Binomial Probabilities" applet, available at `http://www-stat.stanford.edu/~naras/jsm/example5.html`.

There are some statistical routines that rely on the binomial distribution (e.g., `bitest` in Stata), and the binomial distribution can be assumed in generalized linear regression models. Though these tests are common in other fields, they are not used widely in political science.

Some readers might be interested in a more detailed presentation of the binomial distribution. Gill (2006, sections 1.4.3, 6.2, 7.1.3, 7.1.4), Lindsey (1995, pp. 13–14, 99–201), and King (1989, pp. 43–45) are great places to start. A thorough technical overview is available online at `http://mathworld.wolfram.com/BinomialDistribution.html`.

10.6.3 The Multinomial Distribution

The multinomial distribution is an extension of the binomial distribution to cases where more than two mutually exclusive (and collectively exhaustive) outcomes can occur. Whereas the binomial distribution describes the number of times $Y = 1$, where Y is a random variable described by a Bernoulli distribution, the multinomial distribution counts the number of times each one of k different outcomes happens, where each outcome happens with probability p_i, $i \in \{1, \ldots, k\}$. Since the outcomes are mutually exclusive and collectively exhaustive, all these probabilities sum to one. Let Y_i represent a random variable that counts the number of times outcome i occurs. If there are n independent events, then $Y_i \in \{0, 1, 2, \ldots, n\}$ for all i, and $\sum_{i=1}^{k} Y_i = n$. In this case we can write the multinomial PMF for non-negative integers $y_1, \ldots, y_k$ as

$$Pr((Y_1 = y_1) \cap \ldots \cap (Y_k = y_k)) = \begin{cases} \frac{n!}{y_1! \ldots y_k!} \prod_{i=1}^{k} p_i^{y_i} & \text{when} \sum_{i=1}^{k} y_i = n, \\ 0 & \text{otherwise.} \end{cases} \tag{10.6}$$

Though the multinomial distribution is not often invoked in applied statistical work in political science, it can be invoked as one of many possible distributions when using what is called the generalized linear model (GLM) to estimate a regression equation (i.e., a statistical model you will learn about). The GLM has not yet become popular in political science, but see Gill (2001) for an introduction by a political scientist.

Readers interested in more thorough and technical discussions should examine the MathWorld entry at `http://mathworld.wolfram.com/MultinomialDistribution.html` or Zelterman (2004, pp. 8–9).

10.6.4 Event Count Distributions

Many variables that political scientists have created are integer counts of events: the number of bills passed by a legislature, the number of wars in which a country has participated, the number of executive vetoes, etc. Event counts frequently exhibit frequency distributions consistent with those produced by a handful of well-known probability distributions.

10.6.4.1 The Poisson Distribution

The Poisson distribution is named after the French mathematician Siméon Denis Poisson. Its PMF can be written as

$$Pr(Y = y|\mu) = \frac{\mu^y}{y! \times e^{\mu}}, \tag{10.7}$$

where $\mu > 0$ is the expected number of events, y is a positive integer representing the number of events observed, and the variance, σ^2, is equal to the mean, μ.[35]

[35] You will also see this equation written as $Pr(Y = y|\mu) = e^{-\mu} \frac{\mu^y}{y!}$. Recall that $e^{-\mu} = \frac{1}{e^{\mu}}$.

The Poisson has a location parameter (μ) but it does not have a separate scale parameter.

The graph of the Poisson distribution, displayed in Figure 10.2 above, reveals an asymmetry: these distributions tend to have a long right tail. Note, however, that as the mean of the distribution rises, the asymmetry of the distribution declines.

Whence comes equation (10.7)? The goal is to produce a PMF that describes the number of times one observes zero events, one event, two events, 3 events, etc., over a fixed period of time (e.g., wars per century). The Poisson distribution describes event counts produced by a process that meets three criteria: integer count, independence, and a known mean. We discuss each in turn.

First, the individual events must be countable as whole numbers given a period of time, and it cannot be possible to count the non-events. The inability to count non-events may seem odd, but this is actually quite common. For example, we might want to observe the number of wars countries entered into during the twentieth century. We can easily count this using whole numbers. Note, however, that it is nonsensical to count the number of non-wars into which countries entered during the twentieth century.[36] Recall that the Bernoulli and binomial distributions involve events with binary countable outcomes: we can count the events *and* the non-events. When we can only count the events, and not the non-events, the Poisson distribution might be useful.

Second, the events must be produced independently from one another over the period of time one is counting them. Consequently, the probability that the count is, say, two, is computed independently of the probability that the count is, say, five.[37] Third, the average frequency of events in a given period (μ in equation (10.7)) must be known. When used in statistical analyses one can determine μ from one's data, but this requirement explains why we use the notation $Pr(Y = y|\mu)$ in equation (10.7).

A classic example of an event count generated by a Poisson process is the number of traffic accidents at a given intersection over time (e.g., the number of accidents per quarter year). Five years of quarterly data on the number of accidents at a given intersection will often prove to be Poisson distributed. Yet a large number of accidents in any three month period (say four or five) could lead people to conclude that the intersection is dangerous—which is to say that the accidents are *not* independent. The Poisson distribution—which assumes independence of events—shows that even when we assume that events

[36]Though one can, and scholars do, count the number of years that contain no wars between the countries in a particular pair of countries (a dyad).

[37]This implies that one adds all the probabilities of each count's occurring to get the probability that some number occurs (i.e., the probability that some count in the sample space occurs, which is 1). In fewer words, if S is the event that greater than or equal to 0 events occur, $1 = Pr(S) = \frac{\mu^0}{0!e^{\mu}} + \frac{\mu^1}{1!e^{\mu}} + \frac{\mu^2}{2!e^{\mu}} + \ldots = e^{-\mu}\sum_{i=0}^{\infty}\frac{\mu^i}{i!}$. Since the sum is the definition of (and Taylor series for) e^{μ}, we see that the RHS of this does equal 1. This is actually why the e^{μ} is present in the PMF: it is a normalization factor, to ensure that when one sums over all possible outcomes (i.e., all possible counts), one gets 1, as one must for a PMF.

are independent of one another we will still randomly get clusters of relatively large numbers of events. Such clustering of events will be unusual (i.e., have a low frequency), but we should be reluctant to accept a single large cluster as sufficient evidence to infer that the events were not independent. Thus, if one conducts a data analysis and finds that the data fit the Poisson distribution, one can conclude that the accidents were likely produced randomly. If they are not Poisson distributed, then perhaps the light at the intersection or the speed limit needs to be evaluated to determine what systematic factor is producing the accidents.

That said, many integer counts of events that interest political scientists are expected to be related to one another by theory. For example, it seems unlikely that bills passed in a legislature, unanimous court decisions, wars, or executive vetoes are independent of one another. And if we assume that the presence of one event either raises or lowers the probability of another event in a given period of time, then a variable measuring that event type would not be produced by a Poisson process.

For a lucid and detailed discussion of the Poisson distribution, visit Bruce Brooks's entry at his "Acquiring Statistics" site: `http://www.umass.edu/wsp/statistics/lessons/poisson/`.

10.6.4.2 The Negative Binomial Distribution

The Poisson distribution describes the distribution of event counts for rare random events. The negative binomial, on the other hand, provides one with the expected event count prior to the occurrence of a set number of non-events. Because it is built on the binomial distribution, the DGP is one where events have binary countable outcomes (i.e., once we know how many non-events occurred, we can determine the number of events by subtracting the number of non-events from the total number of trials).

The PMF for the negative binomial distribution can be written as

$$Pr(Y = y|r, p) = \binom{y + r - 1}{y} p^y (1 - p)^r, \tag{10.8}$$

where y is the number of observed events (typically called "successes"; e.g., presidential vetoes), r is the number of observed non-events (typically called "failures"; e.g., presidential signatures on bills) over $y + r$ opportunities (or Bernoulli trials), and p is the probability of any particular event ("success"; e.g., veto). The distribution describes the number of events (successes, vetoes), y, prior to observing the rth non-event (failures, signed bills).[38] We should note that what one calls an event (success) or non-event (failure) is arbitrary, and one can frame this distribution as describing the number of successes (vetoes) before a set number of failures (signed bills), as we have done, or the number

[38] Zelterman (2004, pp. 13–14) provides a proof that this PMF sums to 1.

of failures (vetoes) before a set number of successes (signed bills). To switch to the alternative formulation, swap p and $1-p$ in equation (10.8).

The combination in the PMF, $\binom{y+r-1}{y}$, arises because the negative binomial distribution represents the probability of observing r observations of one outcome (call it "signs the bill") and y observations of the alternative outcome (call it "veto") in $y+r$ observations, given that the $(y+r)$th observation has the value "signs the bill." That is a mouthful, so let's break it down.

The negative binomial distribution is built from the binomial distribution, which was built on the Bernoulli distribution. As you know, the Bernoulli distribution concerns the probability of the outcomes for a binary variable, and in our example the binomial distribution describes the number of "veto" outcomes in a series of independent Bernoulli trials. The negative binomial describes a variable that counts the number of vetoes prior to the rth signed bill, which could be interpreted as the number of successes before the rth failure or the number of failures before the rth success. Lethen[39] offers the following succinct description, which employs the second of these interpretations:

> The negative binomial distribution is used when the number of successes is fixed and we're interested in the number of failures before reaching the fixed number of successes. An experiment which follows a negative binomial distribution will satisfy the following requirements:
>
> 1. The experiment consists of a sequence of independent trials.
> 2. Each trial has two possible outcomes, S or F.
> 3. The probability of success, $\Pi = P(S)$, is constant from one trial to another.
> 4. The experiment continues until a total of r successes are observed, where r is fixed in advance.

When would a political scientist suspect that a variable she is studying was produced by a negative binomial DGP? One possibility is the veto example considered above. Another possibility is a study of international conflict focused on the decision to use force in the presence of international disputes. Students of international politics often study event counts of international uses of force. Imagine that we know that the incidence of uses of force over the past two centuries is .01 (i.e., the probability that any given country uses force in any given year is .01). We can now use equation (10.8) to calculate the PMF for various counts of the use of force. That is, once we select a year in which to begin our observations we can use it to determine the probability that a given country will use force for the first, second, third, etc. time, in the first, second, third, etc. year of observation.

[39] "The Negative Binomial Distribution," available online at `http://stat.tamu.edu/stat30x/notes/node69.html`.

To be concrete, let's calculate the probability that the second use of force occurs in the sixth year, so that there are four years without force. If we call war a failure and peace a success, then equation (10.8) states

$$\begin{aligned} P(Y=4|2,0.01) &= \binom{4+2-1}{4} 0.01^2(1-0.01)^4 \\ &= \binom{5}{4} \times .0001 \times 0.9606 \\ &= \frac{5!}{(4!)(1!)} \times 0.00009606 \\ &= 5 \times 0.00009606 \\ &= 0.00048. \end{aligned}$$

We could perform the same calculations for the probability that the third use of force occurs in the seventh year, etc., but that would get tedious very quickly. And since we have the PMF defined, there is no need to do such calculations as we can instruct a computer to do them if we ever need to calculate several.

The PMF for the negative binomial distribution looks similar to the PMF for the Poisson distribution, as we see in Figure 10.5.

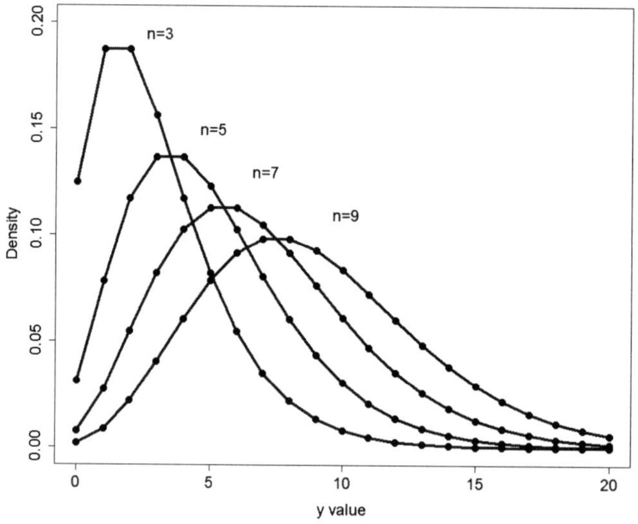

Figure 10.5: PMF of Negative Binomial Distribution, $n = 3, 5, 7, 9; p = 0.5$

One of the key features of the negative binomial distribution relative to the Poisson distribution is that the mean and variance are not constrained to equal one another. The variance is greater than the mean for the negative binomial distribution, and because the negative binomial distribution has two independent parameters, one can set both the mean and the variance separately.

King (1989) brought the negative binomial distribution to the widespread attention of political scientists, and it has been primarily used as a model of event counts when the mean and the variance of the sample data are not equal.[40] In this sense, Poisson regression models tend to be viewed as special cases of negative binomial statistical models.[41]

For a thorough technical overview of the negative binomial distribution, see `http://mathworld.wolfram.com/NegativeBinomialDistribution.html`.

10.6.5 Why Should I Care?

Political scientists are often interested in concepts that can be represented as binary outcomes, ordinal scores, or event counts. Even if one does not intend to use statistics to test hypotheses it is still useful to have an understanding of the difference between these types of distributions. In other words, thinking about distributions leads one to invest in theoretical speculation about what might lead a concept or variable to hold different values in different cases. Another way of saying the same thing is that theory building for the purpose of explaining why different outcomes occur in different cases is equivalent to speculating about a DGP. Further, if one does plan to use statistical hypothesis testing in one's empirical work, then knowledge of discrete distributions and their DGPs is critically important.

10.7 EXPECTATIONS OF RANDOM VARIABLES

We opened this chapter by discussing what a random variable is, but thus far we have mostly discussed how these variables are distributed. This, as we hope we have made clear, is undoubtedly important, but there are still many occasions when one desires more specific knowledge regarding a random variable. For instance, its expected value, its variation around its mean, and one's expected utility when it is a function of the variable are all useful to political scientists. To obtain this knowledge, we must deal with the expectation of the random variable.

The **expectation** of a random variable X, denoted $E_X[X]$ or simply $E[X]$ when no confusion (in the presence of more than one variable) is possible, is the weighted average value that the random variable can take, where the weights are given by the probability distribution.

Let's consider a common example one encounters in game theory and expected utility theory. As noted in the previous chapter, game theorists denote a

---

[40] Or when a dispersion parameter in Poisson regression models suggests that it is unlikely that the dependent variable was drawn from a Poisson distribution with equal mean and variance.

[41] The log-gamma distribution is an alternative to the negative binomial that is used widely in other fields.

lottery any outcome that is uncertain, including a lottery of the kind US states advertise, such as Powerball. A lottery consists of a set of values of outcomes and a corresponding set of probabilities that each outcome might occur. In other words, it is a probability distribution over values of outcomes, and the outcome of the lottery is a random variable.

We compute that variable's expectation by weighting (multiplying) each value by the chance that it occurs, and summing over all values. So, if the lottery has potential outcomes \$0, \$1,000, and \$1,000,000, and these occur with probabilities 0.9998999, 0.0001, and 0.0000001 respectively, then the expectation of the lottery's outcome is $(0.9998999 \cdot \$0) + (0.0001 \cdot \$1,000) + (0.0000001 \cdot \$1,000,000) = \$0 + \$0.1 + \$0.1 = \$0.20$, or twenty cents.

This is known as the **expected value** of the lottery. In general, if a discrete random variable X takes on values x_i, then the expected value is calculated for X according to the formula[42]

$$E_X[X] = \sum_i x_i(Pr(X = x_i)). \tag{10.9}$$

Note that the complex part of equation (10.9) is the $Pr(X = x_i)$ term. For the example we just did that term was provided for each value x_i. Let's try a slightly more complicated example before moving on, one in which the probabilities are dependent on, and specified for, the values. To do this, we'll use the Poisson distribution we introduced in the previous section.

Imagine that you are interested not in the distribution of event counts but rather in how many events one should expect to see. Recalling that $Pr(X = x_i|\mu) = \frac{\mu^{x_i}}{x_i! \cdot e^{\mu}}$, we can compute this via equation (10.9)

$$E_X[X] = \sum_i i(Pr(X = i)) = \sum_{i=0}^{\infty} i \frac{\mu^i}{i! \times e^{\mu}} = e^{-\mu} \sum_{i=0}^{\infty} \frac{\mu^i i}{i!}.$$

This doesn't give us an answer yet, but we can get there by expanding out the sum

$$\sum_{i=0}^{\infty} \frac{\mu^i i}{i!} = 0 + \sum_{i=1}^{\infty} \frac{\mu^i i}{i!} = \sum_{i=1}^{\infty} \mu \frac{\mu^{i-1}}{(i-1)!} = \mu \left(\sum_{i=0}^{\infty} \mu \frac{\mu^i}{i!} \right) = \mu e^{\mu}.$$

In the first step we pulled the $i = 0$ term in the sum out, which is zero. In the second step we divided top and bottom by i, recalling that $\frac{i!}{i} = (i-1)!$, and pulled a μ out, recalling that $\mu^i = \mu \times \mu^{i-1}$. In the third step we noted that the sum from one to infinity of $i - 1$ is the same as the sum from zero to infinity of i, since the first index of both is zero and they both go on forever. Finally, in the fourth step we used the definition of the exponential function. Plugging this back into the sum in the equation for the expected value produces our answer,

$$E_X[X] = e^{-\mu} \mu e^{\mu} = \mu.$$

[42] A very similar equation is true for continuous random variables, and we provide it in the next chapter. Basically, one replaces the sum with an integral and the PMF with a PDF.

Thus the expected value of the event count—which is also the mean—is equal to the parameter μ. Though we technically knew that, it's nice to be able to derive it ourselves, right? More important, this technique can be used for other PMFs, as well to find expected values of other distributions.

10.7.1 Expected Utility

Expected values are useful, but they are limited in that they consider only the (weighted) average value of the variable itself, and not the average value of more complex functions of that variable, which are what we typically care about in political science. In statistics, expectations of functions of the random variable allow the computation of moments of the distribution, which we consider in the next subsection. In this subsection we turn to the expectation of utility functions, or expected utility for short. **Expected utility**, typically denoted $EU(x)$, is much like expected value, except that rather than specifying a weighted average of the variable it specifies a weighted payoff, under a few assumptions on the utility function about which you will learn in your game theory class.[43] Its expression even looks much the same as that for the expected value:

$$EU(X) = \sum_i u(x_i)(Pr(X = x_i)). \tag{10.10}$$

In equation (10.10), the small u (a Bernoulli utility function) gives the payoffs for the *known* values that the random variable can take, while the $EU(x)$ (von Neumann–Morgenstern utility) provides the weighted average utility one can expect to get, given the probability distribution of the values of the random variable.

Let's start with a concrete example that has the character of the example of the lottery above. This relates to the game matching pennies we introduced in the previous chapter. We'll vary it a little and insert some payoff values to make the calculation clearer. Let two people each toss a penny. If the pennies turn up the same (both heads or both tails), then player 1 keeps player 2's penny. If they turn up mixed (one head and one tail), then player 2 keeps player 1's penny. The payoff (or utility) each player receives from a round of play is 1 cent if she wins and -1 cent if she loses.[44] The difficulty is that we do not know whether she will win or lose. That is, we are uncertain about the outcome. Of course, probabilities help us analyze uncertain situations, and an expected utility calculation is nothing more than a means of determining what utility a person should expect to receive in an uncertain situation.

[43] For a brief introduction see Shepsle and Bonchek (1997, pp. 15–35)

[44] Note that this game has the character of a lottery rather than of a strategic interaction. When matching pennies is introduced in game theory it typically involves the decision to play heads or tails, and is an example of a game in which there are no pure strategy equilibria (e.g., Osborne, 2004, pp. 19–20). That is to say, the optimal strategy is to play each of heads and tails half the time, using what is known as a mixed strategy, as noted in the previous chapter. The optimal strategy produces the lottery we analyze here.

We can use equation (10.10) to compute the expected utility for this game for each player. This is

$$
\begin{aligned}
EU(MP_{1,2}) &= (p_{HH} \times u_{1,2}(HH)) + (p_{HT} \times u_{1,2}(HT)) \\
&+ (p_{TH} \times u_{1,2}(TH)) + (p_{TT} \times u_{1,2}(TT)), \qquad (10.11)
\end{aligned}
$$

where MP is the matching pennies game (or lottery); the subscripts 1 and 2 indicate the player; p denotes the probabilities of each joint outcome; H indicates a coin landing heads and T indicates a coin landing tails; and u indicates the utility (or payoff) associated with an outcome.

> One reads equation (10.11) as follows: *the expected utility of playing matching pennies for players 1 and 2 is the probability of heads-heads times the utility of heads-heads plus the probability of heads-tails times the utility of heads-tails plus the probability of tails-heads times the utility of tails-heads plus the probability of tails-tails times the utility of tails-tails.*

We can replace the variables with values and calculate the expected utility of this game (really lottery) for each player. We identified the utilities (or payoffs) to each player above (player 1: HH or TT is $+1$, HT or TH is -1; player 2: HH or TT is -1, HT or TH is $+1$), but where do the probabilities come from? The sample space has four outcomes that are equally likely: HH, HT, TH, or TT. Therefore, the probability of each outcome is $\frac{1}{4}$ or 0.25. Because the players have different payoffs we must calculate two expected utility equations, one for each player

$$
\begin{aligned}
EU(MP_1) &= 0.25(1) + (0.25)(-1) + (0.25)(-1) + 0.25(1) \\
&= 0.25 - 0.25 - 0.25 + 0.25 \\
&= 0. \\
EU(MP_2) &= 0.25(-1) + (0.25)(1) + (0.25)(1) + 0.25(-1) \\
&= -0.25 + 0.25 + 0.25 - 0.25 \\
&= 0.
\end{aligned}
$$

This demonstrates that the expected utility of playing this game is zero for each player. Perhaps that explains why this is not a very popular gambling game.

You may wonder what the point was of presenting this game, given that the utilities were each nothing more than values of the lottery, implying that an expected value computation would be entirely appropriate. Though true in this case, it needn't be: one could assume that both players were risk averse in the realm of gains but risk seeking in the realm of losses, as in prospect theory (Kahneman and Tversky, 1979). In this case, we might assign $u_1(TT) = u_1(HH) = u_2(HT) = u_2(TH) = 2$, and $u_2(TT) = u_2(HH) = u_1(HT) = u_1(TH) = -4$, so that winning is not as good for either player as losing is bad.

One can still use equation (10.10) even though the utilities do not equal the values of the lottery, and verify that $EU(MP_1) = EU(MP_2) = -1$ in this case. Risk-averse players not only get no benefit from playing the game, assuming their utility from doing nothing is zero, but actively prefer not to play the game at all.

We discuss risk preferences a bit more at the end of this section and consider more complex expected utility computations in the next chapter, but first we illustrate further with a more complex and more interesting example, taken from Stokes (2001), which we discussed in the context of Bayes' rule in the previous chapter.

Stokes asks us to consider a voter who places himself on the left side of a left-right ideological continuum (pp. 16–17). The election offers four candidates, none an incumbent, who are vying for the candidacy of two parties. The voter has beliefs about where on the ideological scale both parties sit and can thus identify the party whose policies are closest to (and farthest from) his own. However, he also believes that there are two types of politicians: *ideologues*, who will pursue the policies they campaign on, and *power seekers*, who will lie during the campaign when they know their preferred policy is unpopular, and then switch once in office. The voter's problem is trying to determine how to vote given that though he is confident about the policy the candidates for each party should adopt, he is uncertain whether the candidate for each party is a *power seeker* type or an *ideologue*. One can represent Stokes's voter's decision using the following expected utility equation:

$$EU(v_i) = (p_{iL} \cdot u(L)) + ((1 - p_{iL}) \cdot u(R)) \tag{10.12}$$

where v_i represents a vote for candidate i, p_{iL} represents the probability of a government under i enacting a set of leftist policies and $1 - p_{iL}$ a set of rightist policies,[45] respectively, u represents utility associated with a policy outcome, and L and R represent the leftist and rightist set of policies, respectively. Since i represents the candidate who wins and since several candidates are competing, i is drawn from the set of all candidates.

> A conventional way to read equation (10.12) is: *the expected utility of voting for candidate i is equal to the product of the probability that candidate i adopts leftist policies and the utility derived from leftist policies plus the product of the probability that i does not adopt leftist policies and the utility derived from rightist policies.*

Stokes specifies values for the variables in the equation, thus making it possible to perform calculations and compare the candidates. For her left-leaning voter she assumes that the value of leftist policies is 10 and that the value of rightist policies is −10. If the politician is an ideologue then she will remain faithful to her announced platform with a probability of 1.[46] However, if the

[45] Note that $p_{iR} = 1 - p_{iL}$. One could rewrite the equation using p_{iR} instead of $1 - p_{iL}$.

[46] Since probabilities must sum to 1 and the politician will either remain faithful or switch, the probability that an ideologue switches is $1 - 1 = 0$.

politician is a power seeker, then he will switch policies after the election with probability 0.3.[47]

We can now consider different scenarios. Let us simplify and assume that there are only two candidates, l and r, standing for the left and right party, respectively. Assume further that our voter believes that both candidates are ideologues. To calculate the expected utility for voting for each candidate, place the relevant values from the paragraph above into the equation. In this case, i can take two values, l and r, for each of the two candidates. Because the payoffs to the voter are different for each candidate, we need to calculate the expected utility to the voter for each candidate

$$\begin{aligned} EU(v_l) &= (p_{iL} \cdot u(L)) + ((1 - p_{iL}) \cdot u(R)) \\ &= 1.0(10) + (1 - 1.0)(-10) \\ &= 10. \\ EU(v_r) &= (p_{iL} \cdot u(L)) + ((1 - p_{iL}) \cdot u(R)) \\ &= 0(10) + (1 - 0)(-10) \\ &= -10. \end{aligned}$$

Thus, under the specified assumption, $EU(v_l) > EU(v_r)$. In words, the expected utility of voting for an ideologue leftist candidate is greater than the expected utility of voting for an ideologue rightist candidate: the voter should cast a ballot for the leftist party. There is nothing surprising here.[48] Nonetheless, it illustrates how one can construct an expected utility model.

For practice, let us consider another scenario that Stokes does not evaluate. Let's assume that the voter believes that both candidates are power seekers

$$\begin{aligned} EU(v_l) &= (p_{iL} \cdot u(L)) + ((1 - p_{iL}) \cdot u(R)) \\ &= 0.7(10) + (1 - 0.7)(-10) \\ &= 7 + 0.3(-10) \\ &= 4. \\ EU(v_r) &= (p_{iL} \cdot u(L)) + ((1 - p_{iL}) \cdot u(R)) \\ &= 0.3(10) + (1 - 0.3)(-10) \\ &= 3 + 0.7(-10) \\ &= -4. \end{aligned}$$

Thus, $EU(v_l) > EU(v_r)$. In words, the expected utility of voting for a power seeking leftist is greater than the expected utility of voting for a power seeking rightist, so the voter will again vote for the left party candidate.

[47] Again, the probabilities must sum to 1 and there are only two options. Thus we can use the probability that a power seeker will switch (0.3) to determine the probability that the power seeker will remain faithful: $1 - 0.3 = 0.7$.

[48] If you are wondering why Stokes (2001) would analyze such a simple equation, it turns out that she does not. We made it up as an illustration based on her model, and explain below why she builds her model.

Since this example suggests that the voter's decision is not affected by his beliefs about whether the candidates are ideologues or power seekers, you might be wondering what use Stokes's model has. Had she constructed it for the purpose of determining vote choice it would not have been very interesting (at least not using these values).[49]

10.7.1.1 Expected Utility and Risk Preferences

When we discussed utility functions in Chapter 3 we were really discussing the little u in our expected utility equation. We talked a bit there about what different functional forms for the utility implied substantively, but didn't go into a lot of detail. We can say a little more by bringing expected utility into the mix. Recall our discussion of concave and convex functions in Chapter 8 (or flip there for a moment if you skipped that section or chapter). A function is concave if the secant line joining two points is below the curve, and convex if it is above it. Assume the curve is one's small-u utility function. Equation (10.10) is a linear combination of the utility function u evaluated at several points.

Let's consider two such points for clarity, so that our actor may realize one of two possible utility outcomes. This means that equation (10.10) specifies a point on the secant line joining these two utility outcomes. For a concave function, this secant is below the curve, implying that the expected utility for any lottery over utilities is less than the utility the actor would obtain by receiving with certainty the corresponding combination of the outcomes that produced these utilities. In other words, an actor with a concave utility function prefers the sure thing to the gamble. We call such actors **risk averse**. Conversely, should an actor have a convex utility function, then the secant is above the utility curve, and the actor prefers the gamble to the sure thing. Such actors are said to be **risk seeking**. Finally, if an actor has a linear utility function, then the secant is coincident with the utility function, and the actor is indifferent between the gamble and the sure thing. We call such actors **risk neutral**.

This is a bit complex, particularly in such a small space, but an example will help clarify. Consider the following gamble: you get 0 with probability one-half, and 4 with probability one-half. The expected value of this gamble is $\frac{1}{2}0 + \frac{1}{2}4 = 2$. We'll look at how three different types of people would treat this gamble. First, consider a risk-averse person with the concave utility $u(x) = \sqrt{x}$. Equation (10.10) states that the expected utility for this person is $\frac{1}{2}u(0) + \frac{1}{2}u(4) = \frac{1}{2}0 + \frac{1}{2}2 = 1$. If she were instead to receive with certainty the combination of outcomes that are possible in the lottery (0 and 4), weighted by the same chance that each occurs ($\frac{1}{2}$), then she would receive the expected value of the lottery, 2. Her utility for the expected value of the lottery is $u(2) = \sqrt{2}$,

[49] As an aside, a common exercise in such modeling is to set the expected utility of the options equal to each other and solve for the values of a given variable that make the actor indifferent between the choices. Among other things, this allows computation of what are known as mixed strategy equilibria. Those of you who go on to study formal models and game theory will learn how to do this.

which is *greater* than her expected utility for the lottery itself. Not only would she prefer to get the expected value of the lottery for certain, she'd actually take *less* then the expected value if she could be guaranteed that amount. This is what risk averse means: one is willing to give away potential gains or pay extra to avoid risk. As an example, Feddersen, Sened, and Wright (1990) offer a model of candidate entry that assumes risk aversion.[50]

Next consider a risk-seeking person with the convex utility $u(x) = x^2$. Equation (10.10) states that the expected utility for this person is $\frac{1}{2}u(0) + \frac{1}{2}u(4) = \frac{1}{2}0 + \frac{1}{2}16 = 8$. This is more than her valuation of the expected value of the lottery, $u(2) = 4$; she is so interested in the gamble itself that one would need to pay her to get her to accept the sure thing over the lottery, even though the sure thing here is what the lottery is expected to pay off.

Finally, consider a risk-neutral person with the linear utility $u(x) = x$. Such a person has expected utility equal to her valuation of expected value and so is indifferent between the gamble and the sure thing (e.g., Gradstein, 2006). Risk neutrality is the most common assumption seen in the game theoretic literature in political science for the simple reason that it is easier to deal with mathematically; however, risk aversion is more prevalent substantively, and many models account for this.

10.7.1.2 Why Should I Care?

Expected utility and expected value are central concepts not only in game theory but also in rational choice theory, expected utility theory, and many behavioral and boundedly rational models of politics. Even the theoretical portions of papers and books that are primarily empirical in focus will often use these concepts, and if you want to understand the theory, you will need to understand these concepts.

10.7.2 The Moments of a Distribution

If we replace $u(x)$ with more general functions, equation (10.10) can apply to the expected value of any function of the values of a random variable. One class of such expectations of particular use in statistics is the moments of a distribution. The **moments of a distribution** are an important set of parameters one can use to describe a distribution. They involve the expected values of particular functions of the random variable across the distribution, such that the kth moment of a variable X can be represented as $E[X^k]$, where $E[]$ indicates the expectation of the function of the variable inside the brackets.[51] The expected value of a variable is the sum of the possible values it might take weighted by the probability that each value will turn up, i.e., $E[X]$. The mean (or average)

---

[50] For a useful introduction to this model, see Gelbach (2013, pp. 16–20).

[51] More explicitly, the kth moment of a variable is the kth derivative of the moment-generating function evaluated at zero. See `http://mathworld.wolfram.com/Moment-GeneratingFunction.html` for more detail.

is the expected value of the variable X^1, and so it is also the first moment of a variable. The first moment is a location parameter and is also one measure of the central value of a distribution.[52]

One can define moments about zero and moments about the mean.[53] The equation for moments about zero of discrete variables is

$$\sum_i x_i^k (Pr(X = x_i)), \tag{10.13}$$

where k is the kth moment about zero. As noted, the first moment (i.e., where $k = 1$ in equation (10.13)) is the central tendency or mean. For a variable, X, that takes with equal probability the values x_i, $i = 1, 2, 3 \ldots N$,[54] the first moment of equation (10.13) is $\frac{1}{N}\sum_{i=1}^{N} x_i^1$. Note that this is the (unweighted) average of the values. Make sure it is apparent to you that when $k = 1$, equation (10.13) produces an average for variable X.

In statistical analyses we are often interested in the second, third, and fourth moments about the mean as they can provide useful information about the scale and shape of a distribution (and thus are known as scale and shape parameters). The second moment about the mean is of interest by itself, and the third and fourth moments about the mean are useful components of other indicators. Moments about the mean are defined by the equation

$$\sum_i (x_i - \mu)^k (Pr(X = x_i)). \tag{10.14}$$

The second moment about the mean (i.e., $k = 2$ in equation (10.14)) is the variance and it measures the variation of the distribution about its mean value.

Two other measures of interest are the skewness and kurtosis of a distribution. Skewness involves the third moment about the mean, and it is usually weighted by the standard deviation, though some people use the third moment without a denominator.[55] A common measure of skewness is the third moment divided

[52]You will learn in your statistics courses that there are three common measures of central tendency: the mean, median, and mode. We focus on the mean here.

[53]The equations are different for discrete and continuous variables, and we focus on discrete variables here. One takes the integral, rather than the sum, for continuous variables.

[54]This is known as a uniform distribution; it is more commonly observed in political science as a continuous distribution and so is covered in the next chapter.

[55]Some authors (e.g., Kmenta, 1986, p. 67) define skewness as the third moment alone. Further, most authors refer to *a* measure of skewness rather than *the* measure. The skewness entry at the MathWorld website offers this observation: "Several types of skewness are defined, the terminology and notation of which are unfortunately rather confusing" (`http://mathworld.wolfram.com/Skewness.html`).

by the standard deviation cubed:[56]

$$E\left[\left(\frac{X-\mu}{\sigma}\right)^3\right] = \frac{\mu_3}{\sigma^3}, \tag{10.15}$$

where μ_3 represents the third moment around the mean and σ represents the standard deviation. Skewness measures the symmetry of the distribution about its central value. When skewness is zero, the distribution is symmetric.

Like skewness, kurtosis is used as the label for a number of specific measures.[57] Kurtosis always involves the fourth moment about the mean, and it is often weighted by the standard deviation. A common measure of kurtosis is the fourth moment about the mean divided by the fourth power of the standard deviation:

$$E\left[\left(\frac{X-\mu}{\sigma}\right)^4\right] = \frac{\mu_4}{\sigma^4}. \tag{10.16}$$

The kurtosis measures the flatness or peakedness of the distribution relative to a standard normal distribution.

10.7.2.1 Why Should I Care?

The moments of a distribution are often parameters in the functions one can use to describe them (e.g., the PMF/PDF, CDF, etc.). As such, they show up repeatedly in the study of statistics, and they are often of use when constructing formal models or otherwise trying to discipline one's thinking about a political process. You are likely quite familiar with thinking about the first and second moments of a distribution: in large undergraduate courses you probably paid close attention not only to your own score on a test but also to the average score (i.e., the first moment about zero) and, if your professor made it available, the dispersion or variance of the scores (i.e., the second moment about the mean). And regardless of the extent to which you use formal theory or statistics in your own research, you will need to be familiar with the moments of distributions to do competent grading in large lecture courses.

The third and fourth moments are frequently used to determine whether a given empirical sample deviates from a normal distribution. More generally, using the moments of a sample of data as estimates of the population moments is known as the method of moments in statistics.[58]

[56]The standardized moment is one that is divided by the standard deviation raised to the power of the moment. For example, the standardized second moment is the second moment divided by the standard deviation squared. This measure of skew, then, is the standardized moment. The standardized moments are of interest because they are the moments for a standardized normal distribution (i.e., a normal distribution with a mean of zero and a standard deviation of one). The standardized normal distribution is invoked for a number of hypothesis tests in statistics.

[57]The MathWorld entry observes: "There are several flavors of kurtosis commonly encountered, including the kurtosis proper" (`http://mathworld.wolfram.com/Kurtosis.html`).

[58]The method of moments is the most common way to teach statistics in political science.

10.8 SUMMARY

This chapter argues that political scientists need to think about the (likely) distributions of concepts when developing theories and that a working familiarity of specific distributions is important for (1) developing formal theories of politics and (2) applying statistical inference to hypothesis testing. We introduced several ways one can represent the distribution of a variable (e.g., frequency counts, relative frequency counts, and several functions) and then briefly described several commonly used distributions for discrete variables.

10.9 EXERCISES

1. Write down a research question that interests you. Try to state some assumptions, and then deduce one or more hypotheses from your assumptions. Write them down and bring them all to class.

2. How is the relative frequency distribution different from a frequency distribution?

3. Why can't one create a PDF by plotting the graph of the relative frequency distribution of each value in the sample?

4. What is the difference between a PMF and a CDF?

5. Write down an example where a contingency table would be useful for examining the joint distribution of two variables. Bring it to class.

6. Write down a political process that you think might be drawn from the following discrete distributions: Bernoulli or binomial, Poisson or negative binomial (you should have two political processes).

7. Visit the "Distributions" page of the Virtual Laboratory website at the University of Alabama, Huntsville (`http://www.math.uah.edu/stat/dist/index.xhtml`) and select the "Random Variable Experiment" link under "Applets." Go to the bottom of the Random Variable Experiment applet and select the "Applet" link. Under the label "Bernoulli Trials" you will find applets for the binomial and negative binomial distributions, and under the "Poisson Process" label you will find links to applets for the Poisson distribution (click on "Poisson Experiment"). Investigate the distributions covered in this chapter. More explicitly, select a distribution and note the scale and location of the density function. Adjust one of the parameters using the scroll bar. If there is more than one parameter, adjust it. Write down what happens when you adjust each parameter for the following distributions: Bernoulli, binomial, Poisson, negative binomial.

Wonnacott and Wonnacott (1977) is a good example, and in Chapters 18 and 19 they contrast the method of moments with two other techniques: maximum likelihood estimation and Bayesian inference.

8. Visit the Public Data site at Google (`http://www.google.com/publicdata/`). Select a dataset that is of interest to you (they have many from which to choose). Select the Explore the Data link and plot some univariate distributions (as of this writing, note the options for plotting to the upper right of the page; try different options—you won't break anything). Summarize two things of interest that you learn (or "confirm") by doing so. Now select some other variables that you believe might covary with the first one you selected, and plot some joint bivariate distributions. Again, summarize two or three things of interest that you learn from doing this.

9. If the mean number of wars is three per year, what is the probability that there will be four wars in any given year?

10. A person persuaded a friend to meet her at a concert. Her boss droned on forever at a meeting, and she is running late. To make matters worse, she accidentally dropped her cell phone down the elevator shaft and cannot recall what concert she had said they should attend. She recalls that an orchestra is playing Bach on the north side of town, but a Stravinsky concert is being performed on the west side. She prefers Bach to Stravinsky, such that seeing the former is worth ten units of utility and the latter only five. However, she prefers going with her friend to going alone such that being together yields eight units of utility and being alone yields minus two. We can depict her utility using the following matrix:

		Friend	
		Bach	Stravinsky
Woman	Bach	18	8
	Stravinsky	3	13

Now assume that she recalls wanting to select the concert her friend would prefer, but she has no idea whether her friend likes Bach better than Stravinsky, and therefore assigns a probability of 0.5 that her friend is waiting for her at the Bach concert. Calculate the expected utility of going to the Bach concert and the expected utility of going to the Stravinsky concert. Which should she choose?
Now assume that she knows her friend prefers Stravinsky to Bach and assigns the probability 0.3 that her friend is at the Bach concert. Recalculate the expected utilities of her choices. Which concert should she go to now?

11. Two nations, A and B, face off at the brink of war. A knows B is either strong or weak, and that it would win any war with certainty if B were weak, but lose with certainty if B were strong. A has a prior belief of 40% that B is strong, and observes manuevers that a strong B would do 60% of the time but a weak B would do only 30% of the time. If A gets 1 for winning, −1 for losing, and 0 for not starting a war, should A start a war

after observing the maneuvers? (*Hint:* You'll have to use material from Chapter 9 as well.)

10.10 APPENDIX

Our presentation has been relatively informal, and one can find more formal treatments in Gill (2006) and online (e.g., the various MathWorld we entries we noted throughout). Those interested in studying methods as a subfield will want a more thorough treatment. Another place to look is King (1989, chaps. 2 and 3). The National Institute of Standards and Technology *Engineering Statistics Handbook*, section 1.3.6, "Probability Distributions" (`http://www.itl.nist.gov/div898/handbook/eda/section3/eda36.htm`) is also a good source. Finally, Zelterman (2004) provides a thorough discussion of discrete distributions.

Chapter Eleven

Continuous Distributions

In the previous chapter we covered the concepts of random variables and their distributions, but used only discrete distributions in our discussion and examples. We did this to keep a chapter very important for the development of both empirical and theoretical political science free of calculus, for those readers who might want to skip over Part II of the book. However, there is little in the previous chapter specific to discrete distributions. Indeed, as we show below, replacing sums with integrals gets you much of the way toward representing the distributions of continuous random variables.

In this chapter we make this replacement, as well as discuss the few other concepts necessary to get us all the way there. Section 1 tackles this job, and presents the changes to the conceptual edifice we built in the last chapter necessary for understanding the properties of continuous random variables. We also discuss joint distributions here, both empirically and theoretically. Section 2 makes explicit the comparison between discrete and continuous random variables via more complex examples of expected utility than were presented in the previous chapter. We also introduce the uniform distribution, probably the most common one used in applied game theory and one you've undoubtedly seen before in the discrete case. We discuss the notion of stochastic dominance here as well. Finally, Section 3 presents examples of continuous density functions useful for statistical analysis.

11.1 CONTINUOUS RANDOM VARIABLES

In the preceding chapter we limited the discussion to the probability distributions of discrete concepts or variables. In this chapter the focus is continuous concepts and variables. Though this is a bit loose, the difference can be thought of as similar to countability. If you can list each value the random variable can take and assign an integer and a probability to each, then you have a discrete random variable, represented by a discrete distribution. If you can't, you may have a continuous random variable, represented by a continuous distribution.

Most of the same concepts we introduced in the previous chapter for discrete variables and their distributions apply to continuous ones as well. In particular, like the PMF of a discrete variable, the **probability density function** (PDF) of a continuous variable is related to the relative frequency distribution of that variable. More specifically, the PDF is a function that describes the smooth curve that connects the various probabilities of specific (ranges of) values for a sample.

However, there is a difference between the PMF and the PDF, and the terminology we've used hints at it. Note that we said ranges of values, rather than values, and used the word density, rather than mass, in the name. A PDF differs from a PMF in that it does *not* describe the chance that any particular value of the random variable is drawn at random from the distribution. Rather, it describes the *relative likelihood* of drawing any specific value, and the *exact probability* of drawing a value within some range. This is why it is called a density function rather than a mass function: it describes the density of the probability within some range of values that the random variable may take, rather than the explicit "mass" of probability at a particular value.

We unpack this and make it a bit more formal below, but let's first consider why this difference exists. We'll start by assuming that some random variable X can take all the values between 0 and 1, inclusive. In other (fewer) words, $x \in [0, 1]$ for all values x that X might take.[1] Since all random variables have to take some value, the probability that $x \in [0, 1]$ is 1. Now consider the range $x \in [0, 0.5]$. The chance of being here is probably less than 1, so we've reduced the probability from 1. Now shrink it to $x \in [0, 0.25]$. Again, we've likely shrunk the probability of being in that region.[2] If we keep doing that over and over again we keep shrinking the probability. And because X is continuous, there is no point at which we can stop: X is defined over $[0, 0.001], [0, 0.00001]$, and so on, forever. The probability at a point would be the probability at the limit of this shrinking process, but that's ill-defined. So we don't define it, and don't in general speak about probabilities at specific points, even though the PDF will take non-zero values at these points.

This is likely confusing, and may remain so until we look at some examples. That's fine. But it might help to think about it another way. Specifically, another way to think about this is that the PDF is a function that allows one to sum a series of probability weights to produce the likelihood of drawing a value less than some value; the CDF described in the previous chapter gives this likelihood for all points. These weights do not directly correspond to true probabilities of drawing particular points, though; if they did, we would have to sum over an infinite number of finite probabilities to get the CDF at any point (because the range of values is continuous), and that would give an infinite probability. So the weights are instead the relative likelihoods that each value will be randomly drawn from the population. We show below that the PDF makes it possible to identify different probability distributions for continuous variables, and being able to do so turns out to be very important for developing statistical models that can produce valid hypothesis tests (more on why you care below).

---

[1]Recall that we're using capital letters for the random variables and lowercase letters for the values (realizations) of the random variables.

[2]As with most things in this book, this argument is loose, but the rough idea is what's important.

11.1.1 The PDF of a Continuous Variable

Now that we've discussed the main difference between a PDF and a PMF verbally, let's formalize it. Recall that for a PMF $f(x)$ describing the probability distribution of the random variable X, $Pr(X = x) = f(x)$. The equation is similar but a bit more complex for a PDF $f(x)$. In general, the probability that X takes values in some region B is $Pr(X \in B) = \int_B f(x)dx$. This might be too general, so we consider an application in which the values of the random variable are real numbers, as they typically are in statistical (and most formal) applications in one dimension:

$$Pr(X \in [a, b]) = \int_a^b f(x)dx. \tag{11.1}$$

Recalling Chapter 7, equation (11.1) states that the probability of the variable's being in the range $[a, b]$ is given by the definite integral of the PDF from a to b. As with the PMF, $f(x) \geq 0$ for all PDFs, so this definite integral is the area under the PDF curve between a and b. You might have heard the phrase linking probability and inference to "the area under the curve" in a research methods course; this is the origin of that phrase. Unlike a PMF, though, the PDF function is not limited to taking values no greater than 1, since it does not directly describe probability, only relative likelihood. So, for a PDF, $f(x) \in [0, \infty)$. A value of 0 for the PDF at some point x still means that we can't randomly draw that x, however.

If the probability of being in some interval $[a, b]$ is the integral of the PDF over that region, then we are saying that the probability is computed by summing lots of $f(x)dx$. If we replace the dx with a Δx, we're back to the area under a rectangle, or, in other words, the area under the relative frequency histogram. This is the connection between relative frequency and the PDF. A PDF in this sense is like a smoothed-out histogram.

This covers most of the properties of a PDF of importance for our purposes. But as with a PMF, the probability that some value in the sample space is drawn must be 1, so that $Pr(-\infty < X < \infty) = \int_{-\infty}^{\infty} f(x)dx = 1$.

To this point we have described the PDF only in generic terms. We do this so that you understand where it comes from (especially its connection to the [relative] frequency distribution). In Section 3 below we identify a number of specific PDFs commonly used in political science (and other fields). First, though, we discuss a few more topics: CDFs, parameters, joint distributions, and expectations.

11.1.1.1 Why Should I Care?

Many of the variables that interest political scientists have discrete, not continuous, distributions. Nevertheless, continuous distributions are important. First, some processes of interest—especially measures of time, such as how long a coalition government survives, the timing of sending legislation to the floor for a vote, the length of a war, etc.—are continuous. Second, it turns out that several continuous distributions are remarkably flexible and useful for modeling noncontinuous variables. Further, it is common to assume that various portions of statistical models (e.g., error terms) take a continuous distribution, and a number of statistical hypothesis tests are constructed using continuous distributions. Last, as we demonstrate below, continuous distributions are used in formal theory, and regardless of whether you seek literacy (i.e., the ability to read and understand formal models) or competence (i.e., the ability to create and use formal models), familiarity and comfort with continuous distributions is important. To that end, below we discuss several commonly used continuous distributions and identify their PDFs.

11.1.2 The CDF for a Continuous Variable

Not surprisingly the nettlesome difficulty of an infinite and uncountable number of potential values rears its head again when we think about the CDF of a continuous variable. As with the PDF, the solution lies in thinking about ranges of values instead of discrete ones.

Since we cannot write the CDF for a continuous variable as the sum of the probabilities of each discrete value below the specified value, we have to write it as the sum of all the value ranges below the specified value. Luckily, this is a more straightforward translation from the discrete case. In fact, we merely replace the sum with an integral to get the equation for the CDF:

$$P(X \leq x) = F(x) = \int_{-\infty}^{x} f(t)dt. \tag{11.2}$$

Equation (11.2) states that the probability that a variable drawn randomly from the sample has a value less than or equal to x is the sum of all of the probabilities of all ranges of values less than or equal to x. Because the CDF, unlike the PDF, is a probability, it is constrained to take values between 0 and 1.[3]

Equation (11.2) also introduces, or rather reintroduces, a piece of notation for the CDF: $F(x)$, if the PDF is $f(x)$. This notation is common in both game theory and statistics and arises from the relation between the PDF and the CDF: the latter is the antiderivative of the former.

11.1.3 The Parameters of Continuous Density Functions

Like the PMFs of discrete distributions, the PDFs of many continuous distributions have defined parameters. The most common are the location and scale (dispersion) parameters introduced in the preceding chapter, but some continuous distributions have a **shape parameter**. The shape parameter identifies a point of inflection in a PDF whose graph changes shape. Most distributions

[3]For this reason, CDFs are commonly used to model processes that are constrained between two values, or that involve binary choice.

we use do not have a shape parameter: their central location might change as a function of a parameter and the spread of their values might change as the function of a parameter, but the general shape of the PDF remains the same. Some distributions, however, can also change shape. This is easiest to see in a graph of a distribution that contains a shape parameter, like that of the beta distribution, seen in Figure 11.1.

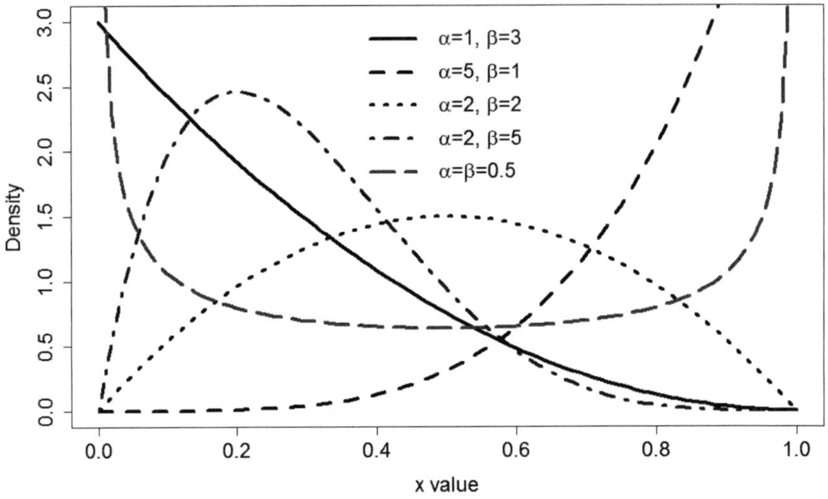

Figure 11.1: Beta PDF with Various Parameter Values

The beta distribution has two parameters, α and β, both of which are shape parameters. As the graph of the PDF demonstrates, the shape of the distribution changes as the parameters change values.

That said, we reiterate that the distributions most frequently used in our discipline do not have shape parameters. As such, you will encounter distributions with location and scale (dispersion) parameters in the political science literature with considerably greater frequency than distributions with a shape parameter.

11.1.4 Joint Distributions

As discussed in the preceding chapter on discrete distributions, we are often interested in the joint distribution of two variables. We discussed empirical joint distributions there. We extend that discussion here for continuous variables, and also introduce theoretical joint distributions of both discrete and theoretical variables.

11.1.4.1 Empirical Joint Distributions

Unlike for discrete variables, contingency tables are not useful for plotting the joint distribution of continuous variables. The tabular (or matrix) format of the contingency table limits its usefulness for looking at the joint distribution of continuous variables (or integer variables with more than a handful of values). The problem is that there are too many potential values that the variable may take. Thus, when we want to examine the joint distribution of continuous variables or one discrete and one continuous variable, rather than a contingency table we use a **scatter plot**.

A scatter plot is a graph with one variable's values listed on the vertical axis (typically the dependent or caused variable) and the other variable's values listed on the horizontal axis (typically the independent or causal variable). As examples, we have produced two scatter plots.

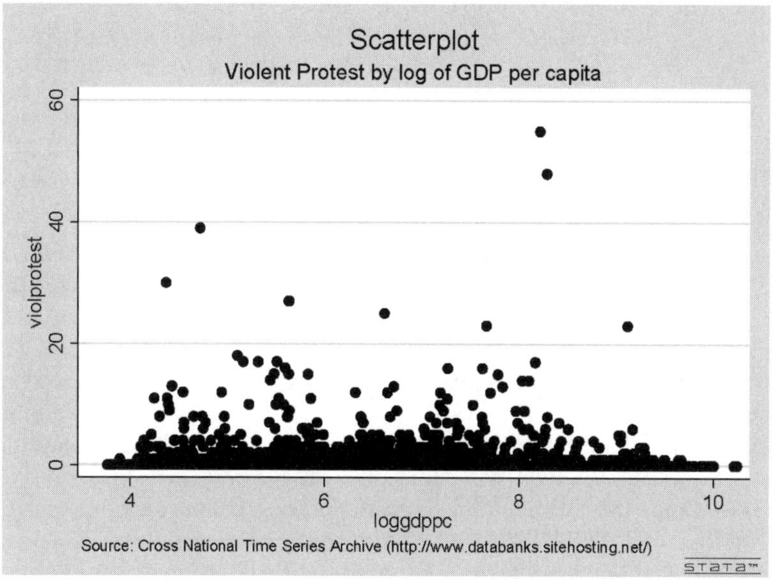

Figure 11.2: Is Violent Protest Related to Macroeconomic Output?

The first plot, in Figure 11.2, is composed of one discrete and one continuous variable. It seems to indicate a slight positive relationship between the size of the economy and the number of violent protest events.

The second plot, in Figure 11.3, depicts two continuous variables and suggests that the number of votes cast in national parliamentary elections is not strongly related to the size of government expenditures.

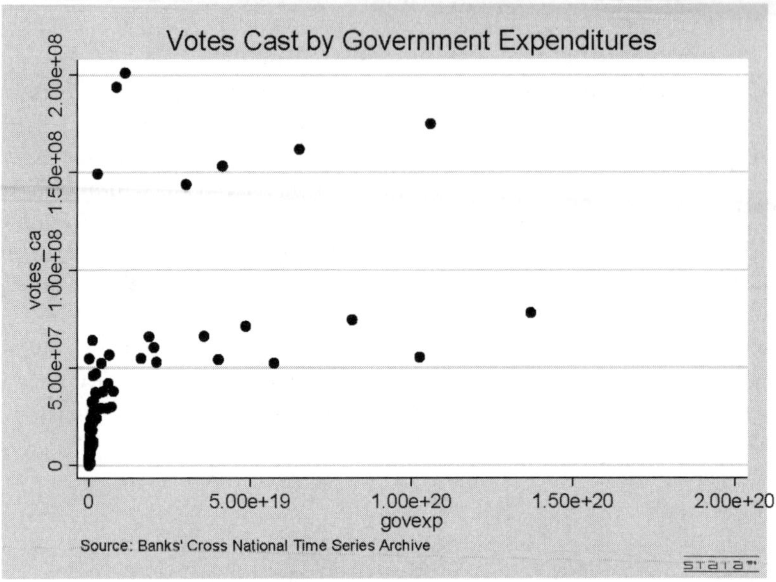

Figure 11.3: Is the Size of the Popular Vote Related to Government Expenditure?

11.1.4.2 Theoretical Joint Distributions

Whenever the values on the y-axes of these scatter plots are correlated with those on the x-axes, we might think there is a relationship between the two variables represented on the axes. Since we think of the variables on both axes as random variables, we can describe the joint variation of the variables, whether or not there is a conditional relationship between them, with joint probability distributions. These are theoretical constructs that describe the *simultaneous* realizations of more than one random variable. We stick to examining two random variables here, but everything we write here can be generalized. We also discuss joint distributions for discrete and continuous random variables at the same time.[4]

Joint distributions merely describe probabilities of more than one outcome at once. For two discrete random variables, we can write their joint PMF as $f(x, y) = Pr(X = x \cap Y = y)$. This is simply the chance that both x and y are simultaneously realized. If $x = y = 1$ and X and Y correspond to the values one might roll on two dice, the joint probability is the chance that two ones are rolled, or $\frac{1}{36}$.

For two continuous random variables, we can write their joint PDF the same

[4]We include discrete joint distributions in this chapter rather than in the previous chapter because the continuous version is more commonly observed, and the previous chapter is already relatively lengthy. Also, the logic behind joint discrete outcomes was provided in Chapter 9.

way: $f(x, y)$. "Summing" the small bits of probability $f(x, y)dxdy$ over some region $X \in A, Y \in B$ produces the probability $Pr(X \in A \cap Y \in B)$.

Thus, $f(x, y)$ is a way of writing the probability that two things occur simultaneously. As in Chapter 9, we can expand this "and" statement. For both discrete and continuous distributions, if the random variables X and Y are independent, then $f(x, y) = f(x)f(y)$. However, if X is conditional on Y (or vice versa), then $f(x, y) = f_{X|Y}(x|y)f_Y(y)$ (or with x and y and X and Y switched). Here the subscripts on the PDFs make clear the nature of the distribution. The function $f_Y(y)$ is the marginal distribution of the random variable Y, which averages over X. For the continuous distribution, this means that $f_Y(y) = \int f(x, y)dx$. The conditional distribution of X is given by $f_{X|Y}(x|y) = \frac{f(x,y)}{f_Y(y)}$, which is nothing more than the equation above rearranged.[5]

As with a single variable, something must happen, so that $\int \int f(x, y)dxdy = 1$ for the continuous case and $\sum_i \sum_j f(x_i, y_j) = 1$ for the discrete case. The double integrals (or sums) tell us to integrate (or sum) over first one variable and then the other; we discuss this more in Part V of the book. These same integrals and sums allow us to compute the joint CDFs: $Pr(X \leq x \cap Y \leq y)$ is $F(x, y) = \int_{-\infty}^{x} \int_{-\infty}^{y} f(x', y')dx'dy'$ for the continuous case[6] and $F(x, y) = \sum_{x_i<x} \sum_{y_i<y} f(x_i, y_i)$ for the discrete case.

This may have seemed complicated, but most of the complication is notational. Basically, joint probability distributions work in the same way as those for a single random variable: they tell you either the probability of two events happening at the same time (for two discrete random variables) or the relative likelihood of two events happening at the same time (for two continuous random variables). When the distributions for each variable are independent you can treat them entirely separately, but when one is conditional on the other you cannot do so. This is no different from how one treats any conditional probability, as we discussed in Chapter 9.

11.2 EXPECTATIONS OF CONTINUOUS RANDOM VARIABLES

In the previous chapter we discussed expectations of random variables. This discussion was general, but all our examples were of, and all our equations were for, discrete random variables. Here we discuss expectations of continuous random variables. Most of what we said in Chapter 10 holds here as well. In fact, all we're going to do is rewrite some of the equations used in the previous chapter for the continuous case, using the substitutions (integrals for sums, etc.) we introduced in the previous section. We'll start by presenting expectations

[5] This rearrangement should look familiar from our discussion of Bayes' rule in Chapter 9.

[6] You will often see integrals over x' and y' when the bounds of the integral contain x or y. Since it doesn't matter what letter or expression we use when we integrate over it (recall Chapter 7), using the "primed" versions of x and y enables us keep track of which integral corresponds to which bounds.

in general, then discuss expected utility and moments of distributions in subsequent subsections. The one new concept here is the uniform distribution, which is commonly used in game theory (and which serves as an uninformative prior in Bayesian statistics); we present this when discussing expected utility.

Recall from the previous chapter that we write the expectation of a random variable X as $E_X[X]$, or $E[X]$ when there is no confusion about other variables. In words, the expectation is the weighted average of the values that a variable can take, where the weights are given by the probability distribution of X. Also recall that the equation for an expectation is $E_X[X] = \sum_i x_i(Pr(X = x_i))$. So you multiply each value x_i by the weight on that value, and add all these up. Since the PMF $(f(x_i))$ of a discrete distribution provides the relevant probability weights $Pr(X = x_i)$, we can also write the expectation as $E_X[X] = \sum_i x_i f(x_i)$. This is for a discrete random variable; for a continuous one we translate the sum to an integral and the PMF to a PDF:

$$E_X[X] = \int_{-\infty}^{\infty} xf(x)dx. \tag{11.3}$$

The bounds on the integral ensure that the expectation includes all possible values of X that might be drawn. Equation (11.3) has the same interpretation as for the discrete case: it's a weighted average of the values of a continuous random variable, and so provides the mean of the distribution.

11.2.1 Expected Utility

As we noted in the previous chapter, we need not limit ourselves to the expectation of the variable itself; we can also consider functions of that variable. When we call these functions $u(x)$, we get expected utility $EU(X) = \sum_i u(x_i)(Pr(X = x_i))$ in the discrete case. Again, replacing the probability with the PMF produces the equation $EU(X) = \sum_i u(x_i)f(x_i)$. And changing to an integral and to the PDF provides the expected utility for the continuous case:

$$EU(X) = \int_{-\infty}^{\infty} u(x)f(x)dx. \tag{11.4}$$

Again, the bounds on the integral ensure that the expectation includes all possible values of X, and so all possible $u(x)$ that might be drawn. Equation (11.4) also has the same interpretation as for the discrete case: it's the weighted average utility one can expect to get, given the probability distribution of the random variable. Everything we said in the previous chapter about risk preferences and why one would need to understand and be able to compute expected utilities continues to be true in this case as well, and we won't repeat it. Instead we'll move right to an example. To do this, though, we'll first introduce the uniform distribution.

11.2.2 The Uniform Distribution

You've undoubtedly seen the uniform distribution before, even if you haven't heard it called that. It is the distribution that assigns equal probability or likelihood to all possible events in the sample space. The discrete case is so straightforward that we didn't even bother mentioning it in the previous chapter. If there are n possible outcomes, then the uniform distribution assigns probability $\frac{1}{n}$ to each outcome. For example, there are two outcomes in a coin flip, so the probability of getting either heads or tails is $\frac{1}{2}$, while there are six outcomes in the roll of a (fair) die, so the probability of getting any number between one and six is $\frac{1}{6}$.

Though this discrete distribution applies commonly outside the social sciences, it is not often used in political science. Few empirical scenarios place equal weight on every possible event. In game theory, whenever the probability is discrete, one typically wants to let it vary as a parameter, and so one assigns probabilities p_i to all events $i \in 1, \ldots, n$, as in the examples we used in the last chapter.[7] And even when the chances of all events are equal, the argument is more typically made in the context of classical probability, as in Chapter 9, than with a uniform distribution.

The continuous uniform distribution, however, is commonly used in game theory, and also as an uninformative prior in Bayesian statistics. We show below that the form of its PDF is somewhat more complicated than $\frac{1}{n}$, yet both the discrete and the continuous uniform distribution share the same fundamental property: the chance of drawing any value is the same.

To get a continuous PDF that satisfies this, let's begin with the easiest option and make the PDF constant at 1 throughout some range.[8] Let's call this range the interval $[\alpha, \beta]$. Then we could let the PDF be 1 from α to β, and 0 for all other values of X. It turns out this is almost good enough. The only problem is that when you integrate the PDF over all X, you need to get 1, and here you don't. Instead you get $\int_{-\infty}^{\infty} f(x) \cdot dx = \int_{\alpha}^{\beta} 1 \cdot dx = x|_{\alpha}^{\beta} = \beta - \alpha$.[9] But this is a constant, so we can readily fix this problem: we divide the PDF by $\frac{1}{\beta-\alpha}$ to cancel out the $\beta - \alpha$ and leave the integral as 1.[10] This yields the expression for the PDF of the uniform distribution:

$$f(x) = \begin{cases} \frac{1}{\beta-\alpha} & \text{if} \quad x \in [\alpha, \beta], \\ 0 & \text{otherwise.} \end{cases} \tag{11.5}$$

Note that α is a location parameter (i.e., it determines the center of the

[7] Some researchers will also assume a discrete uniform distribution as an uninformative prior in Bayesian statistical models.

[8] We need a finite range here, since otherwise we'd have to make an infinite number of subranges equally likely, so that the chance of getting any one of them (such as $[0, 1]$) would effectively be zero.

[9] The second step is true because the PDF is 0 outside $[\alpha, \beta]$, so the definite integral is 0 outside this range.

[10] The $\frac{1}{\beta-\alpha}$ is known as a normalization constant (or factor) for this reason.

distribution), and β is a scale parameter (i.e., it determines the dispersion of the distribution). If X is distributed according to a uniform distribution with these parameters we can write $X \sim U[\alpha, \beta]$. Figure 11.4 plots the PDF of the uniform distribution.

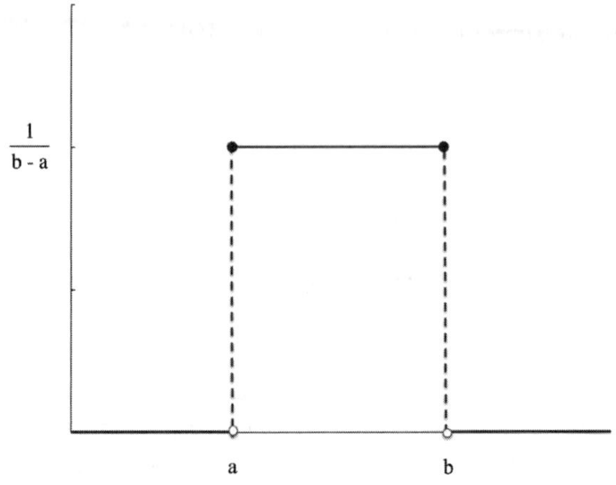

Figure 11.4: Uniform PDF

We are generally interested not in the uniform distribution's PDF but rather in its CDF. Recall that the CDF is the integral of the PDF from negative infinity up to some value x. For the uniform distribution, this function is $F(x) = \int_{-\infty}^{x} \frac{1}{\beta-\alpha} dt = \frac{1}{\beta-\alpha} \int_{\alpha}^{x} dt = \frac{1}{\beta-\alpha} t|_{\alpha}^{x} = \frac{x-\alpha}{\beta-\alpha}$ for any $x \in [\alpha, \beta]$. For smaller values of x the CDF is 0, and for larger values it is 1, since there is no chance of drawing an x less than α or more than β. Putting this together produces the CDF of the uniform distribution:

$$F(x) = \begin{cases} 0 & \text{if} \quad x < \alpha, \\ \frac{x-\alpha}{\beta-\alpha} & \text{if} \quad x \in [\alpha, \beta], \\ 1 & \text{if} \quad x > \beta. \end{cases} \tag{11.6}$$

The important thing to note about this CDF is that it is linear in x. Since the CDF represents the chance of drawing a value from the distribution less than or equal to x, the linearity of the CDF means that this chance increases proportionally with x. We illustrate this in Figure 11.5.

11.2.3 A Game Theoretic Example

In game theory, we often need to know the probability that a given variable has a value below some cutoff point. For instance, we might want to know what the chance is that one's expected utility for things like contesting an election or launching a military expedition would exceed the payoff from taking the alternative option (e.g., doing nothing).

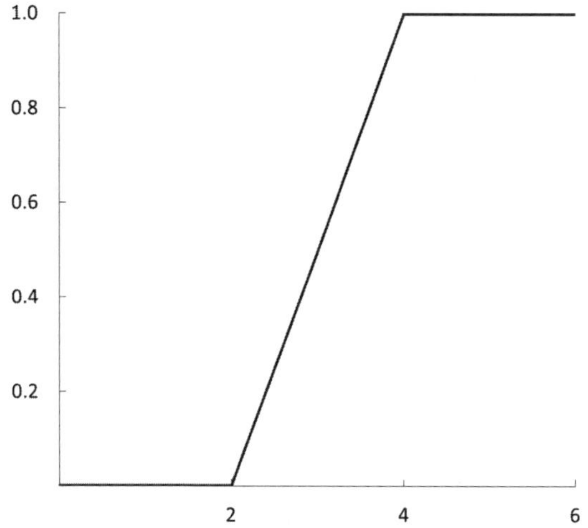

Figure 11.5: Uniform CDF

If one's utility is a function of a random variable, the CDF of the distribution of that variable determines the probability that any realization of the variable is less than some value. If the distribution is continuous, we need to integrate the PDF to get the CDF. In most cases this can be difficult, particularly if we care about obtaining closed-form solutions of the game (i.e., one can write down the equation for the answer). The simplicity of the uniform PDF allows us to readily compute the CDF, and its linearity implies that subsequent calculations will be easier to deal with than would be the case for nearly all alternative distributions.

Further, in game theory we often don't have strong beliefs about the distribution of some parameter of the model, and the uniform distribution allows one to assume no preference for any particular value in the distribution. For both these reasons, game theorists typically assume the uniform distribution when a particular distribution must be chosen.[11]

To see how this works, we'll consider an example of the type you may very well see in a game theory class in political science. Flip back to (or merely recall) Chapter 3, wherein we used as an example of a utility function the quadratic loss function $u(x) = -(x-z)^2$, where z is the ideal policy, x is the enacted policy, and the utility function indicates that z is the most preferred policy, with policies less preferred the further they are from z. We said there that this was a common form of utility for modeling voting behavior. Having now completed Chapter 8, we might recognize that the reason for this is that this utility function is

[11] Of course, it is always better *not* to have to choose a distribution, as then your model doesn't rely on an assumption (of the uniform distribution) that may be wrong, but this option isn't always optimal!

both differentiable and concave everywhere, and so has a maximum at the ideal policy $x = z$ that can be computed comparatively easily.

Assume there exists a pivotal voter who will determine the outcome of an election or the vote in a legislature,[12] and assume that she has a quadratic loss utility function with $z = m$. This means that the pivotal (aka median) voter prefers policies closer to m than those further away. However, while we can assume that the median voter knows her own most preferred option (aka ideal point) m, no one else does. In particular, neither of the two candidates contesting an election does. The candidates do realize, however, that in order to win, they must secure her vote.

Let's say that the policy space—a line over which all policies can be aligned in order, such as a left–right continuum—is $[0, 1]$, which means these are all the available policies from which to choose. Let's also say that neither candidate has any clue where the median voter's ideal point might be within this range. We represent this cluelessness formally by saying $M \sim U[0, 1]$, where M is the random variable corresponding to the median voter's ideal point.

Game theorists are often interested in where the candidate faced with such circumstances will place policy along the continuum. If the candidates knew where m was located, they would select m, as doing so is the best way to win the election. However, they do not know the location of m. This is often referred to as a location game, and it has two or more candidates each choosing a position in the policy space for their platforms. They run on these platforms, then voters (in this case, the median voter only) vote for the candidate they like best, and the winning candidate (the one with the most votes) typically must enact her platform.[13]

In the real game the candidates either enter simultaneously or in some sequence. For our purposes, we'll simplify things. Let's say that there is already an incumbent in office—Sunhee—who has cleverly staked out the position $x_A = \frac{1}{2}$. Sunhee reasons, quite correctly, that occupying the mean of the distribution of M gives her the best chance of winning, which is what her primary interest is in the election. Ryan seeks to challenge Sunhee, but unlike Sunhee he is extreme in his views and only cares about enacting a far left policy of 0, or as close to that as he can get. Specifically, his utility is $u_B = -(x - 0)^2$. If he wins the election with platform x_B, Ryan gets utility $-x_B^2$, and if he loses the election he gets utility $-(\frac{1}{2} - 1)^2 = -\frac{1}{4}$, as Sunhee's platform of $\frac{1}{2}$ gets enacted.

To figure out where Ryan should locate his platform, we need to maximize

[12] In rational choice theories one often appeals to what is called the median voter theorem, which, speaking loosely, says that the voter whose ideal policy in one dimension is in the middle (the median) of all voters' ideal policies is a pivotal voter who determines the outcome of the vote. See Shepsle and Bonchek (1997) for an introduction.

[13] This ignores what are called credible commitment problems (e.g., the candidate once in office can do what she wants), but we need not concern ourselves with this here. There is, however, quite a broad literature on this topic (e.g., McCarty and Rothenberg, 1996), and our discussion of Stokes (2001) in previous chapters is one example.

the expected utility

$$EU(x_B) = Pr(win|x_B)(-x_B^2) + Pr(lose|x_B)(-\frac{1}{4}). \tag{11.7}$$

In Chapter 8 we learned how to maximize this, but before doing so we need to know $Pr(win|x_B)$. (Note that $Pr(lose|x_B) = 1 - Pr(win|x_B)$ since there are only two candidates.) But how does one go about finding this?

Consider the median voter. Her utility function implies she will always vote for the candidate closer to her. So when is she closer to Sunhee's platform of one-half, and when is she closer to Ryan's platform of x_B? Well, the midpoint of x_B and one-half is $\frac{x_B+\frac{1}{2}}{2} = \frac{x_B}{2} + \frac{1}{4}$. Assuming (safely) that Ryan locates to the left of Sunhee, whenever the realized m is less than this midpoint, Ryan wins, because the median voter's ideal point is closer to his position than to Sunhee's, and whenever the realized m is greater than this, Sunhee wins. Stated mathematically, $Pr(win|x_B) = Pr\left(m \leq \frac{x_B}{2} + \frac{1}{4}\right)$.

This probability is the CDF of M evaluated at $\frac{x_B}{2} + \frac{1}{4} \in [0, 1]$. For $U[0, 1]$, $\alpha = 0$ and $\beta = 1$, so the CDF is $F(m) = m$. Thus, $F(\frac{x_B}{2} + \frac{1}{4}) = \frac{x_B}{2} + \frac{1}{4}$. The probability of the median voter's ideal point being less than $\frac{x_B}{2} + \frac{1}{4}$ is therefore equal to $\frac{x_B}{2} + \frac{1}{4}$, and this is the probability that Ryan wins. Plugging this into equation (11.7) yields $EU(x_B) = -x_B^2\left(\frac{x_B}{2} + \frac{1}{4}\right) - \frac{1}{4}\left(1 - \frac{x_B}{2} - \frac{1}{4}\right)$. Simplifying gives $EU(x_B) = -\frac{x_B^3}{2} - \frac{x_B^2}{4} + \frac{x_B}{8} - \frac{3}{16}$.

Maximizing this utility entails first taking the first-order condition (see Chapter 8), which is $-\frac{3x_B^2}{2} - \frac{x_B}{2} + \frac{1}{8} = 0$. We can multiply through by -8 to get rid of the fractions, which yields $12x_B^2 + 4x_B - 1 = 0$. Then we use the tools of Chapter 2 to solve this; e.g., the quadratic equation gives us $\frac{-4\pm\sqrt{16+48}}{24} = \frac{-1\pm2}{6}$. Only one of these is in the range $[0, 1]$, so we'll choose that one. This gives the candidate extremum $x_B^* = \frac{1}{6}$.

We next check whether this is a local maximum by computing the second-order condition at that point. This is $-3x_B^* - \frac{1}{2} < 0$, so x_B^* is a local maximum. Finally, we compare the utility at this point to that at the bounds. At $x_B = 0$, we get $EU(0) = -\frac{3}{16}$. At $x_B = \frac{1}{2}$, which is the furthest Ryan can go and still be to the left of Sunhee, we get $EU(\frac{1}{2}) = -\frac{4}{16}$. Finally, at $x_B^* = \frac{1}{6}$, we get $EU(x_B^*) = -\frac{1}{432} - \frac{1}{144} + \frac{1}{48} - \frac{3}{16} = -\frac{19}{108}$, which is the biggest of the three values. So the global maximum occurs at $x_B^* = \frac{1}{6}$.

Ryan thus locates at a position considerably to the left of one-half, and so accepts that he will lose more often than Sunhee; he's just willing to take the extra risk of losing so as to enact his ideal policy when he wins.

This was a pretty involved example (though it is a simplification of Calvert (1985)), but it illustrated, we hope, the way in which the CDF can be used in game theory. For those interested, we discuss this at a bit higher level in the last part of this section, which introduces the notion of stochastic dominance. Before getting there, though, we briefly illustrate the moments of continuous distributions, again using the uniform distribution as an example.

11.2.4 Moments of Continuous Distributions

In the preceding chapter we discussed moments of distributions and why they are important; since that discussion continues to apply to the case of continuous distributions we do not repeat it. Rather, we will merely present the definitions for the kth moment in the continuous case. The kth moment about zero is

$$\int_{-\infty}^{\infty} x^k f(x)dx, \tag{11.8}$$

and the kth moment about the mean is

$$\int_{-\infty}^{\infty} (x-\mu)^k f(x)dx. \tag{11.9}$$

We can see how this works with the uniform distribution. First we compute its first moment, the mean μ. This is

$$\begin{aligned}\mu &= \int_{-\infty}^{\infty} xf(x)dx \\ &= \tfrac{1}{\beta-\alpha}\int_{\alpha}^{\beta} xdx \\ &= \tfrac{1}{\beta-\alpha}\tfrac{1}{2}x^2|_{\alpha}^{\beta} \\ &= \tfrac{\beta^2-\alpha^2}{2(\beta-\alpha)} \\ &= \tfrac{\beta+\alpha}{2}.\end{aligned} \tag{11.10}$$

You may recognize this as the midpoint of the line segment $[\alpha, \beta]$. When $\alpha = 0$ and $\beta = 1$, $\mu = \frac{1}{2}$.

We can also compute the variance, which is the second moment about the mean. This is

$$\begin{aligned}\sigma^2 &= \int_{-\infty}^{\infty}(x-\mu)^2 f(x)dx \\ &= \tfrac{1}{\beta-\alpha}\int_{\alpha}^{\beta}(x^2-2\mu x+\mu^2)dx \\ &= \tfrac{1}{\beta-\alpha}\left(\tfrac{1}{3}x^3-\mu x^2+\mu^2 x\right)|_{\alpha}^{\beta} \\ &= \tfrac{(\beta^3-\alpha^3)/3-\mu(\beta^2-\alpha^2)+\mu^2(\beta-\alpha)}{\beta-\alpha} \\ &= (\beta^2+\alpha\beta+\alpha^2)/3-(\beta^2+\alpha^2+2\alpha\beta)/2+(\beta^2+\alpha^2+2\alpha\beta)/4 \\ &= (\beta^2+\alpha\beta+\alpha^2)/3-(\beta^2+\alpha^2+2\alpha\beta)/4 \\ &= \tfrac{\beta^2+\alpha^2-2\alpha\beta}{12} \\ &= \tfrac{(\beta-\alpha)^2}{12}.\end{aligned} \tag{11.11}$$

When $\alpha = 0$ and $\beta = 1$, $\sigma^2 = \frac{1}{12}$. Higher moments may be computed similarly. The same procedure may be followed for other, more complex distributions, though their integrals are likely to be more difficult.

We can also use our earlier discussion of theoretical joint distributions to compute a different sort of second moment that will prove convenient for work in statistics. When there are two variables distributed jointly we can discuss their **covariance**. Like the variance, the covariance considers variation around the mean, but now around the means of two variables. It is computed according to the equation $\sigma_{xy} = \int_{-\infty}^{\infty}\int_{-\infty}^{\infty}(x-\mu_x)(y-\mu_y)f(x,y)dxdy$, with μ_x and μ_y the means of the random variables X and Y, respectively.

The form of the covariance implies that when both variables exceed their means or both are below their means the integrand is positive, while if one is above and one is below its mean the integrand is negative. Thus, the covariance measures the degree to which two random variables "move together" in their joint distribution. If one often tends to be large (small) when the other is large (small), then their covariance will be positive, while if one tends to be large while the other is small, then their covariance will be negative. A covariance of zero implies that the two variables are not correlated in this fashion; this often happens when the two variables are drawn from independent distributions, but this is not necessary for a covariance of zero.

Like the variance, the covariance can become large in magnitude. When interested in the relative degree of correlation between two variables, we can instead form a **correlation coefficient**. This is computed from the covariance and the variances of each variable and varies between -1 and 1. A 0 means no correlation, and a 1 (-1) means perfect positive (negative) correlation. It has the form $\rho_{xy} = \frac{\sigma_{xy}}{\sigma_x\sigma_y}$, where the two components of the denominator are the standard deviations (the square roots of the variances) of X and Y, respectively.

11.2.5 Stochastic Dominance

In game theory and expected utility theory, expected utilities for actions are compared together and the action that produces the highest expected utility is chosen. It is natural to compare payoffs for individual outcomes, but when actions produce stochastic payoffs one can also compare the distributions directly. This leads to the important notion of stochastic dominance.

Before giving the formal definitions, let's consider an intuitive example. Consider a type of lottery in which you can receive either nothing or a million dollars. Now compare two specific lotteries, one in which you have a 0.0000000001 chance of getting the million dollars and one in which you have a 0.3 chance of getting the million dollars. Which one would you prefer?

The answer is obvious, of course, but it does illustrate the concept of explicitly comparing probability distributions to each other. The notion of stochastic dominance formalizes this comparison. If we let $f(x)$ and $g(x)$ be two different PDFs, then we say $f(x)$ first-order stochastically dominates (FOSD) $g(x)$ if their CDFs obey this relation: $F(x) \leq G(x)$ for all x.

This is very abstract, so we break it down. The CDF tells you the chance of drawing a value from that distribution below a certain value x. If the CDF for $g(x)$ is always greater than that for $f(x)$ for all x, then the chance of drawing a lower value from the distribution is always greater in $g(x)$ than in $f(x)$. This implies, since $1 - F(x) \geq 1 - G(x)$, that the chance of drawing a value higher than x is always greater in $f(x)$ than in $g(x)$ for all x. So, in a sense, $f(x)$ can be expected to produce higher values than $g(x)$, and we say the former dominates the latter. For our example this is true: the first lottery is more likely to produce the lesser value than the second, and less likely to produce the higher value. So the second FOSD the first.

How does this relate to preference? Another way of writing $f(x)$ FOSD $g(x)$ is $\int_{-\infty}^{\infty} u(x)f(x)dx \geq \int_{-\infty}^{\infty} u(x)g(x)dx$ for all increasing functions $u(x)$.[14] In other words, if you place higher value on obtaining greater levels of some random variable X, perhaps because it corresponds to revenue, shares in a government's cabinet, or your piece of the division of land in a cease-fire bargain, then you always prefer that the distribution $f(x)$ be the one that determines levels of revenue, cabinet shares, or land distributions, as opposed to $g(x)$. And this is for the reason we stated above: it is more likely to produce higher values of these things.

FOSD is thus a useful concept because it lets you state preference over distributions without having to go to the trouble of figuring out expected utilities; you can just compare CDFs. Further, as it works for any increasing utility function, one need not even specify a particular function. This is particularly useful when employing techniques in game theory such as monotone comparative statics (e.g., Ashworth and Bueno de Mesquita, 2005).

It does, however, require a pretty strong assumption on the distributions that may be hard to justify substantively in some cases. We can also define a lesser form of dominance, second-order stochastic dominance (SOSD). It has a very similar definition: $f(x)$ SOSD $g(x)$ if $\int_{-\infty}^{\infty} u(x)f(x)dx \geq \int_{-\infty}^{\infty} u(x)g(x)dx$ for all increasing *concave* functions $u(x)$. Recalling from the previous chapter that concave utility functions represent risk-averse actors, if $f(x)$ SOSD $g(x)$ then it is preferred by all risk-averse individuals.

We can compare two uniform distributions to illustrate these concepts. Let $f(x) \sim U[1,3]$ and $g(x) \sim U[0,2]$. Then $f(x)$ FOSD $g(x)$. *Anyone* who wants higher values of x prefers $f(x)$, because it yields consistently higher values of x. Now let $f(x) \sim U[1,3]$ and $g(x) \sim U[0,4]$. Then $f(x)$ SOSD $g(x)$, but not FOSD. A risk-neutral person (i.e., one with a linear utility function) is indifferent between two lotteries represented by these distributions; they each have the same mean value, and the potential for higher values in $g(x)$ is balanced out by the potential for lower values. However, a risk-averse person prefers the first lottery to the second because she is less likely to draw a really low value.

11.3 IMPORTANT CONTINUOUS DISTRIBUTIONS FOR STATISTICAL MODELING

In this section we introduce several commonly used continuous distributions: the Gaussian family, the logistic distribution, some duration distributions, and three distributions used frequently in statistical hypothesis tests. The Gaussian family includes the normal distribution and the power transformed normal

[14] An equivalent equation holds for discrete distributions, using sums instead of integrals: $\sum_i u(x_i)f(x_i) \geq \sum_i u(x_i)g(x_i)$. Note that if we were to call $u(x)$ utility, then both formulations express expected utility under different probability distributions.

distribution. Duration distributions include the exponential distribution, the Pareto distribution, the gamma distribution, and the Weibull distribution. Finally, we review the chi squared (χ^2), the F, and the (Student's) t distributions.

11.3.1 The Gaussian Family

Families of distributions have the same basic parameter structure. The Gaussian distribution is named after one of the first scholars to use it, Johann Carl Friedrich Gauss.

11.3.1.1 The Normal Distribution

The normal distribution is the best known of all continuous distributions. It may be written as $N(\mu, \sigma^2)$, so that if X is distributed normally, $X \sim N(\mu, \sigma^2)$. As seen in this notation, the distribution admits two parameters, the mean (or average) value, represented by μ, and the variance (or dispersion) of values around the mean, represented by σ^2. The PDF of the normal distribution is given by

$$f(x; \mu, \sigma^2) = \frac{1}{\sqrt{2\pi\sigma^2}} e^{-\frac{1}{2\sigma^2}(x-\mu)^2}. \tag{11.12}$$

Note that while π is often used as a symbol to indicate the probability of observing an event, in equation (11.12) it represents the value 3.14159.... The normal distribution is so commonly used, particularly the standard normal, in which $\mu = 0$ and $\sigma^2 = 1$, that the standard normal PDF has its own symbol: $\phi(x)$. The standard normal CDF is denoted $\Phi(x)$. We can use equation (11.12) to graph some normal distributions, and we have done so in Figure 11.6.

One interesting (and unusual) property of the normal distribution is that the parameters (μ and σ^2) are independent of one another.[15] The mean (μ) determines the central location of the distribution and the variance (σ^2) determines the scale of the distribution. A second property of interest is the symmetry of the normal distribution: the graph of the function to the right of the mean is the mirror image of the graph of the function to the left of the mean.[16]

You have likely heard that the normal distribution has a bell-shaped curve. This is true for many (but as Figure 11.6 demonstrates, not all) values of μ and σ^2, but it is not very meaningful. We have already seen a discrete distribution (the binomial) that frequently produces a bell-shaped curve, and we introduce many others below. It follows that one cannot draw the inference that sample data with a bell-shaped relative frequency distribution were drawn from a normal distribution. In fact, it is a good idea to unlearn the habit of referring to distributions as bell-shaped, as that observation provides precious little information about the distribution. In your statistics courses you will learn some

---

[15] Note that we could have replaced "μ and σ^2" with "the first and second moments." You will sometimes encounter that usage.

[16] Its skewness is zero.

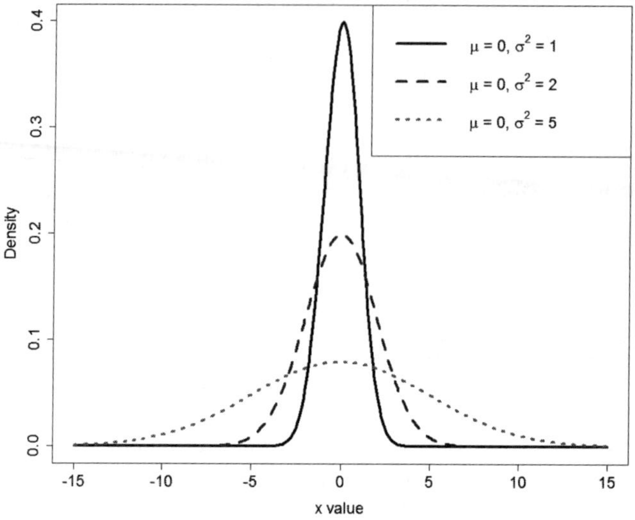

Figure 11.6: Three Normal PDFs, μ, $\sigma^2 = 0, 1; 0, 3; 0, 10$

formal tests one can conduct to determine the probability that a given sample of data was drawn from a normal distribution (e.g., the Jarque-Bera test).

What kinds of processes are likely to produce a normal distribution? Lindsey (1995, p. 113) observes that the "normal distribution describes a continuous response variable, taking any real value, positive or negative, which is the result of a large number of small accumulating, unknown, additive factors." We suspect that description does not strike you as likely to represent the process that produces a majority of the variables political scientists want to explain.

To better understand the limits of the usefulness of the normal distribution in empirical political science, try this as an exercise. Compile a list of variables that come to mind that political scientists use to measure concepts in their theories. How many of them are continuous measures (especially those with both negative and positive values)? Our expectation is that your list will be dominated by integer variables (e.g., the number of seats in parliament or the number of militarized disputes) and ordinal or nominal variables (e.g., an attitudinal scale or party identification).

This does not mean that the normal distribution is useless in statistics. Far from it! In fact, the central limit theorem states that in sufficiently large samples, sampling distributions approximate the normal distribution. We note that the central limit theorem does not imply that the normal distribution is the most appropriate distribution available. However, it does suggest that if one does not have a positive case to make for why the concept for which one has developed hypotheses fits a particular distribution, then the normal distribution is the best choice (though a better decision would be to go back to develop a stronger theory and develop a positive case for the likely distribution of one's concept).

That said, as Lindsey (1995, p. 113) observes, "the normal distribution is primarily important for its nice mathematical properties, and is much overused in many areas of research for this reason." To elaborate, owing to the nice mathematical properties of the normal distribution, it was relatively easier to develop techniques (and, later, software) for inferential statistics assuming a normal distribution than it was to assume other distributions. As such, the practice and teaching of applied statistical work focused on models that invoked the normal distribution. However, the past forty years have witnessed a dramatic increase in computing power, and software that can implement models that invoke different distributional assumptions has become commonplace. Thus, while it is important to use models that invoke a normal distribution when using a dependent variable that is normally distributed, the general point is that *it is important to use a model that invokes the appropriate distribution.* Political scientists became widely aware of this in the 1990s, and the overuse of models that assume a normal distribution has been declining ever since (Krueger and Lewis-Beck, 2008). Nevertheless, the appeal to the central limit theorem remains an important counterargument, but proper consideration will have to wait for your statistics courses.

Interestingly, though the normal distribution may be overused in empirical political science, it is perhaps underused in formal theoretical political science. Parameters that might in fact be distributed normally are rarely modeled as such. The reason is that, as we have noted, CDFs of distributions are important in formal theory, and the CDF of the normal distribution admits no closed-form expression. In other words, to compute the CDF at some value x, one must numerically approximate the integral defining the CDF. This is not an issue in statistics, as statistical computing software can use numerical approximation to do so. You either have already seen or will soon see tables of z-scores;[17] these are computations of the standard normal CDF $\Phi(x)$ or transformations of these. However, in game theory one often wants a closed-form expression that one can maximize, and the normal distribution does not admit this. Computational modeling, lacking this constraint, is more likely to take up use of the normal distribution when appropriate (e.g., Siegel, 2009).

Before turning our attention to the distributions of other variables, we briefly consider some variations of the normal distribution.

11.3.1.2 The Power-Transformed and Log-Normal

Power transformations can be used to make variables whose distributions deviate from the normal more closely approximate the normal. We can write a power-

[17] A z-score is obtained by transforming x to $z = \frac{x-\mu}{\sigma}$, and such tables list the corresponding values of $\Phi(z)$ or transforms thereof.

transformed normal distribution as follows:

$$f(x;\mu,\sigma^2,\lambda) = \frac{\lambda x^{\lambda-1}}{\sqrt{2\pi\sigma^2}} e^{-\frac{1}{2\sigma^2}(x^\lambda-\mu)^2}. \tag{11.13}$$

Like all Gaussian distributions, it has the location and scale parameters μ and σ^2, but it also has a shape parameter λ. Equation (11.13) is quite similar to equation (11.12); to see how this transformation operates, observe that when $\lambda = 1$, equation (11.13) reduces to equation (11.12). However, when $\lambda \leq 1$, the right side of the distribution will be longer than the left side, and when $\lambda \geq 1$, the left side will be longer than the right. An asymmetry where one side (or tail) of the graph of the function is longer than the other is called skewness.[18] Figure 11.7 is a plot of the power-normal.[19]

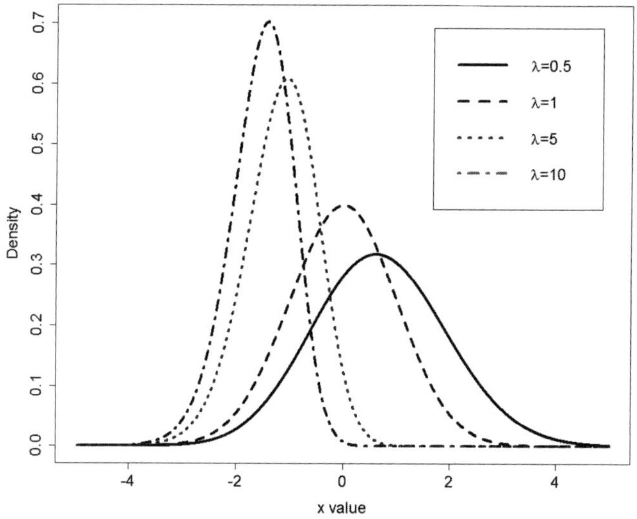

Figure 11.7: Power-Transformed Normal PDF

In the 1970s and 1980s, when political scientists recognized that much of their sample data had a skewed distribution, they frequently sought to transform the variable to make it more closely approximate the normal distribution. The most common transformations are power transformations, and the log transformation is the most widely used of the power transformations. We can write the log-normal distribution as

$$f(x;\mu,\sigma^2) = \frac{1}{x\sqrt{2\pi\sigma^2}} e^{-\frac{1}{2\sigma^2}(\ln(x)-\mu)^2}. \tag{11.14}$$

[18] Recall from the previous chapter that skewness is the third moment about the mean.

[19] This figure is from the National Institute of Standards and Technology's online *Engineering and Statistics Handbook*. Note that their p is our λ. To see their full entry (including the equation that they use), see `http://www.itl.nist.gov/div898/handbook/eda/section3/eda366d.htm`.

The graph of this function looks like Figure 11.8.

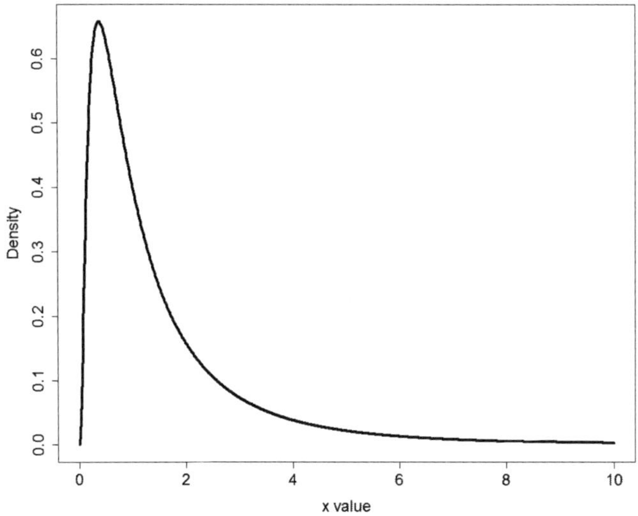

Figure 11.8: Log-Normal PDF, $\mu = 0$, $\sigma^2 = 1$

11.3.1.3 Why Should I Care?

As noted, it used to be fairly common practice to transform skewed distributions. When theory suggests that a given variable has a log (or other power) normal distribution, then these transformations are entirely appropriate. However, in the past ten to fifteen years, increasing numbers of political scientists have come to recognize that skewed sample data might well imply a non-Gaussian distribution. And the popularity of these transformations has declined as computers have made it easier to estimate models that assume non-Gaussian distributions. However, it is critical to keep in mind that some political processes may well produce log-normal or other power variants of the normal distribution, and that it is entirely appropriate to perform such transformations in such cases. For further reading, Mills (1991, pp. 40–50) provides a useful discussion of transforming sample data so that they approximate the normal distribution. For a more thorough overview of the normal distribution, please see the webpage at `http://mathworld.wolfram.com/NormalDistribution.html`.

11.3.2 The Logistic Distribution

The PDF for the logistic distribution can be written as

$$f(x;\mu,\sigma^2) = \frac{\pi e^{-\frac{\pi(x-\mu)}{\sigma\sqrt{3}}}}{\sigma\sqrt{3}\left(1+e^{-\frac{\pi(x-\mu)}{\sigma\sqrt{3}}}\right)^2}. \tag{11.15}$$

Like the normal distribution, the PDF for the logistic distribution is defined by two moments, the mean and the variance. Further, as Figure 11.9 indicates, the logistic distribution's PDF is symmetric. However, the logistic distribution is not part of the Gaussian family.

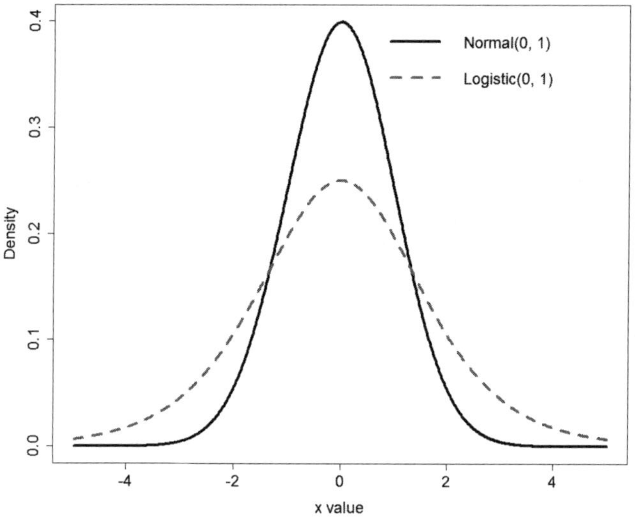

Figure 11.9: Logistic and Normal PDFs, $\mu = 0$, $\sigma^2 = 1$ (the logistic distribution has a lower peak and wider tails)

Figure 11.9 depicts two distributions with the same mean and variance. One is the logistic distribution and the other is the normal. Their similarity is remarkable, yet they have a clear difference: the logistic distribution displays considerably thicker tails, implying that values further from the mean are more likely to be drawn from a logistic distribution than a normal one. If you recall our discussion of moments in the last chapter, the fourth moment, or kurtosis, describes the thickness of the tails of distributions. The logistic has a non-zero kurtosis, while the normal distribution's is zero.

The logistic distribution is primarily used by political scientists modeling *binary outcomes* (e.g., voted/didn't vote, participated in a militarized dispute/did not participate). You will learn more about this in your statistics classes. You can find a more thorough technical overview of the logistic distribution at `http://mathworld.wolfram.com/LogisticDistribution.html`.

11.3.3 Duration Distributions

How long do legislators typically serve in a lower (e.g., provincial or state) house before seeking election to a higher (e.g., federal) legislature? How long do different types of coalition governments (e.g., majority vs. minority) survive? How long do different types of polities (e.g., democracies vs. autocracies) persist? How long do different types of military alliances (e.g., defensive vs. offensive) last? Political scientists are increasingly interested in concepts that are measured in units of time. Variables that measure such concepts can usually be modeled as if they were drawn from a duration distribution.

If you take advanced statistics courses you will likely encounter these models. A widely used model, the Cox proportional-hazards model, makes no distributional assumptions. That sounds too good to be true: we can use this model to employ statistical inference regardless of the distribution of our measure of duration. Yet, as Box-Steffensmeier and Zorn (2001) show, the Cox model can produce misleading inferences when hazards are not proportional. The articles by Zorn (2000) and Box-Steffensmeier and Zorn (2002) may also be consulted to learn more about the statistical models available for duration analysis.

11.3.3.1 The Exponential Distribution

The PDF for the exponential distribution is

$$f(x;\mu) = \frac{1}{\mu}e^{-\frac{x}{\mu}}, \tag{11.16}$$

where $\mu > 0$ is the mean duration between events. If we define $\lambda = \frac{1}{\mu}$, then we can rewrite equation (11.16) as $f(x;\lambda) = \lambda e^{-\lambda x}$, and this is common notation that you may encounter. We provide some representations of the exponential distribution's PDF in Figure 11.10.

The exponential distribution describes events produced by a process with a constant risk to failure. That is, the probability of failure does not change over time. "Failure" is a generic term that indicates the presence of a new state and should not be taken literally. That is, we can use the exponential distribution to study processes where it would be awkward to speak of "failure."

What sort of political process might produce a variable with an exponential distribution? The duration of cabinet governments seems an unlikely candidate since if the government persists long enough, elections are required after a given period of time (e.g., Cioffi-Revilla, 1984; King et al., 1990). So, if elections are required after five years in office and the government survives 1,824 days (i.e., one day less than five years), then we know with certainty that the government will dissolve the next day. So the duration of cabinet governments is not constant over time. Further, one likely expects a government's prospects for failure to be low in the first months in office, then rise some, etc. And lots of political processes seem likely to have risks of failure that change over time. For example, would you be willing to assume that the probability that a person votes is

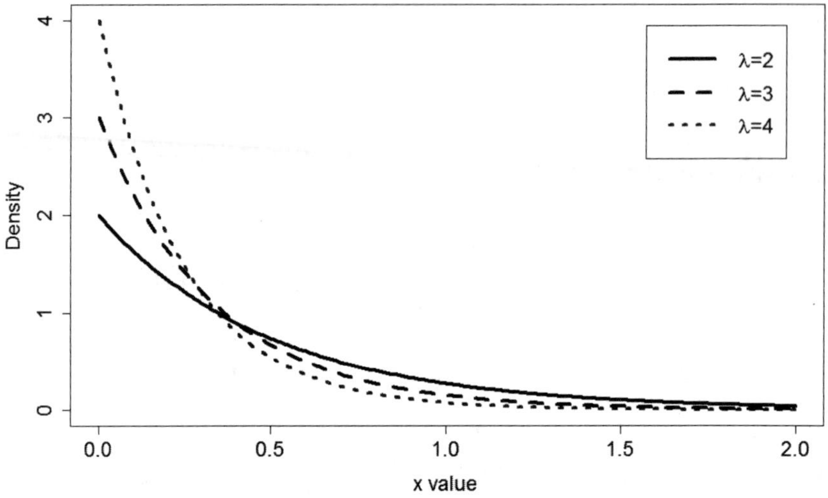

Figure 11.10: Exponential PDF, $\lambda = 2, 3, 4$

independent of her age? Do you suspect that the probability that a country goes to war is independent of the time that has passed since it last went to war?

That we can think of variables that interest political scientists which have failure risks that probably vary over time does not, however, suggest that the exponential distribution is useless. In fact, it is quite useful for processes with a constant risk of failure. Whether there are many duration processes of interest to political scientists that have a constant risk of failure over time is something for scholars to determine as the literatures that explore durations continue to grow.

You can find a thorough technical overview of the exponential distribution at `http://mathworld.wolfram.com/ExponentialDistribution.html`.

11.3.3.2 The Pareto Distribution

The PDF for the Pareto distribution can be written

$$f(x;\kappa,\beta) = \begin{cases} \frac{\kappa\beta^{\kappa}}{x^{\kappa+1}} & \text{for} \quad x \geq \beta, \\ 0 & \text{for} \quad x < \beta, \end{cases} \tag{11.17}$$

where $\kappa > 0$ is a shape parameter and β is a scale parameter such that $x \geq \beta > 0$.

Midlarsky (1988) uses the Pareto distribution to both theoretically and empirically model land inequality in Latin America. He argues that land was settled sequentially over time such that those who first claimed rights to land were able to secure larger properties than those who came later. Unlike many probability density functions which assume independence among observations,

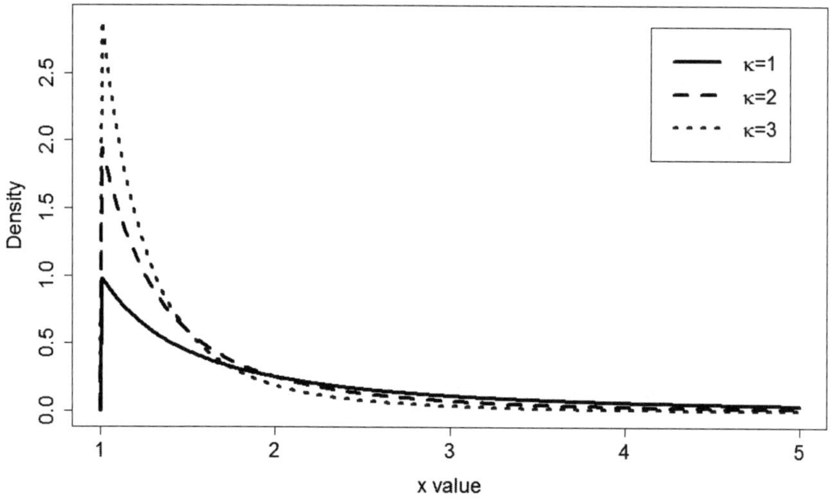

Figure 11.11: Pareto PDF, $\beta = 1, \kappa = 1, 2, 3$

the Pareto distribution assumes that the initial values have an effect on the size of subsequent values. As Midlarsky (1988, p. 494) puts it, "the assumption of a progressive sequential inequality leads to the Pareto distribution."

A thorough technical overview of the Pareto distribution is available at `http://mathworld.wolfram.com/ParetoDistribution.html`.

11.3.3.3 The Gamma Distribution

The PDF for the gamma distribution can be represented for $x \geq 0$ and $\alpha, \beta > 0$ as

$$f(x; \alpha, \beta) = \frac{x^{\alpha-1} e^{-\frac{x}{\beta}}}{\beta^{\alpha}\Gamma(\alpha)}, \tag{11.18}$$

where α is a shape parameter, β is a scale parameter (the mean is $\alpha\beta$), and $\Gamma(\alpha)$ is a named integral function that is equal to $(\alpha - 1)!$ if α is a positive integer.[20] As Figure 11.12 demonstrates, the gamma distribution produces rather different PDFs, depending on the values of α and β.

When $\alpha = 1$, the gamma distribution reduces to the exponential distribution, with $\mu = \beta$. That is because we can think of α as the number of distinct periods of constant risk. The exponential assumes that risk is constant over the entire time, so there is only one period (i.e., $\alpha = 1$). However, imagine that a political scientist were to argue that there are four distinct periods that cabinet

[20] You can find a definition of the gamma function at `http://mathworld.wolfram.com/GammaFunction.html`.

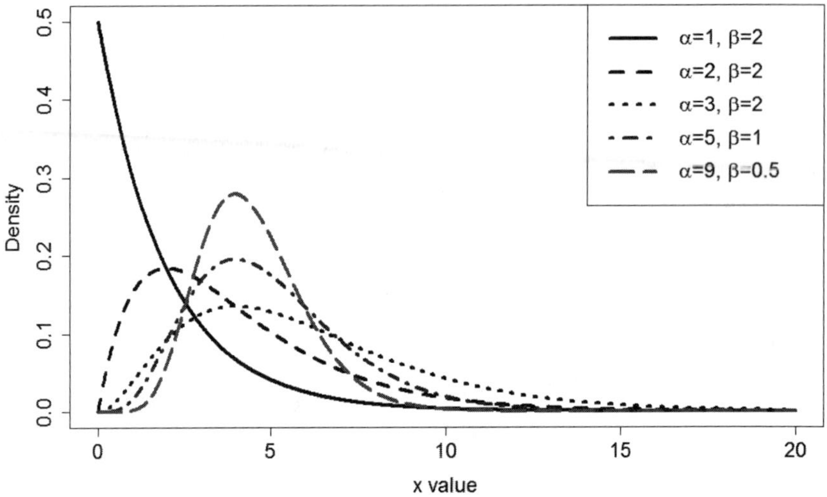

Figure 11.12: Gamma PDFs

governments experience with respect to risk of failure: (1) a honeymoon period with low risk, (2) a period of risk (when the honeymoon is over), (3) a mature period with reduced risk (for those governments that survive), and (4) a high-risk period (because the government is approaching the constitutional limit of its life). If that is a reasonable theory, then one would expect that if a variable measuring the life of a government is produced by a gamma distribution, then $\alpha = 4$.

You can find a thorough technical overview of the gamma distribution at `http://mathworld.wolfram.com/GammaDistribution.html`.

11.3.3.4 The Weibull Distribution

One can represent the PDF for the Weibull distribution as follows:

$$f(x;\alpha,\beta) = \begin{cases} \frac{\alpha x^{\alpha-1}}{\beta^{\alpha}} e^{\left(-\frac{x}{\beta}\right)^{\alpha}} & \text{for} \quad x \geq 0, \\ 0 & \text{for} \quad x < 0, \end{cases} \tag{11.19}$$

where α is a shape parameter and β is the scale parameter. The Weibull PDF is very similar to the gamma PDF; the differences are that Γ is not in the denominator, α is in the numerator, and the exponent of the e is raised to the power α (compare equations (11.18) and (11.19)). As such, note that, as with the gamma distribution, when $\alpha = 1$ the Weibull reduces to the exponential distribution, again with $\mu = \beta$.

We provide some graphs of the Weibull distribution PDF using different values of the parameters in Figure 11.13. You can see that the risk changes over time, and further, that the changes depend on the values of the parameters α and β.

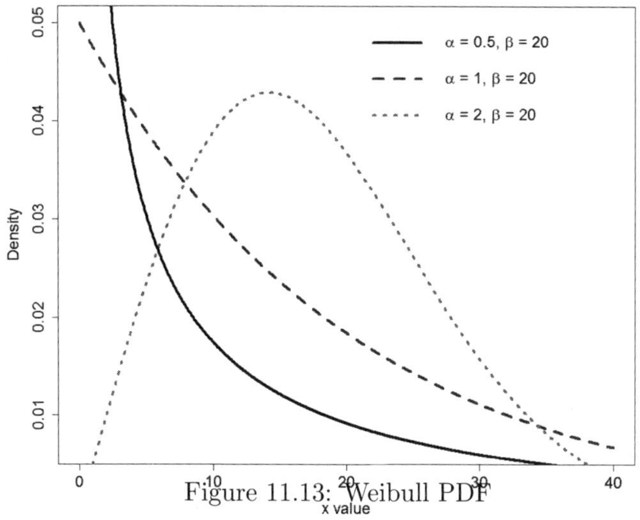

Figure 11.13: Weibull PDF

Lindsey (1995, p. 133) offers the following explanation of the processes that produce a Weibull distribution:

> It can be interpreted as if several processes are running in parallel, with the first to stop ending the duration. This is a weakest link mechanism, as when the failure of some part causes a machine to break down and the total operating time of the machine is the duration.

In other words, imagine that a number of "things" were required for a cabinet government to persist, such that if any one of those "things" were no longer present, the government would fall. The Weibull distribution should be useful when we want to study a variable measuring duration and we believe that the political processes that affect the duration are each necessary conditions to continuation. Though the hypotheses that are frequently tested using models built on the Weibull distribution are not often stated as sets of necessary conditions, it is probably the most widely used duration distribution in political science.

As a concrete example, we may consider Bennett (1997), who studies the duration of international military alliances. He estimates a statistical model that assumes that the duration of alliances have a Weibull probability distribution. In other words, the probability that an alliance is broken at any given moment in time, t, depends on (1) how long the alliance has lasted and (2) a number of other factors that Bennett specifies (e.g., changes in the power of the allies, regime change, etc.; see the article for details).

Those interested in a thorough technical overview of the Weibull distribution should visit `http://mathworld.wolfram.com/WeibullDistribution.html`.

11.3.4 Distributions Used Frequently in Statistical Hypothesis Tests

These distributions are of interest primarily in testing statistical hypotheses and are not much used to structure theoretical expectations. They will also be discussed at length—or at least the tests based on them will be—in your statistics classes. Accordingly, we will introduce them only briefly here.

11.3.4.1 Chi-squared (χ^2) Distribution

The sum of the squares of n independent variables each distributed according to a standard normal distribution is distributed according to a chi-squared (χ^2) distribution. We write a variable distributed in this way as $Q \sim \chi^2(n)$, where n is the number of degrees of freedom. Its PDF is

$$f(x;n) = \begin{cases} \frac{x^{n/2-1}e^{-x/2}}{2^{n/2}\Gamma(n/2)} & \text{for} \quad x \geq 0, \\ 0 & \text{for} \quad x < 0. \end{cases} \tag{11.20}$$

The chi-squared distribution is actually a special case of the gamma distribution, as one can see by using the parameters $\alpha = \frac{n}{2}$ and $\beta = 2$ in equation (11.18).

11.3.4.2 The (Student's) t Distribution

The Student's t distribution is the distribution of a random variable that is proportional to the ratio of a variable that is distributed according to the standard normal distribution and the square root of a variable that is distributed according to a chi-squared distribution. Such ratios arise when normalizing differences in means by the standard deviation. The distribution looks much like a normal distribution but with thicker tails, is particularly useful for small sample sizes, and approaches the standard normal distribution as the sample size approaches infinity. Its PDF, where n is the number of degrees of freedom, is

$$f(x;n) = \frac{\Gamma\left(\frac{n+1}{2}\right)}{\sqrt{n\pi}\,\Gamma\left(\frac{n}{2}\right)} \left(1 + \frac{x^2}{n}\right)^{-\frac{n+1}{2}}. \tag{11.21}$$

11.3.4.3 The F Distribution

The F distribution is the distribution of a random variable that is equal to the ratio of two random variables, each distributed according to a chi-squared distribution and each scaled according to its number of degrees of freedom. It is used commonly in the analysis of variance and in testing the hypothesis that several parameters are not jointly null. If its two degrees of freedom are n_1 and n_2 and if $x \geq 0$, then the F distribution's PDF is

$$f(x;n_1,n_2) = \frac{\sqrt{\frac{(n_1 x)^{n_1} n_2^{n_2}}{(n_1 x + n_2)^{n_1+n_2}}}}{x B\left(\frac{n_1}{2}, \frac{n_2}{2}\right)}, \tag{11.22}$$

where $B(x, y) = \frac{\Gamma(x)\Gamma(y)}{\Gamma(x+y)}$ is the beta function.

11.4 EXERCISES

1. Why can't one create a PDF by plotting the graph of the relative frequency distribution of each value in the sample?

2. What is the difference between a PMF and a PDF?

3. What is the difference between a PDF and a CDF?

4. Write down an example where a scatter plot would be useful for examining the joint distribution of two variables.

5. Why can't we eyeball a probability distribution and determine whether it is normal?

6. What is the difference between a relative frequency distribution of a sample and a PDF?

7. Why does a PDF require calculating an integral?

8. Show that $\text{Var}(X) = E[(X - \mu)^2] = E[X^2] - \mu^2$.

9. An individual benefits from an action whenever $X > 10$. If X is a random variable distributed uniformly on $[0, 25]$, what is the probability that the individual will benefit?

10. Annual country budget deficits (surpluses) are distributed normally, with a mean of $-\$100$ million and a standard deviation of $\$300$ million. What do both of these parameters tell us substantively about this distribution? Explain.

11. Write down a political process that you think might be drawn from the following distributions: normal or log-normal; logistic; or exponential, Pareto, gamma, or Weibull (you should have three political processes).

12. Visit the "Distributions" page of the Virtual Laboratory Website at the University of Alabama, Huntsville (`http://www.math.uah.edu/stat/dist/index.xhtml`). Click on the "Random Variable Experiment" link under "Applets." You can do experiments changing the parameters of a number of distributions (the normal, gamma, chi-squared, Student's *t*, F, beta, Weibull, Pareto, logistic, and log-normal are available). Investigate the distributions covered in this chapter. More explicitly, select a distribution and note the shape and location of the density function. Adjust one of the parameters using the scroll bar. If there is more than one parameter, adjust it. Write down what happens when you adjust each parameter for the following distributions: normal, log-normal, logistic, beta, gamma,

Pareto, and Weibull. Note the distributions that can be made to have a bell shape given some parameter values.

13. Using the random variable experiment applet you used in the previous exercise, run the simulation 1,000 times (set Stop to 1,000) with an update frequency of 10 (use the Update tab), and note the apparent convergence of the empirical density to the true density. What does this imply, in your opinion, for the shape of the distribution of real data relative to the shape of a theoretically derived PDF?

11.5 APPENDIX

Our presentation has been relatively informal, and one can find more formal treatments in Gill (2006) and online (e.g., the various MathWorld entries we noted throughout). Those interested in studying methods as a subfield will want a more thorough treatment. Another place to look is King (1989, chaps. 2 and 3). The National Institute of Standards and Technology Engineering Statistics Handbook, section 1.3.6, "Probability Distributions," available at `http://www.itl.nist.gov/div898/handbook/eda/section3/eda36.htm`, is also a good source.

Part IV

Linear Algebra

Chapter Twelve

Fun with Vectors and Matrices

You first encountered mathematics as whole numbers and learned how to add, subtract, multiply and divide them. That field of inquiry is arithmetic. A loose distinction between arithmetic and algebra is that the latter replaces numbers with variables. So, for example, we can let the variable x take any value within the set of whole numbers. What you may not know is that the algebra you learned in middle school (in the United States) could actually be called scalar algebra, which is to say algebra concerning single variables. Scalar algebra in this usage is a subset of vector algebra, which is itself a subset of matrix algebra. Loosely speaking, vector algebra deals with one-dimensional collections of variables, while matrix algebra deals with two-dimensional collections of variables. Now, instead of a single *scalar* x, we might have the *vector* $\mathbf{x}$, which contains the set of scalars $x_1, x_2, \ldots, x_i, \ldots, x_n$, with each x_i a different element of the vector $\mathbf{x}$. Or we might have the *matrix* X, which contains the set of vectors $\mathbf{x}_1, \ldots, \mathbf{x}_m$ or, equivalently, the set of scalars $x_{11}, \ldots, x_{1m}, x_{21}, \ldots, x_{2m}, \ldots, x_{n1}, \ldots, x_{nm}$. A scalar variable thus has one element, a vector n elements, and a matrix $n \times m$ elements.

One useful way to think about the three sets of algebras is that they are nothing more than different means of representing variables that we want to manipulate. A matrix holds more information than a vector, which holds more information than a scalar. It is often quite sufficient to represent variables in scalar form, and when that is all we need to conduct our business, we turn to scalar algebra. At other times, however, it turns out that there is an advantage to writing down a problem using vectors rather than limiting ourselves to a scalar representation of variables, and in such circumstances we need to understand vector algebra. Finally, for some problems matrix representations of variables are superior to vector or scalar representations. The key point to appreciate is that scalar, vector, and matrix algebra define the rules for manipulating variables in different representations.

Another way to think about this is to make an analogy to the summation operator, $\sum$. We can represent the sum of three values of a variable, x, by writing: $x_1 + x_2 + x_3$. Alternatively, we can more compactly express the same operation as $\sum_{i=1}^{3} x_i$. In this case the gain from using the sum operator is minimal, but there are circumstances in which there are considerable efficiency gains to using one notation versus another. In the case of vector and matrix algebra it turns out that there are gains beyond notational convenience: by studying problems expressed using vector or matrix representation, mathematicians have devised a number of operations and outcomes that make it possible to solve problems such as determining whether three different equations composed of the same variables have a linear relationship across them. So the first task in studying vector and matrix algebra is to learn the notation and then the operations one can perform. That is the focus of this chapter. With that foundation one can then study how to put this knowledge to use to solve problems.

As noted, some problems are easier to solve using vectors or matrices than scalars. This will most commonly be observed in statistics, and here many students of political science need only become familiar with the representation of variables in vector and matrix form (you already know scalar representation—you have been using it all along) so that they can read textbooks and understand research that uses vectors and matrices to present results. This chapter provides the nuts and bolts these students need. Each of Sections 1, 2, and 3 defines in turn scalars, vectors, and matrices, introduces common notation, and briefly discusses some contexts in which they are used in political science. Sections 2 and 3 also detail the rules of vector algebra and matrix algebra. Section 4 provides some additional properties of vectors and matrices that may be useful. Section 5 is a brief illustrative example of the use of matrices in linear regression and ordinary least squares (OLS) estimation. In Chapter 13 we provide a deeper introduction to vector spaces, including such important topics as linear independence and spanning vectors, and illustrate how to use these concepts to solve systems of linear equations. We also discuss the rank of a matrix and provide a sketch of a proof for why OLS estimation is so widely used. For those students who want to pursue more advanced but increasingly common topics in statistics and formal theory useful to political scientists, we offer Chapter 14, which introduces eigenvalues, eigenvectors, and Markov processes. These help us model time series data, networks, and dynamical systems.

12.1 SCALARS

We refer to a **scalar** in this book as any single element of some set. The most common examples we use are elements of the real numbers, $\mathbb{R}$. For example, the number 3.5 is a scalar, as is any variable $x \in \mathbb{R}$.[1] We dealt with the rules for manipulating scalars—arithmetic and algebra—in Chapter 2, and would imagine that most readers are comfortable with these rules by this point in the book. We'll continue using lowercase letters, both Latin and Greek, to refer to scalars. So, if not otherwise specified, you may assume that x, y, and θ are all scalars.

[1]Elements of other sets can also be scalars, e.g., an integer variable ($q \in \mathbb{Z}$) or a complex variable ($r \in \mathbb{C}$). In the latter case the scalar can even have two components of its own, a real part and a complex part.

12.2 VECTORS

There are several ways to conceptualize a **vector**. Perhaps the most intuitive calls on the n-tuples we introduced in Section 2 of Chapter 1. Recall (or just flip back to that section) that $\mathbb{R}^3$ is a space that contains three-dimensional objects such as $(3, 1, 2), (100, 500, 10)$, and (x, y, z). One way of specifying a vector in this space is as an arrow from the origin (the zero) in three dimensions, $(0, 0, 0)$, to the point in question, such as from $(0, 0, 0)$ to $(3, 1, 2)$. The arrow points to $(3, 1, 2)$ in this case. The arrow notes that the vector indicates motion, in general from zero to some (x, y, z). Figure 12.1 displays a graphical representation of the vector from $(0, 0)$ to $(5, 2)$.

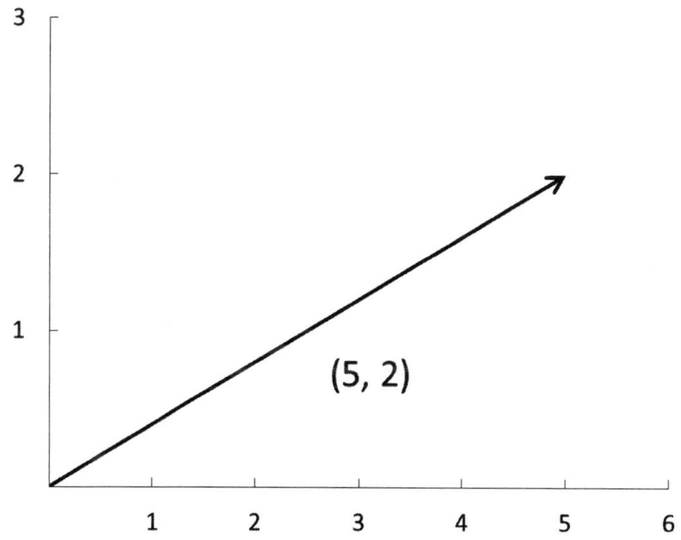

Figure 12.1: Vector $(5, 2)$

Though a graphical representation is often useful to develop intuition, we will by and large stick to a simpler manner of expression for ease of manipulation, in which the vector is simply the n-tuple itself. So we would write $(5, 2)$ for the vector in Figure 12.1, or $(3, 1, 2)$ for our three-dimensional example. Thus, coordinates in any three-dimensional spaces (and any n-dimensional space beyond that) can be seen as vectors.

In both cases, the notation is similar. A vector will either be a lowercase letter in a bold font, such as $\mathbf{x}$ or $\mathbf{a}$, or a lowercase letter with an arrow over it, such as $\vec{x}$. Each **element**, or **component**, of the vector will be denoted by a subscript signifying its place in the vector. So, if $\mathbf{x} = (3, 1, 2)$, then $x_1 = 3$, $x_2 = 1$, and $x_3 = 2$. The **dimension** of a vector is the number of components in the vector.

In general, a vector is an element of a vector space, a concept we will explore in more depth in the next chapter. For now we will focus on simple properties of vectors: a vector's length, vector addition, scalar multiplication, and the dot (or inner or scalar) product.

12.2.1 Vector Length

The **length of a vector**, not to be confused with a vector's dimension, tells us how big it is. In one dimension this is straightforward: the scalar 5 has "length" 5.[2] We could see this by drawing an arrow from 0 to 5 on the x-axis and measuring its length. The same idea is true in more dimensions but requires a bit more effort. You may remember how it works in two dimensions from high school geometry. If $\mathbf{a}$ is two-dimensional, then the length of $\mathbf{a}$, denoted $||\mathbf{a}||$, is $\sqrt{a_1^2 + a_2^2}$. This generalizes: for any vector of dimension n, its length is given by $||\mathbf{a}|| = \sqrt{a_1^2 + a_2^2 + \ldots + a_n^2}$. For example, the length of the vector $(2, 4, 4, 1)$ is $\sqrt{2^2 + 4^2 + 4^2 + 1^2} = \sqrt{4 + 16 + 16 + 1} = \sqrt{37}$. We call a vector **normalized** if it has length 1. Often we will divide a vector by its length (using scalar multiplication, defined below) to normalize it. Finally, note that $||\mathbf{a}|| \geq 0$, and is equal to 0 only if the vector $\mathbf{a} = \mathbf{0}$, where $\mathbf{0} = (0, 0, \ldots, 0)$. That is, a vector has length 0 only if all of its elements are zero. We call $\mathbf{0}$ the **zero vector**.

12.2.2 Vector Addition

Vectors add just like scalars (i.e., numbers). To add (or subtract) vectors, they must have the same dimension. The only trick beyond that requirement is to keep track of which numbers add to which, and where the sums go. In every case, the first component of the first vector adds to the first component of the second vector to form the first component of the added vector, and so on. Using our notation, $\mathbf{a}+\mathbf{b} = (a_1+b_1, a_2+b_2, \ldots, a_n+b_n)$ for two n-dimensional vectors. The same is true for subtraction, so that $\mathbf{a}-\mathbf{b} = (a_1-b_1, a_2-b_2, \ldots, a_n-b_n)$. Here are a few concrete examples: $(1, 2)+(5, 8) = (6, 10)$, $(5, -3, -6)+(1, 8, 7) = (6, 5, 1)$, and $(5, 1, 4, 1) - (1, 2, 3, 4) = (4, -1, 1, -3)$.

We can use this and the concept of vector length to compute the lengths of sums of vectors:[3]

$$||\mathbf{a} + \mathbf{b}|| = \sqrt{(a_1 + b_1)^2 + (a_2 + b_2)^2 + \ldots + (a_n + b_n)^2}.$$

One could also replace the + with a − to get the length of vector differences:

$$||\mathbf{a} - \mathbf{b}|| = \sqrt{(a_1 - b_1)^2 + (a_2 - b_2)^2 + \ldots + (a_n - b_n)^2}.$$

[2] The length of a negative number is the absolute value of that number.

[3] If you do remember your high school geometry, you can check that this works for right triangles, which can be seen as adding a vector on the x-axis to one on the y-axis. The sum is the hypotenuse of the triangle. So, for example, the length of the hypotenuse of a right triangle with sides 3 and 4 is $\sqrt{(3 + 0)^2 + (0 + 4)^2} = \sqrt{9 + 16} = \sqrt{25} = 5$.

This gives us a scalar measure of the distance between two vectors. The **triangle inequality** relates the length of the sum of two vectors to the sum of the lengths of the vectors, and can be useful in proofs. It states that $||\mathbf{a}+\mathbf{b}|| \leq ||\mathbf{a}|| + ||\mathbf{b}||$. That is, the length of a sum of vectors is never greater than the sum of the lengths of the individual vectors.

Graphically, one can add vectors by moving the second one so that its tail (i.e., the end at 0) now touches the head (i.e., the end with the arrow) of the first, and then drawing a line from 0 to the head of the second. This is done for the vectors $(5, 2)$ and $(1, 1)$ in Figure 12.2. We see that the sum is $(6, 3)$. Subtracting works similarly, as long as one swaps the head and the tail of the second vector.

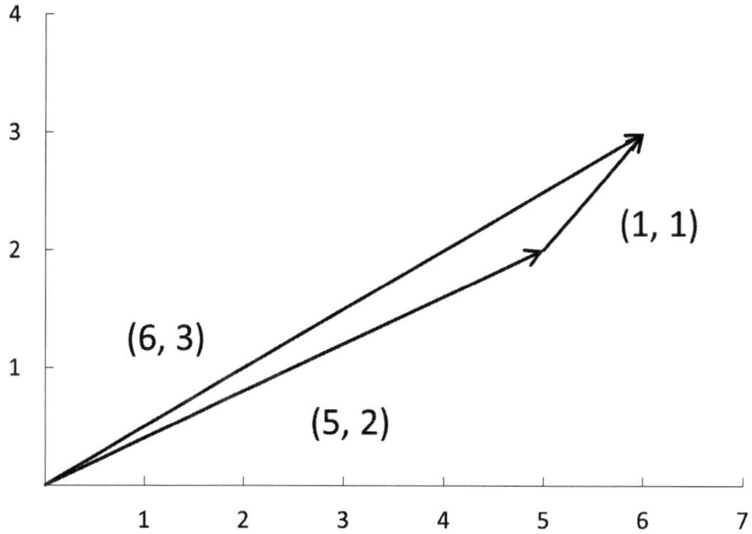

Figure 12.2: Vector Addition: $(5, 2) + (1, 1)$

12.2.3 Scalar Multiplication

Multiplication is a bit more complex for vectors, as it can have different meanings. The most familiar form of multiplication is **scalar multiplication**, which is multiplication of a vector by a scalar. To accomplish this, one needs to multiply each element in the vector by the scalar. So, in general, if $\mathbf{x}$ is an n-dimensional vector and c is a scalar, then $c\mathbf{x} = (cx_1, cx_2, \ldots, cx_n)$. A more concrete example would be $5\mathbf{a}$, where $\mathbf{a} = (2, 1)$. The multiplied vector is $(10, 5)$, where each of 2 and 1 have been multiplied by the scalar 5. Dividing by a scalar works the same as multiplying by one over that scalar (i.e., multiplying each element by $\frac{1}{c}$), so we do not cover it separately. The length of a scalar multiple of a vector is just the absolute value of the scalar times the length of the vector, or $||c\mathbf{x}|| = |c|||\mathbf{x}||$.

One can represent scalar multiplication graphically by stretching (for scalars greater than 1) or contracting (for scalars less than 1) the original vector according to the size of the scalar, switching the head and the tail if the scalar is negative.[4] Figure 12.3 illustrates this for $5\mathbf{a}$, where $\mathbf{a} = (2, 1)$.

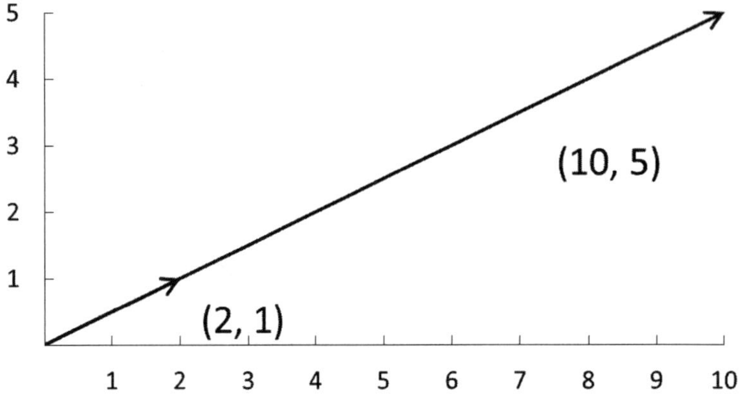

Figure 12.3: Scalar Multiplication: $5\mathbf{a}$, where $\mathbf{a} = (2, 1)$

12.2.4 Vector Multiplication

One can guess from the scalar rule that multiplying two vectors might be tricky. In fact, there are many ways in which one can do so. Luckily, we'll only need to consider two in typical applications, and one more commonly than the other. We'll discuss the less common one, the Kronecker product, when talking about matrices below. Here we'll cover the more common one, the scalar product.[5]

The **scalar product** is a way of multiplying vectors that results in a scalar. It is also known as a **dot product** owing to its common notation, $\mathbf{a} \cdot \mathbf{b}$, read "a dot b."[6] Its computation is more straightforward that its nomenclature. In general, if $\mathbf{a}$ and $\mathbf{b}$ are both n-dimensional vectors, then $\mathbf{a} \cdot \mathbf{b} = a_1 b_1 + a_2 b_2 + \ldots + a_n b_n$. Or, using summations, $\mathbf{a} \cdot \mathbf{b} = \sum_{i=1}^{n} a_i b_i$. In words, one multiplies each component of the first vector by the corresponding component of the second, and then adds up these products for all n components. This requires that the two vectors be of equal dimension, just as does addition. Here are a couple of concrete examples: $(3, 1) \cdot (2, 3) = 6 + 3 = 9$ and $(6, 5, 4) \cdot (9, 8, 7) = 54 + 40 + 28 = 122$.

[4] This last rule is whence the rule for subtraction arises.

[5] Another one that might arise infrequently in the context of permutations is the cross product, written as $\mathbf{a} \times \mathbf{b}$. This returns a vector. For more detail, see the Wikipedia or Wolfram MathWorld entries.

[6] It is also called in some contexts an **inner product**, but this usage is more rare in political science.

The dot product also has a geometrical interpretation that is particularly intuitive in two dimensions. Consider any two vectors $\mathbf{a} \cdot \mathbf{b}$ in the plane $\mathbb{R}^2$. Since both start at the origin, there is some angle between them. Call this angle θ. This is shown graphically in Figure 12.4. The dot product in two dimensions is $\mathbf{a} \cdot \mathbf{b} = ||\mathbf{a}||||\mathbf{b}|| \cos(\theta)$.[7]

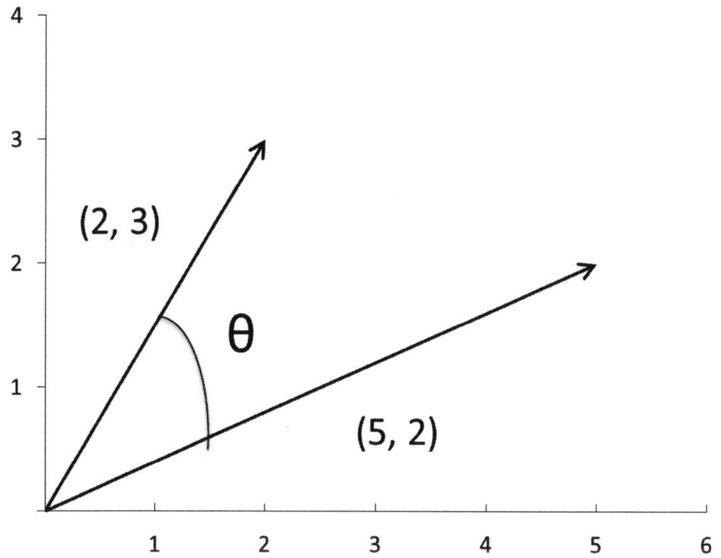

Figure 12.4: Dot Product

When $\theta = 0$, so that the vectors point in exactly the same direction, the cosine is equal to 1, and the dot product is just the product of the lengths. When $\theta = \frac{\pi}{2}$ (or 90°), so that the vectors are perpendicular (also known as orthogonal) to each other, then the cosine is zero and the dot product is zero. Thus, the dot product can also tell you when the two vectors are perpendicular to each other.[8] Finally, dot products obey the **Cauchy-Schwartz inequality**: $|\mathbf{a} \cdot \mathbf{b}| \leq ||\mathbf{a}||||\mathbf{b}||$ This can be arranged to yield $\frac{|\mathbf{a} \cdot \mathbf{b}|}{||\mathbf{a}||||\mathbf{b}||} \leq 1$, and we can define

[7]Geometrically, the dot product considers the "pieces" of the two vectors that point in the same direction. To find them, we need to project one vector onto the other; the length of the projection of **a** onto **b** is $||\mathbf{a}|| \cos(\theta)$ (and of **b** onto **a** is $||\mathbf{b}|| \cos(\theta)$). To see why, draw a line in Figure 12.4 from $(5, 2)$ to $(2, 3)$ that is perpendicular to $(5, 2)$ and connects to the head of $(2, 3)$. This forms a right triangle with $(2, 3)$ as the hypotenuse. In this right triangle, $||(2, 3)|| \cos(\theta)$ is the length of the side of the triangle that lies along the vector $(5, 2)$. Hence the $\cos(\theta)$ in the dot product in two dimensions.

[8]More generally, the dot product gets at the concept of *projection*, indicating the degree to which one vector overlaps another. This overlap is what we computed in the previous footnote for two two-dimensional vectors. One can frame certain concepts in statistics in terms of projections, but that is beyond the scope of this book.

the LHS of this equation as the cosine of the angle between any two vectors. This can be useful for some proofs.

12.2.5 Why Should I Care?

Vectors are common in both statistics and formal theory. In statistics, data are stored in vectors: typically each variable has an associated vector, and each component of this vector is an observation or data point. If there are n data points, the vectors are n-dimensional. Parameters are also stored in vectors. For example, logistic regression returns a vector of coefficients that tells the analyst what effect varying each of the independent variables has on the dependent variables: $y \in (0, 1) = e^{\alpha+\beta X+\epsilon}$, where β is the coefficient vector.

In formal theory vectors play a similar role, allowing one to analyze problems with more than one dimension of choice or more than one player. For example, in the bargaining model we discuss in Chapter 4, each player's strategy involves a bargaining offer comprising what each player would receive in the bargain. These strategies are represented as vectors: $\mathbf{x} = (x_1, x_2)$, where each x_i is the offer to player i. As a result, you'll be dealing with vectors quite often!

12.3 MATRICES

A **matrix** is a rectangular table of numbers or variables that are arranged in a specific order in rows and columns. Just like any other table, matrices consist of columns and rows, and they can vary in size from a few columns and rows to hundreds of thousands of rows and columns. Any dataset you might use in political science is nothing more than a large matrix, and as you will learn in statistics coursework, when you estimate OLS regressions using such datasets, the computer actually uses matrix algebra to calculate coefficient estimates and standard errors. You get to do this (for a very small dataset) in the problems at the end of this chapter, and we provide an extended example in the last section of this chapter.

The size of a matrix in mathematics is known as its dimensions and is expressed in terms of how many rows, n, and columns, m, it has, written as $n \times m$ (read "n by m"). Matrices are often represented by capital letters rather than writing out the whole matrix, and sometimes their size is subscripted as well. A matrix $A_{n\times m}$ thus is a matrix with n rows and m columns. Individual elements of the matrix are scalars, and are typically represented by the lowercase letter corresponding to the matrix, with subscripts indicating their position in the matrix. The row number is always first in this notation. So, If we are considering matrix A, then the element in row i and column j is given by a_{ij}.[9] For example, a_{32} is used to refer to the number or variable that is in the third row and second

[9]Be sure not to confuse the subscript on A, which tells you how big the matrix is, with the subscript on a, which tells you where the element is.

column of the matrix. We can write a general three-by-two matrix as

$$A_{3\times 2} = \begin{pmatrix} a_{11} & a_{12} \\ a_{21} & a_{22} \\ a_{31} & a_{32} \end{pmatrix}.$$

Since matrices may be any shape, we can also consider matrices that have only one row or one column. Both these types have identical properties for our purposes to the vectors we discussed in the previous section. We say a matrix is a **column vector** if it has only one column but two or more rows, and a **row vector** if it has only one row but two or more columns. A **scalar** is a matrix with only one column and one row.

In the following subsections we provide some commonly used matrices, detail matrix transpose, addition, and multiplication operations, and introduce the trace, determinant, and matrix inverse. In the next section we discuss some properties of matrices and vectors.

12.3.1 Some Special Types of Matrices

There are several special types of matrices that are important to know in matrix algebra, either because they allow you to perform certain operations or because they allow you to see at a glance whether a certain operation is possible at all, and what the results should look like if so.

A **square matrix** is a matrix that has an equal number of columns and rows, i.e., $m = n$. For example,

$$A_{3\times 3} = \begin{pmatrix} a_{11} & a_{12} & a_{13} \\ a_{21} & a_{22} & a_{23} \\ a_{31} & a_{32} & a_{33} \end{pmatrix}.$$

A **zero matrix** is a square matrix in which all elements are 0.

The main **diagonal** of a matrix comprises the elements running from the top left to the bottom right of the matrix. A **diagonal matrix** is a square matrix in which all elements other than those on the main diagonal are zero. An **identity matrix** is a diagonal matrix in which all elements on the main diagonal are 1. Identity matrices are used frequently enough to have their own notation, $I_{n\times n}$, or simply I when there is no confusion about its dimension. The identity matrix is special because, when multiplied by another matrix, it produces the original matrix back again (i.e., $AI = IA = A$). Examples of both types of matrices are

$$D_{3\times 3} = \begin{pmatrix} a_{11} & 0 & 0 \\ 0 & a_{22} & 0 \\ 0 & 0 & a_{33} \end{pmatrix}, I_{3\times 3} = \begin{pmatrix} 1 & 0 & 0 \\ 0 & 1 & 0 \\ 0 & 0 & 1 \end{pmatrix},$$

and

$$\begin{pmatrix} a_{11} & a_{12} & a_{13} \\ a_{21} & a_{22} & a_{23} \\ a_{31} & a_{32} & a_{33} \end{pmatrix} \times \begin{pmatrix} 1 & 0 & 0 \\ 0 & 1 & 0 \\ 0 & 0 & 1 \end{pmatrix} = \begin{pmatrix} a_{11} & a_{12} & a_{13} \\ a_{21} & a_{22} & a_{23} \\ a_{31} & a_{32} & a_{33} \end{pmatrix}.$$

A **lower triangular matrix** has non-zero elements only on or below the main diagonal, while an **upper triangular matrix** has non-zero elements only on or above the main diagonal. Examples of these are:

$$L_{3\times 3} = \begin{pmatrix} a_{11} & 0 & 0 \\ a_{21} & a_{22} & 0 \\ a_{31} & a_{32} & a_{33} \end{pmatrix}, U_{3\times 3} = \begin{pmatrix} a_{11} & a_{12} & a_{13} \\ 0 & a_{22} & a_{23} \\ 0 & 0 & a_{33} \end{pmatrix}.$$

A **symmetric matrix** is a square matrix in which the elements are symmetric about the main diagonal, or more formally one in which $a_{ij} = a_{ji}$. For example,

$$A_{3\times 3} = \begin{pmatrix} a_{11} & a_{12} & a_{13} \\ a_{12} & a_{22} & a_{23} \\ a_{13} & a_{23} & a_{33} \end{pmatrix}.$$

A **submatrix** of an element is the matrix that remains when we take out the row and column in which the element is (so it has one fewer column and row than the original). Thus, for example, the submatrix for a_{21} in a three-by-three square matrix is

$$\begin{pmatrix} - & a_{12} & a_{13} \\ - & - & - \\ - & a_{32} & a_{33} \end{pmatrix} \text{ or } \begin{pmatrix} a_{12} & a_{13} \\ a_{32} & a_{33} \end{pmatrix}.$$

A **permutation matrix** is one in which there is only a single value of 1 in any row and column, with all other elements 0. The identity matrix is a trivial permutation matrix that does not permute the elements. Other permutation matrices, such as the example below, flip rows or columns of the thing they are multiplying:[10]

$$P_{3\times 3} = \begin{pmatrix} 0 & 1 & 0 \\ 1 & 0 & 0 \\ 0 & 0 & 1 \end{pmatrix}.$$

An **idempotent matrix** A is a matrix with the property $AA = A$. That is, when you multiply it by itself, it returns the original matrix. A **singular matrix** is one that has a determinant of 0, while a **nonsingular matrix** has a determinant that is not 0 (we describe determinants below, in Subsection 12.3.6). Nonsingular matrices have inverses, as we will see shortly. A **block matrix** or **partitioned matrix** is a matrix composed of other, smaller matrices. For example, if A, B, C, D are all three-by-three matrices, then the block partitioned matrix E below is a six-by-six matrix whose elements are the elements of each of the four component matrices. A **block diagonal matrix** has blocks only on the diagonal, so B and C in E below would each be zero in such a matrix (or, more precisely, the three-by-three zero matrix):

$$E_{6\times 6} = \begin{pmatrix} A & B \\ C & D \end{pmatrix}.$$

[10] Flip back to Chapter 9, Section 2.2, if you forget what a permutation is.

Finally, an **orthogonal matrix** is one in which all the columns of the matrix, treated as vectors, are orthogonal, or perpendicular, to each other. The identity matrix is an orthogonal matrix. An **orthonormal matrix** is an orthogonal matrix in which the length of each column is 1. The identity matrix is orthonormal as well.

12.3.2 Matrix Transposition

The next few subsections detail operations one can perform on a matrix. We begin with the **transpose** of a matrix. The transpose of a matrix is another matrix in which the rows and columns have been switched, i.e., the rows of the first matrix are written as columns in the second and the columns in the first matrix are written as rows in the second. Formally, to find the transpose B of a matrix A, rewrite each element in B so that $b_{j,i} = a_{i,j}$. The typical notation for the transpose of A is either A^T or A'. We will use the former in this book as a rule, but you will see the latter frequently. Here is an example of how to find the transpose of a matrix:

$$
\begin{aligned}
A &= \begin{bmatrix} 1 & 3 & 0 \\ -1 & 6 & 2 \end{bmatrix}, \\
A^T &= \begin{bmatrix} 1 & -1 \\ 3 & 6 \\ 0 & 2 \end{bmatrix}.
\end{aligned}
$$

Note that the transpose of a matrix has the same number of columns as the original had rows, and the same number of rows as the original had columns. So, for the example here, $A_{2\times3}$ becomes $A^T_{3\times2}$. Because of this, as we will see below, a matrix can always be multiplied with its transpose, which is a useful property of the transpose, as we will see. For instance, finding the coefficient estimates in OLS regression requires transposition. Further, transposition can be used on vectors as well. The transpose of a row vector is a column vector, and the transpose of a column vector is a row vector.

12.3.3 Matrix Addition (and Subtraction)

Matrix addition (and subtraction) is straightforward, and follows the exact same rules as vector addition (and subtraction): one adds (or subtracts) the corresponding elements of the two matrices and places the sum (or difference) in the corresponding location in the new matrix. As a precondition, as with vectors, both matrices have to be of the same dimensions. More formally, given two matrices A and B of equal dimensions, the operation $A + B$ will result in a matrix C with the same dimensions where each element $c_{i,j} = a_{i,j} + b_{i,j}$. For subtraction, e.g., $A - B$, each element $c_{i,j}$ in C will equal $a_{i,j} - b_{i,j}$.

Here is an example of addition to illustrate this further:

$$\begin{bmatrix} 1 & -2 \\ 0 & 5 \\ 4 & 3 \end{bmatrix} + \begin{bmatrix} 3 & 9 \\ -1 & 1 \\ 0 & 2 \end{bmatrix} = \begin{bmatrix} 4 & 7 \\ -1 & 6 \\ 4 & 5 \end{bmatrix}.$$

And another example, of subtraction:

$$\begin{bmatrix} 1 & -2 \\ 0 & 5 \\ 4 & 3 \end{bmatrix} - \begin{bmatrix} 3 & 9 \\ -1 & 1 \\ 0 & 2 \end{bmatrix} = \begin{bmatrix} -2 & -11 \\ 1 & 4 \\ 4 & 1 \end{bmatrix}.$$

12.3.4 Matrix Multiplication

As with vectors, there is more than one type of multiplication that can be accomplished with matrices. The two most common mirror scalar multiplication and the dot product for vectors; these are called scalar or matrix multiplication, respectively. We will consider each in turn. We will also consider another useful type of multiplication that may be applied to vectors as well, known as the Kronecker product.

12.3.4.1 Scalar Multiplication

Scalar multiplication is almost as easy as matrix addition and subtraction, and again mirrors exactly the rule for vectors. Multiply each individual element of the matrix by the scalar to find the product. Formally, $C = rA$, where each $c_{i,j} = r \times a_{i,j}$. For example,

$$5 \times \begin{bmatrix} 1 & -2 \\ 0 & 5 \\ 4 & 3 \end{bmatrix} = \begin{bmatrix} 5 & -10 \\ 0 & 25 \\ 20 & 15 \end{bmatrix}.$$

12.3.4.2 Matrix Multiplication

The second kind of matrix multiplication is between two matrices, or between a matrix and a vector. This is where matrix algebra can get confusing and difficult, as you might remember from high school: the product of two matrices looks nothing like the multiplication of scalars (numbers) we are used to. In order to be able to multiply two matrices, the number of columns in the first must match the number of rows in the second matrix, e.g., $A_{n \times m}$ and $B_{m \times p}$. If this is not the case, the matrices cannot be multiplied. This goes for multiplying vectors by matrices as well: B has to be a column vector of length m, or A a row vector of length m, if one of A or B in the example above is a vector.

If A and B can be multiplied, however, the product is a matrix that will have the same number of rows as the first matrix and the same number of columns as the second matrix, i.e., $A_{n \times m} B_{m \times p} = C_{n \times p}$. If A is a matrix and B is a column vector, then C also is a column vector; if A is a row vector and B is a matrix, then C is also a row vector.

Once we have figured out that it is possible to multiply two matrices, A and B, the multiplication process is a little bit complex. Here is the formal definition of how to do it: for $C_{n \times p} = A_{n \times m} B_{m \times p}$, each element is $c_{i,j} = \sum_{k=1}^{m} a_{ik} b_{kj} = a_{i1} b_{1j} + a_{i2} b_{2j} + \ldots + a_{im} b_{mj}$. And here is an example to illustrate this:

$$
\begin{aligned}
AB &= \begin{bmatrix} 1 & -2 \\ 0 & 5 \\ 4 & 3 \end{bmatrix} \times \begin{bmatrix} 3 & 1 & 4 \\ -1 & 2 & 5 \end{bmatrix} \\
&= \begin{bmatrix} (1 \times 3) + (-2 \times -1) & (1 \times 1) + (-2 \times 2) & (1 \times 4) + (-2 \times 5) \\ (0 \times 3) + (\ 5 \times -1) & (0 \times 1) + (\ 5 \times 2) & (0 \times 4) + (\ 5 \times 5) \\ (4 \times 3) + (\ 3 \times -1) & (4 \times 1) + (\ 3 \times 2) & (4 \times 4) + (\ 3 \times 5) \end{bmatrix} \\
&= \begin{bmatrix} 3 + 2 & 1 - \ 4 & 4 - 10 \\ 0 - 5 & 0 + 10 & 0 + 25 \\ 12 - 3 & 4 + \ 6 & 16 + 15 \end{bmatrix} \\
&= \begin{bmatrix} 5 & -3 & -6 \\ -5 & 10 & 25 \\ 9 & 10 & 31 \end{bmatrix}.
\end{aligned}
$$

In words, to multiply two matrices, you have to construct a new matrix with as many rows as the first matrix and as many columns as the second. Then each element in the new matrix will equal the sum of the product of the kth element in the corresponding row of the first matrix and the kth element in the corresponding column in the second matrix.

This may seem arbitrary, but it is very similar in nature to the dot product. In fact, the formula for each element c_{ij} in the product is the same as the formula for a dot product between two vectors. How does this work? Let the vector $\mathbf{a}_i$ be the ith row of A, and the vector $\mathbf{b}_j$ be the jth column of B. Then c_{ij}, the element in the ith row and jth column of C, is the dot product of $\mathbf{a}_i$ and $\mathbf{b}_j$: $c_{ij} = \mathbf{a}_i \cdot \mathbf{b}_j$.[11] This provides a handy way to remember how to multiply matrices: fill out the elements of C by "lifting" each row from the left matrix in the product, "rotating" it ninety degrees clockwise, and taking the dot product of it with a column from the matrix on the right in the product. Take a look at the example above, and see if you can replicate the steps via this method. This method also lets you check quickly to see whether multiplication is possible. A dot product requires two vectors of equal length, so the length of the rows in A (which is the number of columns) must be the same as the length of the columns in B (which is the number of rows).

It's easy to practice multiplying matrices together and worthwhile to do so, given the frequency with which they appear in statistics. We offer some examples at the end of this chapter, but really, populating two matrices with random

[11]Conversely, one can think of a dot product as matrix multiplying a row vector by a column vector.

numbers is sufficient. Start with two square matrices of the same dimensions before branching out to ones that differ.

If you do this exercise, you may note that the order in which the matrices are multiplied matters. That is, matrix multiplication does not *commute*. We'll discuss this more later in this chapter. For now we'll deal with some of its consequences. First, consider two ways of multiplying $A_{3\times 2}$ and $B_{2\times 4}$. $A_{3\times 2}B_{2\times 4}$ is possible, as the number of columns of the first equals the number of rows of the second. However, $B_{2\times 4}A_{3\times 2}$ is not possible, as this is not true. Thus for some cases you can only multiply matrices in a certain order.

Second, even when matrix multiplication is possible in both orders, the product may not be the same. Consider the example above, except in the reverse order. In this case we get the following result:

$$\begin{aligned}
BA &= \begin{bmatrix} 3 & 1 & 4 \\ -1 & 2 & 5 \end{bmatrix} \times \begin{bmatrix} 1 & -2 \\ 0 & 5 \\ 4 & 3 \end{bmatrix} \\
&= \begin{bmatrix} (\ 3\times 1) + (1\times 0) + (4\times 4) & (\ 3\times -2) + (1\times 5) + (4\times 3) \\ (-1\times 1) + (2\times 0) + (5\times 4) & (-1\times -2) + (2\times 5) + (5\times 3) \end{bmatrix} \\
&= \begin{bmatrix} 3+0+16 & -6+\ \ 5+12 \\ -1+0+20 & 2+10+15 \end{bmatrix} \\
&= \begin{bmatrix} 19 & 11 \\ 19 & 27 \end{bmatrix}.
\end{aligned}$$

Because of this, we must be very careful when we multiply matrices. We call **left multiplication** the act of multiplying by a matrix on the left and **right multiplication** the act of multiplying by a matrix on the right. For example, left-multiplying B by A means the product AB, while right-multiplying B by A means the product BA.

As you can see from these two examples, multiplying matrices is very labor-intensive, and the number of calculations needed increases very fast with the size of the matrices involved. You will not have to multiply matrices larger than three-by-three here, although you might have to deal with slightly larger matrices in your methods classes if you calculate OLS regression coefficients by hand. Computers, thankfully, can do this much faster and more effectively than we can. Before computers were commonly used to deal with large matrices, a lot of this type of work had to be done by hand. Imagine a model with one dependent variable and three independent or explanatory variables (plus the intercept term), e.g., $y = a + b_1x_1 + b_2x_2 + b_3x_3 + \epsilon$. Let's say you have thirty observations, in order to get reasonably accurate coefficient estimates. To calculate just the coefficient estimates will then, as part of the process, involve multiplying a four-by-thirty matrix by a thirty-by-four matrix!

12.3.4.3 Kronecker Product

While scalar and matrix multiplication are the most common products of matrices, they are not the only ones. Leopold Kronecker, a nineteenth-century German mathematician, is credited with developing the **Kronecker product**, a particular product of two matrices of different dimensions that proves to be rather useful in some contexts you will encounter. For example, panel corrected standard error (PCSE) models are developed using Kronecker products (Beck and Katz, 1995).

Unlike regular matrix multiplication, one can take a Kronecker product of any two matrices of arbitrary size. Given two matrices A and B with the respective dimensions $m \times n$ and $p \times q$, the Kronecker product, denoted $A \otimes B$, will be a matrix with the dimensions $mp \times nq$. Thus, the Kronecker product can vastly increase the dimensions of a matrix. Fortunately, however, finding each element of this matrix is easier than in regular matrix multiplication. One can think of the Kronecker product as consisting of different blocks: each block consists of the products of an element in the first matrix, say a_{11}, and each individual element in the second matrix, e.g., $b_{11}, b_{12}, \ldots b_{pq}$. Thus the first elements in the Kronecker product would be $a_{11} \times b_{11}$, $a_{11} \times b_{12}$, and so on, until all the elements in B have been multiplied with a_{11}. This would be followed by the second block, which consists of the products of a_{12} and all elements in B, and so on:

$$\begin{aligned}
A \otimes B &= \begin{bmatrix} a_{11} & a_{12} \\ a_{21} & a_{22} \end{bmatrix} \otimes \begin{bmatrix} b_{11} & b_{12} & b_{13} \\ b_{21} & b_{22} & b_{23} \end{bmatrix} \\
&= \begin{bmatrix} a_{11}B & a_{12}B \\ a_{21}B & a_{22}B \end{bmatrix} \\
&= \begin{bmatrix} a_{11}b_{11} & a_{11}b_{12} & a_{11}b_{13} & a_{12}b_{11} & a_{12}b_{12} & a_{12}b_{13} \\ a_{11}b_{21} & a_{11}b_{22} & a_{11}b_{23} & a_{12}b_{21} & a_{12}b_{22} & a_{12}b_{23} \\ a_{21}b_{11} & a_{21}b_{12} & a_{21}b_{13} & a_{22}b_{11} & a_{22}b_{12} & a_{22}b_{13} \\ a_{21}b_{21} & a_{21}b_{22} & a_{21}b_{23} & a_{22}b_{21} & a_{22}b_{22} & a_{22}b_{23} \end{bmatrix}.
\end{aligned}$$

12.3.5 Trace

After that complexity, you will be relieved at the simplicity of the next operator. The **trace** of an $n \times n$ square matrix is the sum of its diagonal elements. Formally, $\text{Tr}(A) = \sum_{i=1}^{n} a_{ii} = a_{11} + a_{22} + \ldots + a_{nn}$. While the trace comes up rarely, you may see it in reference to eigenvalues, discussed in Chapter 14, as the trace of a matrix is the sum of its eigenvalues.

12.3.6 Determinant

The **determinant** of a matrix is a commonly used function that converts the matrix into a scalar. Determinants are defined only for square matrices. One can rapidly memorize the formula for the determinant of a two-by-two matrix, but most students find the function for the determinant of an $n \times n$ matrix, where $n > 2$, a bit more challenging. We will begin with the two-by-two case and then consider the $n \times n$ case.

First, let us introduce some notation. You are familiar with the absolute value sign: $|-1| = 1$. In scalar algebra the vertical bars signify the absolute value of a number. In matrix algebra those same vertical bars are used to identify determinants. Thus, $|A|$ is the determinant of matrix A. Alternatively, some people use $\det(A)$ to signify the determinant of matrix A. We will use both, but be aware that the determinant may be negative, unlike an absolute value!

Consider the two-by-two matrix,

$$A = \begin{bmatrix} a_{11} & a_{12} \\ a_{21} & a_{22} \end{bmatrix}.$$

The determinant of A is the difference of the diagonal products:

$$|A| = (a_{11} \cdot a_{22}) - (a_{12} \cdot a_{21}).$$

Okay, so that's the easier bit. Now let's tackle the general $n \times n$ matrix, where $n > 2$. There are a variety of different ways to find determinants of larger matrices, and we are going to present what is known as a **Laplace expansion.**[12] The basic idea is that we can easily calculate the determinant of two-by-two matrices, so our problem can be reduced to figuring out a general rule for using determinants of the 2×2 submatrices to calculate the determinant of the larger matrix. And it turns out that such a rule exists: informally, the determinant of a three-by-three matrix is the sum of the products of elements in any given row or column, alternating in sign, and the determinants of specific 2×2 submatrices. For bigger matrices, you reduce them by one dimension each time until you get a 2×2 matrix.

Consider the following three-by-three matrix:

$$B = \begin{bmatrix} b_{11} & b_{12} & b_{13} \\ b_{21} & b_{22} & b_{23} \\ b_{31} & b_{32} & b_{33} \end{bmatrix}.$$

Select a given element, say b_{23}. Define the **minor** of element b_{23} as the determinant of the submatrix of b_{23}. Recall that the submatrix of b_{23} is the matrix remaining when we eliminate the elements in the row and column of which b_{23} is the intersection. Thus, the minor of b_{23} is

$$M_{23} = \begin{vmatrix} b_{11} & b_{12} \\ b_{31} & b_{32} \end{vmatrix} = (b_{11} \cdot b_{32}) - (b_{31} \cdot b_{12}).$$

Similarly, the minor of b_{11} is

[12]By introducing permutations one can present the Laplace expansion technique much more formally than we do here. Another common technique for computing determinants is to "decompose" the matrix into a product of matrices with more attractive properties, then use some of the properties of the determinant, provided in the next section, to compute the determinant quickly. We'll touch on this briefly in the next chapter.

$$M_{11} = \begin{vmatrix} b_{22} & b_{23} \\ b_{32} & b_{33} \end{vmatrix} = (b_{22} \cdot b_{33}) - (b_{32} \cdot b_{23}).$$

Good enough, but what does one do with a minor? Well, it turns out that the determinant of a matrix is related to the sum of its signed minors. A **cofactor** is a minor with a prescribed sign. If we define the minor for an element in row i and column j as M_{ij}, then the sign of the cofactor C_{ij} retains the sign of its minor (M_{ij}) when the sum of $i + j$ is even. When the sum of $i + j$ is odd, then the sign of the cofactor C_{ij} is the opposite of the sign of its minor (M_{ij}). It is perhaps useful to illustrate by listing the signs of the cofactors (in parentheses) next to their elements in the B matrix above:

$$\begin{bmatrix} b_{11}(+) & b_{12}(-) & b_{13}(+) \\ b_{21}(-) & b_{22}(+) & b_{23}(-) \\ b_{31}(+) & b_{32}(-) & b_{33}(+) \end{bmatrix}.$$

To recapitulate, the minor of an element is the determinant of the submatrix of that element. The cofactor of a minor is the signed minor of an element such that its sign is the same as the minor when the row and column numbers of the element have an even sum, and its sign is the opposite of the minor when the row and column numbers of the element have an odd sum. If you have trouble remembering the sign of each cofactor, just multiply the minor by the factor $(-1)^{i+j}$. This is negative when the sum is odd and positive when the sum is even, as required.

With that as background we are ready for the main result of Laplace expansion: *the determinant of an $n \times n$ matrix, where $n > 2$, is the sum of the products of each element and its cofactor for any row or column.* This last part is important: one need only "expand" (i.e., calculate the sum of the products of each element and its cofactor) one row or column. And any row or column will do.

Calculating the determinant of a matrix via Laplace expansion is tedious, but it works, and it really isn't too bad for a three-by-three matrix. We will work through an example to illustrate. Consider the following matrix:

$$D = \begin{bmatrix} 1 & 4 & 3 \\ 1 & 2 & 0 \\ 2 & 3 & 1 \end{bmatrix}.$$

We will work with column 2. (In general, you want to pick a column or row with a zero in it to eliminate the need to calculate one of the cofactors, since it gets multiplied by zero. However, we're trying to provide additional practice.) Thus, the determinant of D is $\det(D) = 4(C_{12}) + 2(C_{22}) + 3(C_{32})$. We will concern ourselves with cofactors after we determine the value of the minors. M_{12} is the determinant of the submatrix of the element in the first row and second column:

$$M_{12} = \begin{vmatrix} 1 & 0 \\ 2 & 1 \end{vmatrix} = (1 \cdot 1) - (2 \cdot 0) = 1.$$

Similarly, M_{22} is the determinant of the submatrix of the element in the second row and second column:

$$M_{22} = \begin{vmatrix} 1 & 3 \\ 2 & 1 \end{vmatrix} = (1 \cdot 1) - (2 \cdot 3) = -5.$$

Finally, M_{32} is the determinant of the submatrix of the element in the third row and second column:

$$M_{32} = \begin{vmatrix} 1 & 3 \\ 1 & 0 \end{vmatrix} = (1 \cdot 0) - (1 \cdot 3) = -3.$$

So the minors are $M_{12} = 1$, $M_{22} = -5$, and $M_{32} = -3$. Recall that if the sum of the row and column values is odd, we flip the sign of the minor to get the cofactor. If the sum of the row and column is even, then the cofactor is the minor. The cofactors, then, are $C_{12} = -1$, $C_{22} = -5$, and $C_{32} = 3$.

With the cofactors in hand, we can now calculate the determinant: $\det(D) = 4(-1) + 2(-5) + 3(3) = -4 - 10 + 9 = -5$. You can verify for yourself that expanding any other row or column of D will produce the same determinant value, and it's good practice to do so.

The general formula for this expansion for an $n \times n$ matrix A, expanding around row 1 (other expansions are similar), is

$$\det(A) = a_{11}C_{11} + a_{12}C_{12} + \ldots + a_{1n}C_{1n}.$$

This can get messy, as noted; however, the determinant of certain matrices are easier to compute. For example, the determinant of any diagonal, upper triangular, or lower triangular matrix is just the product of the diagonal elements.

There is also another method for finding the determinants of three-by-three matrices, sometimes called the **butterfly method** (or the **rule of Sarrus**) for the lines drawn around the matrix.[13] Consider a generic three-by-three matrix:

$$\begin{bmatrix} a_{11} & a_{12} & a_{13} \\ a_{21} & a_{22} & a_{23} \\ a_{31} & a_{32} & a_{33} \end{bmatrix}.$$

Start by rewriting the first two columns to the right of the matrix:

$$\begin{bmatrix} a_{11} & a_{12} & a_{13} \\ a_{21} & a_{22} & a_{23} \\ a_{31} & a_{32} & a_{33} \end{bmatrix} \begin{matrix} a_{11} & a_{12} \\ a_{21} & a_{22} \\ a_{31} & a_{32} \end{matrix}.$$

Now each element in the top or bottom rows of the original matrix forms a full diagonal of three elements by going down or up, respectively, and to the right, as depicted in Figure 12.5.

The determinant will then equal the sum of the signed products of each diagonal, as follows:

$$|A| = a_{11}a_{22}a_{33} + a_{12}a_{23}a_{31} + a_{13}a_{21}a_{32} - a_{31}a_{22}a_{13} - a_{32}a_{23}a_{11} - a_{33}a_{21}a_{12}.$$

[13] We offer this method only because we have found that it is a popular shortcut among students. However, we do not believe it to be much faster than the method we provide above, and it is less generalizable.

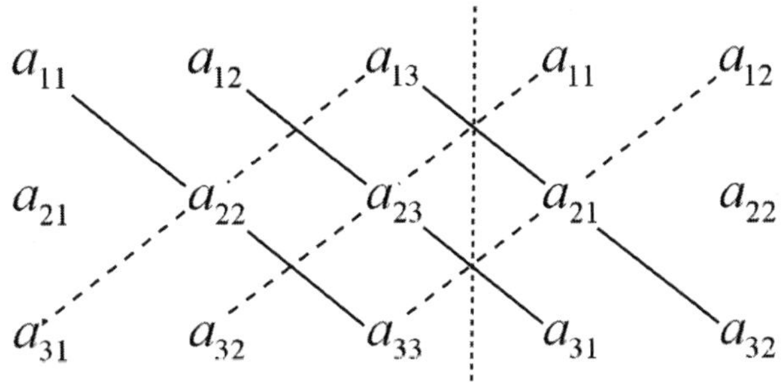

Figure 12.5: Rule of Sarrus for Three-by-Three Determinants

12.3.6.1 Why Should I Care?

The major use of determinants is to determine whether a matrix is invertible (i.e., has a calculable inverse). We discuss inverses and their usefulness in the following subsection. When a matrix has a non-zero determinant, we call it nonsingular, and it is invertible. If the determinant of the matrix is zero, then the matrix is singular and so is not invertible. Determinants are also useful in assessing the linear independence of vectors and in solving systems of equations, discussed in the next chapter; in computing eigenvalues, discussed in the chapter after that; and in some multidimensional computations that get used in game theory, a topic we raise in Part V of this book.

12.3.7 Inverse

Some matrices can be inverted, others cannot. A square matrix that is not invertible is called singular; as noted earlier, it has a determinant of zero. An invertible square matrix is called nonsingular, and its determinant is non-zero. Since one can use the inverse of a matrix to solve the system of equations represented by that matrix, as we show in the next chapter, being able to invert the matrix indicates whether the system has a solution (i.e., a set of linear equations that is noninvertible, or singular, has no solution).

An $n \times n$ matrix, A, is invertible (only square matrices can be inverted) if one can find a second $n \times n$ matrix, B, such that the product AB and the product BA both produce the $n \times n$ identity matrix, $I_{n \times n}$. In such a situation, B is the inverse of A. Put more simply, then, the inverse of a square matrix is the matrix that produces the identity matrix when it is multiplied by it on either the left or the right:

$$A \cdot B = B \cdot A = I$$

One denotes the inverse of a matrix by using a -1 superscript. So A^{-1} is the inverse of A:

$$A \cdot A^{-1} = A^{-1} \cdot A = I.$$

Finding the inverse of a matrix, then, involves solving for one unknown (i.e., A^{-1}) given two knowns (i.e., A and I). Let's start with a two-by-two matrix:

$$E = \begin{bmatrix} 1 & 1 \\ -1 & 2 \end{bmatrix}.$$

We know that $E \cdot E^{-1} = I$. Start by assigning the unknown elements of E^{-1} variables, as follows:

$$E^{-1} = \begin{bmatrix} e_{11} & e_{12} \\ e_{21} & e_{22} \end{bmatrix}.$$

Right-multiply E by E^{-1}, yielding

$$EE^{-1} = \begin{bmatrix} e_{11} + e_{21} & e_{12} + e_{22} \\ -e_{11} + 2e_{21} & -e_{12} + 2e_{22} \end{bmatrix}.$$

We can now set EE^{-1} equal to I, as follows:

$$\begin{bmatrix} e_{11} + e_{21} & e_{12} + e_{22} \\ -e_{11} + 2e_{21} & -e_{12} + 2e_{22} \end{bmatrix} = \begin{bmatrix} 1 & 0 \\ 0 & 1 \end{bmatrix}.$$

This can be rewritten as

$$\begin{aligned} e_{11} + e_{21} &= 1, \\ e_{12} + e_{22} &= 0, \\ -e_{11} + 2e_{21} &= 0, \\ -e_{12} + 2e_{22} &= 1. \end{aligned}$$

This is the more familiar world of scalar algebra. We'll talk about how to solve systems of equations like this in the next chapter, but we'll give you a sneak peak while we derive the inverse of E. First, we can add the first and third equations:

$$\begin{aligned} (e_{11} + e_{21}) + (-e_{11} + 2e_{21}) &= 0 + 1, \\ 3e_{21} &= 1, \\ e_{21} &= \tfrac{1}{3}. \end{aligned}$$

Now we substitute the value of e_{21} back into the first equation and solve for e_{11}:

$$\begin{aligned} e_{11} + (\tfrac{1}{3}) &= 1, \\ e_{11} &= \tfrac{2}{3}. \end{aligned}$$

Next we solve for e_{12} and e_{22}. We begin by adding the second and fourth equations and solving for e_{22}:

$$\begin{aligned} (e_{12} + e_{22}) + (-e_{12} + 2e_{22}) &= 1, \\ 3e_{22} &= 1, \\ e_{22} &= \tfrac{1}{3}. \end{aligned}$$

We can substitute $\frac{1}{3}$ for e_{22} in the second equation and solve for e_{12}:

$$e_{12} + (\tfrac{1}{3}) = 0,$$
$$e_{12} = -\tfrac{1}{3}.$$

We thus have the values for the inverse of E:

$$E^{-1} = \begin{bmatrix} \frac{2}{3} & -\frac{1}{3} \\ \frac{1}{3} & \frac{1}{3} \end{bmatrix}.$$

It's always a good idea to check to make sure that you did your algebra right, so as an exercise, compute both EE^{-1} and $E^{-1}E$. (*Hint*: it should check out.)

It turns out that there is a simple formula for computing the inverses of two-by-two matrices. Let

$$A^{-1} = \begin{bmatrix} a_{11} & a_{12} \\ a_{21} & a_{22} \end{bmatrix}.$$

Then the inverse is

$$A^{-1} = \frac{1}{|A|} \begin{bmatrix} a_{22} & -a_{12} \\ -a_{21} & a_{11} \end{bmatrix}.$$

In words, swap the elements of the diagonal, take the negative of the elements off the diagonal, and divide each element by the determinant of the original matrix to get the inverse.

This rule is a useful special case of the more general way to compute a matrix inverse, given by the following formula:

$$A^{-1} = \frac{1}{|A|} C^T, \tag{12.1}$$

where C^T is the transpose of the matrix of cofactors of A. Each element of C is the cofactor of the corresponding element of A. That is, each element of C is C_{ij}, where C_{ij} is defined as in the previous subsection.[14]

The first step in determining the inverse is thus to find the determinant of the matrix we want to invert. Conveniently, this will also let us know if there is an inverse at all. If the determinant is zero and the matrix is singular, it does not have an inverse. You can see from the formula that if the determinant is zero, the scalar multiplying the matrix will be undefined.

The next step is to construct the cofactor matrix C. The transpose of this matrix, known as an **adjoint matrix**, then gets multiplied by $\frac{1}{|A|}$ to find A^{-1}. Note that in computing the determinant we may have already computed three of the elements of the cofactor matrix, saving some time.

Let's try to find the inverse for the following matrix in this manner:

$$A = \begin{bmatrix} 1 & 2 & 1 \\ 0 & 4 & 3 \\ -6 & -2 & 2 \end{bmatrix}.$$

[14] That is, $C_{ij} = (-1)^{i+j} \times M_{ij}$, where M_{ij} is the minor corresponding to element a_{ij}.

The determinant equals

$$|A| = 1 \cdot M_{11} - 2 \cdot M_{12} + 1 \cdot M_{13} = 1(14) - 2(18) + 1(24) = 2.^{15}$$

Now we need to construct the cofactor matrix. Thus we first have to find all the minors:

$$M_{11} = \begin{vmatrix} 4 & 3 \\ -2 & 2 \end{vmatrix} = (4 \cdot 2) - ((-2) \cdot 3) = 8 + 6 = 14,$$
$$M_{12} = \begin{vmatrix} 0 & 3 \\ -6 & 2 \end{vmatrix} = (0 \cdot 2) - ((-6) \cdot 3) = 0 + 18 = 18,$$

...and so forth, until we find that all the minors are[16]

$$\begin{aligned} M_{11} &= 14, \\ M_{12} &= 18, \\ M_{13} &= 24, \\ M_{21} &= 6, \\ M_{22} &= 8, \\ M_{23} &= 10, \\ M_{31} &= 2, \\ M_{32} &= 3, \\ M_{33} &= 4. \end{aligned}$$

With this information we can now construct our cofactor matrix (remember to multiply each minor by $(-1)^{i+j}$):

$$C = \begin{bmatrix} 14 & -18 & 24 \\ -6 & 8 & -10 \\ 2 & -3 & 4 \end{bmatrix}.$$

Now we can transpose the cofactor matrix to find the adjoint matrix of A:

$$\text{adj}(A) = \begin{bmatrix} 14 & -6 & 2 \\ -18 & 8 & -3 \\ 24 & -10 & 4 \end{bmatrix}.$$

Finally, we multiply this matrix by $\frac{1}{|A|}$ to find the inverse of A:

$$A^{-1} = \frac{1}{|A|}\text{adj}(A) = \frac{1}{2}\begin{bmatrix} 14 & -6 & 2 \\ -18 & 8 & -3 \\ 24 & -10 & 4 \end{bmatrix} = \begin{bmatrix} 7 & -3 & 1 \\ -9 & 4 & -\frac{3}{2} \\ 12 & -5 & 2 \end{bmatrix}.$$

Again, it's good to check to see whether $AA^{-1} = A^{-1}A = I$ (it does).

[15] We compute two of these minors below.

[16] Yes, this is where we tell you to check them yourself for practice.

Table 12.1: Matrix and Vector Properties

Associative property	$(AB)C = A(BC)$
Additive distributive property	$(A + B)C = AC + BC$
Scalar commutative property	$xAB = (xA)B = A(xB) = ABx$

12.3.7.1 Why Should I Care?

The inverse of a matrix can be used to solve systems of linear equations, and in the next chapter we provide several examples. Such systems are important for game theoretic and dynamic models. Linear regression, as in OLS, can also be represented as a system of equations. In Section 4 of Chapter 8 we showed how to derive the regression coefficients via calculus, and then provided them in matrix form: $\boldsymbol{\beta} = (X^T X)^{-1} X^T \mathbf{y}$. Here X is an $n \times m$ matrix of m independent variables each with n associated data points, and $\mathbf{y}$ is an n-dimensional vector corresponding to the dependent variable. Thus, matrix inversion can be used to compute these coefficients, which are the objects of interest in regression. We provide examples of this in the last section of this chapter and in the exercises, and develop a fuller exposition in the following chapter.

12.4 PROPERTIES OF VECTORS AND MATRICES

Here we collect some useful properties of vectors and matrices. Some were stated earlier in the chapter, some we hinted at, and some are new. A few more will be provided in the next chapter; e.g., we'll introduce the rank of a matrix there, after discussing linear independence.

With respect to matrix multiplication, the properties in Table 12.1 are important.

Of course, keep in mind that $AB \neq BA$. That is, matrix multiplication is *not* commutative in general. That said, certain types of multiplication are commutative. For example, multiplication by either the identity matrix or the matrix inverse is always commutative: $IA = AI$ and $AA^{-1} = A^{-1}A$. As is multiplication by the zero matrix, rather trivially. Further, the dot (scalar) product of vectors is commutative: $\mathbf{a} \cdot \mathbf{b} = \mathbf{b} \cdot \mathbf{a}$. Both vector and matrix addition satisfy the normal associative, distributive, and commutative properties.

Transposition has a few useful properties we list in Table 12.2 that you may encounter in proofs in statistics and formal models. Note in particular that transposition flips the order of multiplication.

Recall that symmetric matrices have the same elements above and below the main diagonal. It can also sometimes be useful to write the dot product in terms of the transpose. If $\mathbf{a}$ and $\mathbf{b}$ are both column vectors, then $\mathbf{a} \cdot \mathbf{b} = \mathbf{a}^T \mathbf{b}$.

The trace has a few properties that aren't really worth breaking out: $\text{tr}(A + B) = \text{tr}(A) + \text{tr}(B)$, $\text{tr}(A^T) = \text{tr}(A)$, $\text{tr}(AB) = \text{tr}(BA)$.[17]

[17] Cyclic permutations are in general allowed, but not all permutations.

Table 12.2: Matrix and Vector Transpose Properties

Inverse	$(A^T)^T = A$
Additive property	$(A + B)^T = A^T + B^T$
Multiplicative property	$(AB)^T = B^T A^T$
Scalar multiplication	$(cA)^T = cA^T$
Inverse transpose	$(A^{-1})^T = (A^T)^{-1}$
If A is symmetric	$A^T = A$

Table 12.3: Matrix Determinant Properties

Transpose property	$\det(A) = \det(A^T)$
Identity matrix	$\det(I) = 1$
Multiplicative property	$\det(AB) = \det(A)\det(B)$
Inverse property	$\det(A^{-1}) = \frac{1}{\det(A)}$
Scalar multiplication ($n \times n$)	$\det(cA) = c^n \det(A)$
If A is triangular or diagonal	$\det(A) = \prod_{i=1}^{n} a_{ii}$

Table 12.4: Matrix Inverse Properties

Inverse	$(A^{-1})^{-1} = A$
Multiplicative property	$(AB)^{-1} = B^{-1}A^{-1}$
Scalar multiplication ($n \times n$)	$(cA)^{-1} = c^{-1}A^{-1}$ if $c \neq 0$

There are several useful properties of the determinant that you might encounter. Some of these follow from others, but we list them all in Table 12.3 separately for reference.

Finally, the matrix inverse has a few properties that can be useful, and we list these in Table 12.4.

12.5 MATRIX ILLUSTRATION OF OLS ESTIMATION

In your statistics coursework you will likely be taught OLS regression, and it is conventional to present some of that material using matrix algebra. For example, consider the OLS regression equation $\mathbf{y} = \alpha + \beta\mathbf{x} + \boldsymbol{\epsilon}$, where $\mathbf{x}$ and $\mathbf{y}$ are vectors containing the values of the independent and dependent variables, respectively, for each observation;[18] α is a scalar containing the y-intercept (i.e., the expected value of y when $x = 0$); $\boldsymbol{\epsilon}$ is a vector that holds the errors (i.e., the distance between the regression line and the value of $\mathbf{y}$); and β is a scalar holding the value of average change in $\mathbf{y}$ given a one-unit increase in $\mathbf{x}$. We know the values of $\mathbf{x}$ and $\mathbf{y}$, and our problem is to "estimate" the values of α and β that produce the regression line (aka, a vector containing the predicted

[18]Note that some of this material was presented first in Section 4 of Chapter 8; we repeat it to accommodate readers who skipped ahead to Part IV of the book.

Table 12.5: Per Capita Income and Size of Government in Some Southern US States

State	Per Capita Income	% Gov't Employees
Alabama	\$24,028	19.2
Florida	\$30,446	14.5
Georgia	\$29,442	16.4
Mississippi	\$23,448	21.8
North Carolina	\$28,235	17.3
South Carolina	\$26,132	18.2
Tennessee	\$28,455	15.5

values of $\mathbf{y}$) that best fits the data. OLS regression proposes that the best fit is produced by selecting the values of α and β that minimize the sum of squared errors ($\sum_i \epsilon_i^2$). You will likely learn that the OLS estimator is the best linear unbiased estimator (BLUE) in your statistics coursework, and we sketch a proof of this fact using matrix algebra in the next chapter.

It turns out that we can calculate a vector that contains the values of α and β, call it $\hat{\boldsymbol{\beta}}$, by using the equation

$$\hat{\boldsymbol{\beta}} = (X^T X)^{-1} X^T \mathbf{y}.$$

We briefly discussed this in Section 4 of Chapter 8, and you will learn why this is so in more depth in your statistics coursework. Here we walk through the mechanics of manipulating actual $\mathbf{x}$ and $\mathbf{y}$ vectors to produce an OLS estimate. To keep things manageable, we use a dataset with only seven observations. In applied statistical work you will have considerably larger samples. Table 12.5 records the average per capita income and the percent of workers employed by the government in seven US states.

Let's call $\mathbf{y}$ "size of government" and $\mathbf{x}$ "per capita income." This gives us the vector $\mathbf{y}^T = (19.2, 14.5, 16.4, 21.8, 17.3, 18.2, 15.5)$. Because we are estimating both α and β, we need to add a column of 1s to the $\mathbf{x}$ vector, producing the matrix X.

$$X = \begin{bmatrix} 1 & 24,028 \\ 1 & 30,446 \\ 1 & 29,442 \\ 1 & 23,448 \\ 1 & 28,235 \\ 1 & 26,132 \\ 1 & 28,445 \end{bmatrix}.$$

We can now take the product of the matrices X transpose and X, yielding

$$X^TX = \begin{bmatrix} 7 & 190,356 \\ 190,356 & 5,218,840,922 \end{bmatrix}.$$

The next step is to calculate the inverse of that matrix:

$$(X^TX)^{-1} = \begin{bmatrix} 17.603 & -0.001 \\ -0.001 & 0.000 \end{bmatrix}.$$

Let's now calculate the product of X transpose and $\mathbf{y}$:

$$X^T\mathbf{y} = \begin{bmatrix} 1 & 24,028 \\ 1 & 30,446 \\ 1 & 29,442 \\ 1 & 23,448 \\ 1 & 28,235 \\ 1 & 26,132 \\ 1 & 28,445 \end{bmatrix} \begin{bmatrix} 19.2 \\ 14.5 \\ 16.4 \\ 21.8 \\ 17.3 \\ 18.2 \\ 15.5 \end{bmatrix} = \begin{bmatrix} 122.9 \\ 3,301,785.2 \end{bmatrix}.$$

That completed, we are now ready to calculate the OLS estimates of α and β given the data in Table 12.5 and the equation $\hat{\boldsymbol{\beta}} = (X^TX)^{-1}X^T\mathbf{y}$:

$$\hat{\boldsymbol{\beta}} = \begin{bmatrix} \beta \\ \alpha \end{bmatrix} = \begin{bmatrix} 17.603 & -0.001 \\ -0.001 & 0.000 \end{bmatrix} \times \begin{bmatrix} 122.9 \\ 3,301,785.2 \end{bmatrix} = \begin{bmatrix} 50.228 \\ -0.079 \end{bmatrix}.$$

We thus have an OLS estimate of 50.228 for β and -0.079 for α. There are all sorts of reasons why the OLS estimates from this bivariate regression are not of substantive interest, and in your statistics coursework you will review them. As such, we do not concern ourselves with substantive interpretation of the estimates beyond noting that they suggest that among states in the US South, those with lower per capita income tend to have a higher percentage of their labor force working for the government. We conclude this chapter by observing that this example demonstrates, we suspect, that using a computer to calculate OLS regression estimates is much easier than doing it by hand (once one has learned how to instruct the software program one is using, of course). Still, we provide an opportunity to practice this skill in the exercises, as it may come in handy in statistics coursework.

12.6 EXERCISES

1. Let: $\mathbf{a} = \begin{pmatrix} 10 \\ 2 \\ 5 \\ 2 \end{pmatrix}$, $\mathbf{b} = \begin{pmatrix} 4 \\ 15 \\ 6 \\ 8 \end{pmatrix}$, $\mathbf{c} = (2, 6, 8)$, $\mathbf{d} = (1, 15, 12)$, $\mathbf{e} = (14, 17, 11, 10)^T$, and $\mathbf{f} = (20, 4, 10, 4)^T$. Calculate each of the following, indicating that it's not possible if there is a calculation you cannot perform.

   a) $\mathbf{a} + \mathbf{b}$

   b) $\mathbf{a} + \mathbf{c}$

c) $\mathbf{b} - \mathbf{e}$

d) $15\mathbf{c}$

e) $-3\mathbf{f}$

f) $||\mathbf{b}||$

g) $||\mathbf{c} + \mathbf{d}||$. Show that the triangle inequality holds for this case.

h) $||\mathbf{c} - \mathbf{d}||$

i) $\mathbf{a} \cdot \mathbf{b}$

j) $\mathbf{c} \cdot \mathbf{d}$

2. Identify the following matrices as diagonal, identity, square, symmetric, triangular, or none of the above (note all that apply).

a) $A = \begin{bmatrix} 0 & 1 & 5 \\ 1 & -2 & -1 \\ 5 & -1 & 2 \end{bmatrix}.$

b) $B = \begin{bmatrix} 1 & 0 & 0 \\ 0 & 1 & 0 \\ 0 & 0 & 1 \end{bmatrix}.$

c) $C = \begin{bmatrix} 1 & 1 \\ 3 & -2 \end{bmatrix}.$

d) $D = \begin{bmatrix} 0 & 1 & 2 \\ 5 & 1 & -1 \\ 2 & 4 & 0 \\ 1 & 1 & 0 \end{bmatrix}.$

3. Write down the transpose of matrices A through D from the previous problem.

4. Given the following matrices, perform the calculations below.

$$A = \begin{bmatrix} 5 & 1 & 2 \\ 6 & 2 & 3 \end{bmatrix} \; B = \begin{bmatrix} 3 & 4 & 5 \\ -2 & -3 & 6 \end{bmatrix} \; C = \begin{bmatrix} 1 & 2 \\ -5 & 3 \\ -3 & 1 \end{bmatrix} \; D = \begin{bmatrix} 2 & 1 \\ 4 & 3 \end{bmatrix}$$

a) $A + C$

b) $A - B$

c) $A + 5B$

d) $3A$

e) $2B - 5A$

f) $B^T - C$

g) BA

h) DA

i) AD

j) CD

k) BC

l) CB

5. Find the determinants of the following matrices:

a) $A = \begin{bmatrix} 2 & 2 \\ -2 & 1 \end{bmatrix}$.

b) $B = \begin{bmatrix} 4 & 2 \\ 6 & 3 \end{bmatrix}$.

c) $C = \begin{bmatrix} 4 & 0 \\ 5 & 3 \\ 1 & 11 \end{bmatrix}$.

d) $D = \begin{bmatrix} 3 & 2 & -4 \\ -1 & -5 & 1 \\ 3 & 2 & 3 \end{bmatrix}$.

e) $E = \begin{bmatrix} 3 & 2 & 0 \\ 0 & 2 & 1 \\ 0 & 1 & 0 \end{bmatrix}$.

6. If it exists, find the inverse of the following matrices:

a) $A = \begin{bmatrix} 4 & 2 \\ 6 & 3 \end{bmatrix}$.

b) $B = \begin{bmatrix} 1 & 4 \\ 3 & 2 \end{bmatrix}$.

7. Let: $A = \begin{pmatrix} 3 & -2 & 1 \\ 0 & 4 & 2 \end{pmatrix}$, $B = \begin{pmatrix} 1 & 2 & 1 \\ 5 & -1 & 3 \end{pmatrix}$, $C = \begin{pmatrix} 2 & -3 \\ -1 & 1 \\ 1 & 4 \end{pmatrix}$, $D = \begin{pmatrix} 1 & 2 \\ 3 & 4 \end{pmatrix}$, $E = \begin{pmatrix} 3 & 1 \\ 6 & 2 \end{pmatrix}$, $F = \begin{pmatrix} 1 & 1 & 5 \\ 3 & -2 & -1 \\ 2 & 4 & 2 \end{pmatrix}$, $\mathbf{g} = \begin{pmatrix} 2 \\ 3 \\ 1 \end{pmatrix}$, $\mathbf{h} = (1, 2, 3)$.
Calculate each of the following, indicating that it's not possible if there is a calculation you cannot perform.

a) $A + B$

b) $A - C$

c) $B - A$

d) $3A$

e) $2A - 3B$
f) AB
g) DA
h) CD
i) DE
j) ED
k) $F\mathbf{g}$
l) $\mathbf{h}F$
m) A^T
n) $\mathbf{g}^T$
o) $(ED)^T$
p) Trace F
q) $\det(E)$
r) $\det(F)$
s) D^{-1}
t) E^{-1}

8. A multivariate regression model looks like $\mathbf{y} = \beta_1 x_1 + \ldots + \beta_n x_n + \epsilon$. With only one data point, each of the $\mathbf{y}$, x_j, and ϵ would be scalars, but with many data points, they are each column vectors. With β a column vector and X a matrix containing each x_j in a different column, we can write the model simply as $\mathbf{y} = X\boldsymbol{\beta} + \boldsymbol{\epsilon}$. Minimizing squared error amounts to minimizing $\epsilon \cdot \epsilon$, which results in $\boldsymbol{\beta} = (X^T X)^{-1} X^T \mathbf{y}$. If your independent variable is $\mathbf{y} = \begin{pmatrix} 7 \\ 2 \end{pmatrix}$ and your two dependent variables are $\mathbf{x_1} = \begin{pmatrix} 1 \\ 2 \end{pmatrix}$ and $\mathbf{x_2} = \begin{pmatrix} 3 \\ 2 \end{pmatrix}$, what is your estimate of $\boldsymbol{\beta}$?

9. True or false?

   a) $BA = AB$ for all matrices A, B.
   b) $XX^{-1} \neq I$.
   c) $M_{i \times j} N_{j \times k} = (MN)_{i \times k}$.

10. Why are the following useful?

    a) Determinant
    b) Inverse

11. What does it mean if you have a singular matrix?

12. Fill in the blanks:

    a) AB indicates that you ______________ A by B.
    b) AB indicates that you ______________ B by A.

Chapter Thirteen

Vector Spaces and Systems of Equations

Having read the previous chapter, one could easily have gotten the impression that linear algebra is a bunch of disconnected rules. Sure, they might be useful in some (a lot of) circumstances, but they may not seem to have a lot of coherence to them. That's a perfectly reasonable inference given our presentation thus far, which was geared toward providing the tools needed to do most of the manipulations of vectors and matrices that you'll need to do during your first couple of statistics classes. But, it turns out, this characterization is hardly fair. The rules of vector and matrix manipulation are part of the larger topic of **linear algebra**, the name attached to this part of the book. Linear algebra, loosely speaking (as always), deals with linear mappings, or functions, of just the type we talked about way back in Section 2.2 of Chapter 3. More specifically, it deals with linear mappings of vectors, which are elements of vector spaces.

We go into vector spaces at some length in the first part of this chapter, but put simply, they are collections of vectors, which are loosely defined as objects that can be added together and multiplied by scalars in just the way we defined in the previous chapter. For most purposes in which we use them, vector spaces will also take on the added structure of a *norm*, which is much like the measure of vector length we introduced in the last chapter,[1] and we generally lump the norm in with the definition of the vector space for convenience. These vectors stand for things: for data on dependent or independent variables, estimates of coefficients describing how a dependent variable changes with an independent variable, different aspects of the state of the world or the system, or equilibrium actions taken by an individual in a game theory model.

In this conceptualization, matrices act as linear mappings, taking the vector from one value to another. This is particularly clear in the context of Markov chains, which we cover in the next chapter. There vectors correspond to the state of the system, and matrices describe the manner in which the state of the system changes over time.

The reason for the connection between matrices and linear mappings is that, as we saw in Chapter 3, one can describe via functions how a variable (like the state of the system) changes. In one dimension, a linear function might look like $y = f(x) = 3x$. In more than one dimension, with lots of different y's and x's of interest, we get a series of equations like $y_i = 3x_1 - 2x_2 + \ldots + 5x_m$ for each $y_i, i \in \{1, \ldots, n\}$. As we shall see in the second section below, *any* system

[1] This is related to why we call a vector of length 1 "normalized."

of linear equations can be represented by a matrix that tells you how all the x_j change into the y_i when the linear function is applied. This is helpful, as we can then utilize techniques of matrix manipulation to do things like solve systems of equations or understand properties of dynamical systems. We see how to do the former in this chapter and the latter in the next.

Linear algebra also offers several useful concepts that help us understand what kind of solutions we should expect to see for a given system of equations. Vectors are considered linearly independent if, intuitively, they don't point in the same direction. This ties closely to the problem of multicollinearity in statistics: if two independent variables vary in the same fashion, then they both should not be included as explanatory variables. In fact, it is likely either that one depends on the other or both depend on some third factor that is more important in explaining the dependent variable. You'll learn much more about this in your statistics classes.

Here we cover two broader topics. In the first section below we introduce vector spaces more formally and discuss important, if abstract, concepts in linear algebra such as norms, linear independence, spanning vectors, matrix rank, and quadratic forms. In the second section we move to the primary point of the chapter: to provide several methods to solve systems of linear equations. Such systems arise in statistics and are common in formal theory. We leave until the third section examples of the payoff for learning this material in both statistics and formal theory. This payoff includes being able to show that the ordinary least squares (OLS) estimator is the best linear unbiased estimator (BLUE), and being able to solve for the equilibria of games with more than one actor or action. The next chapter introduces matrix methods useful for dynamical systems and analyzing certain types of data, such as those in network relationships.

13.1 VECTOR SPACES

In abstract mathematics, things like numbers and vectors get defined in a very precise fashion, according to a series of axioms that often uniquely define them. You've seen some of these axioms, such as the distributive property of (scalar) multiplication, in this book, though not presented so formally. We stick to the informal presentation here. First we define a vector space and its associated norm and offer some intuition. Then we discuss linear combinations of vectors, along with geometric interpretations thereof. Next we specify what it means for two or more vectors to be linearly independent, and how vectors can span (that is, generate all vectors in) a vector space, forming what is known as a basis for that space. We also relate these concepts to matrices via the notion of the rank of a matrix. Finally we briefly introduce quadratic forms, the next most complex form to the linear mapping. These prove useful for the last part of the book.

13.1.1 Definitions and Intuition

We defined a vector several ways in the previous chapter, and we won't alter that. We stick to the ordered n-tuple version here. In other words, a vector is nothing more than a list of numbers put in some order. A **vector space** is a set of vectors that satisfy certain properties. Though technically speaking, we need only have vectors in any given vector space satisfy the properties of vector addition and scalar multiplication that we defined in the previous chapter, we assume that they satisfy *all* the properties listed in the previous chapter.

This means that for all vectors in any given vector space we are dealing with, not only can we add them together and multiply them by scalars, we can also compute their lengths and take their dot (or scalar) products with other vectors. The **norm** of a vector $\mathbf{a}$, denoted $||\mathbf{a}||$, provides the length of the vector and can be related to the dot product according to[2] $||\mathbf{a}|| = \sqrt{\mathbf{a} \cdot \mathbf{a}}$. Thus the dot product allows us to easily compute the length of any vector. The same is true for the scalar difference between vectors: $||\mathbf{a} - \mathbf{b}|| = \sqrt{(\mathbf{a} - \mathbf{b}) \cdot (\mathbf{a} - \mathbf{b})}$.

For most purposes we won't need even this level of generality. Typically, the elements of a vector are either real numbers or variables that take on real values. In these cases vectors are ordered n-tuples, such as the ordered pair $(3, 2)$, and the vector space containing them is known as a coordinate space. Given the vector norm we have defined, a coordinate space containing real-valued vectors of dimension[3] d is just $\mathbb{R}^d$, which is the Euclidean space we discussed way back in Chapter 1. Thus, despite all this formalism, vector spaces can and should be intuitive: they contain, for example, the points on a line (one-dimensional space, 1-D), or in a plane (two-dimensional space, 2-D), or in three-dimensional (3-D) space.

So why introduce all this formalism at all? Because in formal theory and particularly statistics social scientists do not frequently work in 2-D or 3-D space, and thus visualization is not feasible. Rather, work is generally done in d-dimensional space, where d may be very, very large. For example, if we have $10,000$ data points, as is common in studies of international conflict, where the unit of analysis is often the country-year, then the vectors in our space have

[2]These definitions are a bit looser than even our usual, so it is worth elaborating a bit. A vector space can be thought of as an ordered set of elements of a *field*, e.g., the real numbers with the usual ordering, that satisfies a series of axioms. We've provided the two main ones, vector addition and scalar multiplication, but strictly speaking, some of the things we took for granted, such as the existence of a vector $\mathbf{0}$ that when added to a vector yields the vector back again (i.e., $\mathbf{a} + \mathbf{0} = \mathbf{a}$), are separate axioms. When you add a norm or an inner product to a vector space, the result is a normed vector space or an inner product space, respectively. Further, strictly speaking, the inner product, written as $< \mathbf{a}, \mathbf{a} >$, is not the same as the dot product in some contexts. This allows different norms to be defined, as the more general relation is between norm and inner product: $||\mathbf{a}|| = \sqrt{< \mathbf{a}, \mathbf{a} >}$. However, as all the vector spaces you are likely to deal with have associated norms that are closely related to the dot product in the manner given in the main text and dot products that are identical to inner products, we try to avoid confusion and refer to vector spaces as automatically having dot products and associated norms.

[3]Recall that the dimension of a vector is its number of components.

dimension 10,000, and using our intuition alone becomes much more difficult. The formalism in this part of the book is designed to allow us to draw insights from even such complex but important cases.

13.1.2 Linear Combinations

As we've seen, we can do things with vectors in vector spaces, such as add or subtract them. A **linear combination** of any number of vectors in general looks like this: $a_1\mathbf{x_1} + \ldots + a_n\mathbf{x_n}$, where the $\mathbf{x_i}$ are all vectors in some vector space and the a_i are all scalars. In words, a linear combination means that we add (or subtract) scalar multiples of vectors to get some new vector. Because of the rules of vector addition, any such linear combination will also automatically be in the vector space, too. Thus, adding vectors is a way to get new vectors.

Some linear combinations have specific geometric interpretations that can be useful. We provide two examples. First, if t is any scalar and $\mathbf{x}$ and $\mathbf{y}$ are any vectors, then $(1-t)\mathbf{x} + t\mathbf{y}$ represents a line on which both points lie. This is a fancy way of saying that a line is specified by any two points.

Second, if $\mathbf{z}$ is also a vector and s is also a scalar, then $(1-t-s)\mathbf{x} + t\mathbf{y} + s\mathbf{z}$ forms a plane on which all three points lie. Again, this is a fancy way of saying that a plane is formed by any three points. A **normal vector**, $\mathbf{n}$, to a plane is perpendicular to the plane, so if $\mathbf{x}, \mathbf{p}$ are any points on the plane, then $\mathbf{n} \cdot (\mathbf{x} - \mathbf{p}) = 0$. This equation is one way to specify a plane. We can do this for any dimensionality as well: a surface with one fewer dimension than the Euclidean space in which it rests can be described according to this equation and is called a **hyperplane**. You will sometimes see the word hyperplane used in the proofs for social choice theory; all it really means is a surface that has one fewer dimension than the space in which it is embedded. This is a line if the space is a plane, a plane if the space is 3-D, and so on.

13.1.3 Linear Independence and Spanning Vectors

We can use linear combinations for another purpose as well: to determine if a set of vectors chosen from some vector space is in a sense minimal.[4] By minimal, we mean that no vector in this set can be written as the linear combination of other vectors in this set. If we cannot write any vector in a set as a linear combination of the other vectors in the set, then we say the set of vectors is **linearly independent**. Formally, a set of vectors $\mathbf{v_i}$ is linearly independent if whenever $a_1\mathbf{v_1} + a_2\mathbf{v_2} + \ldots + a_n\mathbf{v_n} = \mathbf{0}$ then $a_1 = a_2 = \ldots = a_n = 0$. In words, the only way to get all the vectors to add to zero is for all the coefficients on

[4] If you are confused at this point by the relationship between vectors and vector spaces, just think of a vector space as some set of vectors that has something in common. For instance, the collection of all 3-D vectors with real elements that obeys all the vector properties given in the previous chapter is the vector space $\mathbb{R}^3$. As with any set, we can draw elements from it to form smaller subsets. So, for example, a set of vectors in $\mathbb{R}^3$ might be $\{(1,2,3),(4,1,9),(2,-1,-2)\}$. When we talk about sets of vectors in this chapter, that's what we mean: specific subsets of some vector space.

all the vectors to be zero. Why is this the same thing as linear independence? Well, if the sum can add to zero, then we can rewrite one vector in terms of all the others; e.g., if $a_1 \neq 0$, $\mathbf{v_1} = \frac{a_2\mathbf{v_2}+\ldots+a_n\mathbf{v_n}}{-a_1}$.

Linear independence proves to be an incredibly important concept across many applications. Of particular importance to us, it can tell us when systems of equations have solutions and what exactly is the size of a vector space. We cover the former in the next section, but we can do the latter here. To do so, we first need to introduce another two definitions.

First, a set of vectors **spans** a vector space if every vector in that vector space can be written as a linear combination of vectors from that set. You are already familiar with a spanning set of vectors: the coordinate axes in 3-D space. Consider the set of vectors $\mathbf{e_1}, \mathbf{e_2}, \mathbf{e_3}$, where $\mathbf{e_1} = (1, 0, 0), \mathbf{e_2} = (0, 1, 0), \mathbf{e_3} = (0, 0, 1)$. These three vectors are known as **coordinate vectors** or **unit vectors**. Via scalar multiplication, one can get any point on the x-axis from $\mathbf{e_1}$, any point on the y-axis from $\mathbf{e_2}$, and any point on the z-axis from $\mathbf{e_3}$. One can then add these up to get any point in 3-D space. Since vectors in $\mathbb{R}^3$ can be represented by points in 3-D space, these three unit vectors span $\mathbb{R}^3$.

These unit vectors are not the only way to span $\mathbb{R}^3$, however. For example, we could have used $(1, -1, 0), (1, 2, 0)$, and $(0, 0, 4)$, or $(2, -3, 0), (1, 2, 0), (0, 1, -2)$, and $(0, 2, -5)$. In fact there exists an infinite number of possible sets of spanning vectors, some of which contain three vectors, some of which contain more.

We are not interested in most of these, however. To avoid redundancy we prefer to stick to the *minimal* number of spanning vectors, and this number will be the same as the dimension of the vector space. So you need at least three vectors to span $\mathbb{R}^3$, for instance. But how do we determine if our set of three is sufficient? That is where linear independence comes in. If a set of vectors is linearly independent, then one cannot write any one of the vectors as a linear combination of the others. This means that one can't eliminate any vectors in the set and still span the space—at the very least, one wouldn't be able to obtain the vector one eliminated via a linear combination of the remaining ones. Consequently, if you have three vectors that are linearly independent in $\mathbb{R}^3$, for example, they must span $\mathbb{R}^3$ and be the minimal set capable of doing so.

We call a set of linearly independent vectors that span a vector space a **basis** of that vector space. Any basis of a vector space is minimal, and the number of vectors in the space will equal the dimension of the space. This fact is useful for gauging the dimension of a vector space when it is not obvious: the **dimension** of a vector space is equal to the number of vectors in its basis. Despite their minimal quality, though, bases are still not unique. For example, $(1, 0)$ and $(0, 1)$ form a basis (and an orthonormal one, since the vectors are orthogonal and have length 1) for 2-D Euclidean space, but so do $(1, -1)$, $(1, 2)$, or even something more exotic, such as polar coordinates.[5] Still, despite this multiplicity, the most

[5] Polar coordinates are a way of representing the plane in which the two coordinates are r, the distance from the origin, and θ, the angle from the x-axis. You can write any point in standard Cartesian coordinates (x, y) as $(r\cos(\theta), r\sin(\theta))$.

commonly used basis for any Euclidean space is the set of coordinate vectors of appropriate dimension. We use bases in a concrete fashion in the next chapter to diagonalize a square matrix, which allows us to perform operations on that matrix more easily.

13.1.4 Matrix Rank

Matrices prove useful in linear algebra in numerous ways. We introduce their use in dynamical systems in the next chapter; in this chapter we consider their ability to help us solve systems of equations (in the next section) and determine linear independence of vectors quickly (right now). To do that, we need to define the rank of a matrix. Consider a matrix formed by either placing column vectors side by side or stacking row vectors on top of each other. The **rank** of this matrix is the maximum number of linearly independent rows *or* columns; these are always the same number.

We make use of the matrix rank more in the next section, but, for square matrices, the rank is also useful in determining linear independence of vectors. If the rank of a square matrix is equal to its number of rows or columns, then the matrix is nonsingular. Since nonsingular matrices have non-zero determinants, one way of checking whether a set of n n-dimensional is linearly independent is to create a matrix out of the set, as described in the previous paragraph, and then take the determinant of this matrix. If the determinant is non-zero, the vectors are linearly independent; otherwise, the vectors are linearly dependent.

13.1.5 Quadratic Forms

The final topic of this section takes us beyond linear algebra for a moment, but since it uses the same techniques as discussed in this part of the book, we've placed the topic here. We've stated that a linear mapping takes $\mathbf{x}$ to $A\mathbf{x}$. A quadratic form can be written as $Q = \mathbf{x}^T A\mathbf{x}$, where A is a symmetric matrix and $\mathbf{x}$ remains a column vector.[6]

Our interest in quadratic forms is typically whether these are **positive definite** ($Q > 0$), **positive semidefinite** ($Q \geq 0$), **negative definite** ($Q < 0$) or **negative semidefinite** ($Q \leq 0$). If this relation is true for all $\mathbf{x}$, then we call A positive definite, and so on. These facts are used primarily for optimization in more than one dimension, a topic we take up in Chapter 16.

[6]Note that the vector left-multiplying A must be a row vector to make the multiplication work out, which explains the transpose operation.

13.2 SOLVING SYSTEMS OF EQUATIONS

To this point in the text we have by and large limited our attention to functions of a single variable.[7] To the extent that we have addressed functions of more than one variable, we have nevertheless restricted the presentation to single equations (or functions). For the remainder of this chapter we turn our attention to systems of equations, which are situations in which more than one equation exists and there is at least one relationship across the equations. In Chapter 2 we discussed scalar manipulation of single equations. Here we introduce the vector and matrix algebra techniques for solving systems of equations.

It turns out that solving systems of linear equations is one of the more useful things one can do with the techniques of linear algebra. Such systems of equations come up in both statistics and formal theory. We go through a few techniques for doing so: substitution, elimination, matrix inversion, and finally Cramer's rule. We also explore examples in which there are no, one, and infinite numbers of solutions. Huckfeldt, Kohfeld, and Likens (1982, pp. 46–56) do an excellent job of illustrating the use of matrix algebra to solve the system of equations implied by Richardson's model of arms race spending between two countries that are international rivals (e.g., India and Pakistan, US and USSR), and readers interested in a political science application will find it useful to consult that text after reading this section. Blalock Jr. (1969, pp. 100–40) similarly provides a useful illustration of a formal model of influence within groups that has several equations.

To make the different methods more clear, we carry through three different systems of equations, and solve each of them with the different methods provided. You may want to write these down on a piece of paper as you read through the rest of the chapter. Each of x, y, z is a variable for which we will solve; collectively, we term them *unknowns*. Additional examples may be found in the exercises.

Example 1 is

$$\begin{aligned} 2x - y + 3z &= 9, \\ x + 4y - 5z &= -6, \\ x - y + z &= 2. \end{aligned}$$

Example 2 is

$$\begin{aligned} 4x + 3y + 2z &= 3, \\ -3x + 5y - 2z &= 2, \\ -6x + 10y - 4z &= 4. \end{aligned}$$

[7]The major reason that linear algebra occurs at this point in the book is that it serves as a transition from one variable to many, a necessary step before tackling multivariate calculus in Part V.

Example 3 is

$$3x + 6y - 6z = 5,$$
$$x + 2y - 2z = 3,$$
$$x + y - z = 5.$$

13.2.1 Substitution

Perhaps the easiest and most intuitive manner of solving systems of linear equations is **substitution**. In this method, you first choose one variable for which to solve (e.g., x). Then you pick an equation and solve for that variable in terms of the other variables (e.g., solve the third equation for x in terms of y and z). You want to choose the easiest variable to solve for, and the easiest equation in which to solve for it, where easiest in each case means the one with the least associated algebra. Next you plug this expression for the variable into the other two equations you did not yet use (e.g., plug this x into equations one and two). This gives you one fewer equation than before, and one fewer variable as well. You keep repeating these steps until one of three things happen.

One, you might get a numerical solution for the last variable. In this case you can then plug this solution into the expression for the other variables, acquire a numerical solution for at least one of these variables, plug this solution into the other expressions, and so on until you have numerical expressions for all variables. This yields the unique solution to the system. Systems in which all the equations are different (i.e., not multiples of each other or linear combinations of other equations) and there is the same number of equations as unknowns often have unique solutions. We say the system in this case is **uniquely determined**.

Two, you might not be able to obtain a numerical solution for any variables, but the expressions you end up with are not contradictory. In this case you can choose one or more of the variables for which you will not solve, and solve for the other variables in terms of these. There is an infinite number of solutions in this case, with each solution corresponding to a different variable for which you did not solve. Systems in which there are more unknowns than equations often have an infinite number of solutions, as do systems in which one of the equations can be formed from some linear combination of the others, but the equations are not contradictory. We say the system in this case is **underdetermined**.

Three, you might end up with contradictory expressions, such as $x = 2, x = 4$. In this case there is no solution to the system, and there's nothing else you can do. This is often the case when there are more equations than unknowns, or any time in which some of the equations are contradictory, which is more likely to happen the more equations there are. We say the system in this case is **overdetermined**.

One might be wondering at this point how substitution is intuitive, but this is one of those cases where seeing how to do it makes more sense than the general procedure. So let's solve our three examples, in order.

Let's start with the first example, and begin by solving the third equation for

x. We chose this because it involves no real algebra: we isolate the x on the left-hand side (LHS) to yield $x = y - z + 2$.[8] Now that we have x, we can plug this x into equations one and two. Equation one becomes $2(y - z + 2) - y + 3z = 9$, or $y + z = 5$, and equation two becomes $(y - z + 2) + 4y - 5z = -6$ or $5y - 6z = -8$. Now we have two equations and two unknowns (y and z). We repeat the procedure on the new equations, this time choosing y in the new equation one. Solving for y yields $y = -z + 5$. We can plug this into the new equation two to get an even newer equation two: $5(-z + 5) - 6z = -8$ or $-11z = -33$. We now have one equation and one unknown, and see that we can obtain a numerical solution for z by dividing both sides by -11. This yields $z = 3$. Now that we've solved for z, we can plug this into the most recent version of equation one to get $y = -(3) + 5 = 2$. Finally, we can plug the values for y and z into our equation for x to get $x = 2 - (3) + 2 = 1$. Thus, we've obtained a single, unique solution to this system: $x = 1, y = 2, z = 3$.

Now for the second example. It's a little less clear where to begin, so let's solve the first equation for z. There's no way to go wrong here, so this is more a matter of making our lives easier. Solving equation one for z yields $z = -2x - \frac{3y}{2} + \frac{3}{2}$. We can plug this in to equations two and three to get $-3x + 5y - 2\left(-2x - \frac{3y}{2} + \frac{3}{2}\right) = 2$ and $-6x + 10y - 4\left(-2x - \frac{3y}{2} + \frac{3}{2}\right) = 4$, which reduce to $x + 8y = 5$ and $2x + 16y = 10$. Call these the new second and third equations. We can now solve the new second equation for x to get $x = -8y + 5$, which can then be plugged into the new third equation to get $2(-8y + 5) + 16y = 10$, which reduces to $10 = 10$. We were hoping to solve for y here, but it turns out that it dropped out, leaving instead an equation that is always true. So, while we don't have a contradiction, we do have a situation in which we can't solve for any unique numerical solution. We thus have an infinite number of solutions. To find them, let's assume we never solve for y, and leave it as an unknown. Then we know $x = -8y + 5$, and we've written x in terms of the unknown y. All we need to do is solve for z in terms of y too. We can plug our expression for x into our solution for z to get $z = -2(-8y + 5) - \frac{3y}{2} + \frac{3}{2}$ or $z = \frac{29y}{2} - \frac{17}{2}$. Our solution is then any point along the line specified by $x = -8y + 5$, y, and $z = \frac{29y}{2} - \frac{17}{2}$.[9]

Finally, let's tackle the third example. We start with x in equation three, to get $x = -y + z + 5$. We plug that into the other two equations to get our new equations one and two: $3(-y+z+5)+6y-6z = 5$ and $(-y+z+5)+2y-2z = 3$, which become $3y-3z = -10$ and $y-z = -2$. We can then solve the new equation three for y to get $y = z - 2$, which we plug into the new equation two to get $3(z - 2) - 3z = -10$. This simplifies to $-6 = -10$. Not only is there no third

[8] If this is difficult, it is worth revisiting Chapter 2.

[9] Note that we could have used either x or z as the unsolved variable as well (the "known unknown"). The solution line would have been the same. Also, we could have saved ourselves some time by recognizing that the third equation is twice the second equation, implying that they are linearly dependent and so redundant. Solving the system of two equations resulting from removing either the second or third one from the system would have produced the same outcome, as we show below.

unknown for which to solve in this, as there was in the previous example, but we also have a contradiction. Thus the equations can never be simultaneously satisfied, and there is no solution to the system of equations.[10]

13.2.2 Elimination

The benefit of substitution is primarily that, as long as you do the algebra right, you'll find the answer, and you don't have to remember any procedure for doing so. The downside is it is not always very systematic, and you can waste a lot of time on algebra that you didn't need to do. **Elimination** and associated methods like Gaussian elimination provide a more systematized procedure. There are various ways to do this. Rather than go through a large number of them, we present a single example that captures the basic idea.

This basic idea is to manipulate the equations you must solve to produce something that looks not coincidentally like an upper triangular matrix when you stack the equations one on top of the other. That is, the top equation will have all the variables, the next one will have one fewer variable, and so on until the last one only has one variable. You can then do what's called back substitution to work out the values for all variables.

Back substitution you now know how to do, as it is no different from any other form of substitution: you solve for one variable, plug it into other equations, and keep doing this until all the variables have values. But how does one manipulate the equations? The answer is that one can use the rules of logic to perform operations on the equations that keep them true, but possibly make them simpler.

Let's see how this works. Each equation is a logical statement that says one expression equals another. To keep them very general, we write each equation number i like this: $LHS_i = RHS_i$. In words, the left-hand side of equation i equals the right hand side of that equation. We know this is true because the equation tells us it is. But if that's true, then so is $a(LHS_i) = a(RHS_i)$, where a is any scalar, including any variable. After all, if $x + y = 5$, then $2(x + y) = 10$, since one can replace $x + y$ in the second expression by 5 to get $2(5) = 10$. Similarly, one can add (or subtract) any scalar a to (from) each side, too: $LHS_i \pm a = RHS_i \pm a$. Further, since each side of each equation is a scalar, and the LHS must equal the RHS of each equation, we can also add *equations*, or multiples of equations, to each other. For example, we could add 3 times equation two to 2 times equation one to get a new equation: $3(LHS_2) + 2(LHS_1) = 3(RHS_2)+2(RHS_1)$. Finally, since the order in which the equations are listed is arbitrary, we can also change the order without changing the content of the equations.

Now that we know how to manipulate equations, let's solve our examples

[10]Again, we might have noted the contradiction between equations one and two right off the bat: the LHS of equation one is three times the LHS of equation two, but the RHS of one is not three times the RHS of two, implying that they can't both be true, and so there is no solution. We see easier ways to figure this out later.

again using elimination, beginning with example one. We start by getting rid of x from all equations but the first. First, -2 times equation two plus equation one produces an equation with no x, since $2x-y+3z-2(x+4y-5z) = 9-2(-6)$, which reduces to $-9y + 13z = 21$. Second, we can do the same thing with -2 times equation three plus equation one, since $2x-y+3z-2(x-y+z) = 9-2(2)$, which reduces to $y + z = 5$. Thus our equations are now

$$\begin{aligned} 2x - y + 3z &= 9, \\ -9y + 13z &= 21, \\ y + z &= 5. \end{aligned}$$

Next we want to eliminate y from the last equation, which we can do by multiplying 9 times the last equation and adding it to the second equation, since $-9y+13z+9(y+z) = 21+9(5)$, which reduces to $22z = 66$. Now our equations are

$$\begin{aligned} 2x - y + 3z &= 9, \\ -9y + 13z &= 21, \\ 22z &= 66. \end{aligned}$$

This is the triangular shape we need. We complete it by back substitution. The third equation solves to $z = 3$ once you divide both sides by 22. Plugging $z = 3$ into the second equation gives $-9y + 13(3) = 21$, or $-9y = -18$, or $y = 2$. Plugging both $y = 2$ and $z = 3$ into the first equation gives $2x - (2) + 3(3) = 9$, or $2x = 2$, or $x = 1$. This matches our solution using substitution.

This was somewhat easier than using substitution, but the benefit is more pronounced with the second and third examples. As we are now primed to manipulate equations, we should notice quickly that equation three is twice equation two in example two. This means that, were we to add -2 times equation two to equation three we would get the uninformative equation $0 = 0$. This implies that equations two and three are redundant, so we can eliminate equation three entirely and replace it with $y = y$.[11] As before, we could have also chosen $x = x$ or $z = z$, but we're trying to replicate the answers from substitution for comparability, and there we chose not to solve for y. This produces the equations

$$\begin{aligned} 4x + 3y + 2z &= 3, \\ -3x + 5y - 2z &= 2, \\ y &= y. \end{aligned}$$

We want to eliminate x from equation two, which involves adding 3 times equation one to 4 times equation two, since $3(4x + 3y + 2z) + 4(-3x + 5y - 2z) =$

[11] Even if we were not to notice this, going through the elimination procedure would show that we could not eliminate one variable out of the last two equations without eliminating both variables at the same time, yielding $0 = 0$ eventually anyway. Noticing it early saves time; it is not essential for the method.

$3(3) + 4(2)$, which becomes $29y - 2z = 17$. This yields the equations

$$4x + 3y + 2z = 3,$$
$$29y - 2z = 17,$$
$$y = y.$$

Now we're done, so we can use back substitution (which here involves doing nothing, since our third equation is $y = y$) to get $z = \frac{29y}{2} - \frac{17}{2}$. All we did here was solve for z in terms of y. Then we finish by plugging our solutions for y and z into equation one to get $4x+3y+2\left(\frac{29y}{2} - \frac{17}{2}\right) = 3$, which becomes $x = -8y+5$. These are again the same answers as we obtained using substitution, but more quickly.

The answer is even faster to obtain in example three. We should again note while looking for productive ways to eliminate variables that the LHS of equation one is three times that of equation two, but the RHS of equation one is not three times that of equation two, implying a contradiction. We can check this by adding equation one to -3 times equation two, which yields $3x + 6y - 6z - 3(x + 2y - 2z) = 5 - 3(3)$, which reduces to $0 = -4$, a contradiction. Thus the equations can never be simultaneously satisfied, and there is no solution to the system.

13.2.3 Matrix Inversion

Thus far, these techniques have been heavier on algebra than on linear algebra, but they've all been leading up to the next two methods. First, consider the connection noted above of elimination to upper triangular matrices. The connection of systems of linear equations to matrices is not an accident. In fact, one can write any system of n linear equations and m unknowns as an equation involving matrices. In particular, if we label all of our variables $x_1, x_2, \ldots, x_m$, and all of our RHSs as $b_1, b_2, \ldots, b_n$, then we can succinctly call our unknowns the elements of the vector $\mathbf{x}$, and all of the RHS_i the vector $\mathbf{b}$. All we need then is to concatenate the coefficients. A matrix provides the easiest way to do this. Let's label all the coefficients on the variables in equation i: $a_{i1}, a_{i2}, \ldots, a_{im}$. We can view this as a row vector, $\mathbf{a}_i$. If we then stack all these row vectors on top of each other, we get the matrix A. If we write $\mathbf{x}$ and $\mathbf{b}$ as column vectors, the rules of matrix multiplication you learned in the previous chapter tell us that $A\mathbf{x} = \mathbf{b}$. (Multiply it out if you are unsure.)

There are several things we can do with this representation. One, we can create an "augmented matrix" by taking our A and adding the column $\mathbf{b}$ to it as a new column on the right. Using what are called elementary row operations, such as multiplying rows by scalars, adding rows, or interchanging rows, we can attempt to get the identity matrix in all but the last column. If we can do this, then we get what is called reduced row echelon form.[12] More important,

[12] This is for a square matrix. A matrix that is not square, and so has a different number

if we can do this, then we can read off the answers to the system by returning to equation form, since each row will provide the value of one of the variables. However, because these elementary row operations are the logically consistent ones we used in performing elimination, all this framework does is help us keep track of our coefficients as we manipulate the equations. While useful, this is sufficiently similar to elimination that there is little reason to discuss it further.

There are more things we can do with this representation, though. For one, we can use it to identify several important properties of the system. Recall that the *rank* of the matrix is either the number of linearly independent rows or the number of linearly independent columns and that these two numbers are always equal. If the rank of A is equal to its number of rows, then the system has at least one solution. In a nutshell, this is because each equation is linearly independent, and so cannot produce a contradiction, as none of the LHSs of the equations are identical. Further, there can't be fewer unknowns than equations in this case because if there were, the rank would be equal to the (smaller) number of columns.

If the rank of A is equal to its number of columns, then the system has at most one solution. Again in a nutshell, this is because there are enough independent equations to solve for all the unknowns as long as there isn't a contradiction, since the number of independent rows (corresponding to equations) is at least as great as the number of variables (corresponding to columns).[13]

If A is a square matrix, then there is an equal number of rows and columns, and so an equal number of equations and unknowns. If the rank is equal to $m = n$, then there is no more than one solution, but there is at least one solution, implying that there is a unique solution to the system of equations. Since we can discern whether or not the matrix has full rank (i.e., rank equal to $m = n$) by checking its determinant, we are able to test for the number of solutions of the system. If it is square, take the determinant. If the determinant is non-zero, the matrix is nonsingular, and so full rank, and so the system of equations it represents has a unique solution. If the determinant is zero, the matrix is singular. In this case, or in the case of $m \neq n$, we can start checking for linearly dependent columns or rows, removing them until we can identify the rank of the matrix and apply the logic of the last two paragraphs to bound the number of solutions the system has.

When the matrix A is square and nonsingular, we can do more, though: we can *invert* the matrix to figure out what the unique solution is. This is comparably easy to explain. If $A\mathbf{x} = \mathbf{b}$, then left-multiplying both sides by A^{-1} yields $A^{-1}A\mathbf{x} = A^{-1}\mathbf{b}$, which becomes, since $A^{-1}A = I$, $\mathbf{x} = A^{-1}\mathbf{b}$. So all we need to do is find the inverse of A, and then multiply that inverse by $\mathbf{b}$ to get our solution. Since many computational solutions exist for inverting and

of equations and unknowns, might have not have a one in all columns, for example, since not all variables may have a unique numerical solution.

[13]There's a more elegant way of saying this involving the dimensions of the image and the kernel of the linear map specified by A, but this is beyond the scope of the book.

multiplying matrices,[14] this is often the easiest way to solve systems of linear equations. We attempt the method on our examples shortly.

Before doing so, though, it is worth mentioning one other use of this matrix representation. Consider a set of m column vectors denoted $\mathbf{a}_j$, and let the matrix A be formed by placing these column vectors side by side. We can determine whether any vector $\mathbf{b}$ is in the space spanned by these vectors by checking to see if the equation $A\mathbf{x} = \mathbf{b}$ has a solution. If it has a solution for all $\mathbf{b} \in \mathbb{R}^m$, then these vectors span $\mathbb{R}^m$.

Okay, back to equation solving via **matrix inversion**, starting with example one. First we form a matrix A and a vector $\mathbf{b}$:

$$\mathbf{b} = \begin{pmatrix} 9 \\ -6 \\ 2 \end{pmatrix}, A = \begin{pmatrix} 2 & -1 & 3 \\ 1 & 4 & -5 \\ 1 & -1 & 1 \end{pmatrix}.$$

Next we find the determinant of A to make sure it's invertible (we need it for the inverse anyway, as you may recall): $\det(A) = 2(4-5) - 1(-1+3) + 1(5-12) = -2 - 2 - 7 = -11$. As this is not zero, the matrix is nonsingular, and we can invert it. To invert it, we need to compute nine minors. Specifically, $M_{11} = -1, M_{12} = 6, M_{13} = -5, M_{21} = 2, M_{22} = -1, M_{23} = -1, M_{31} = -7, M_{32} = -13, M_{33} = 9$.[15] Now we can use the formula for the inverse from the previous chapter, $A^{-1} = \frac{1}{|A|}C^T$, where $C_{ij} = (-1)^{i+j} \times M_{ij}$, to find the inverse. This is

$$A^{-1} = \frac{1}{-11} \begin{pmatrix} -1 & -2 & -7 \\ -6 & -1 & 13 \\ -5 & 1 & 9 \end{pmatrix}.$$

Finally, we multiply this by $\mathbf{b}$ to get

$$A^{-1}\mathbf{b} = \frac{1}{-11} \begin{pmatrix} -1 & -2 & -7 \\ -6 & -1 & 13 \\ -5 & 1 & 9 \end{pmatrix} \begin{pmatrix} 9 \\ -6 \\ 2 \end{pmatrix} = \begin{pmatrix} 1 \\ 2 \\ 3 \end{pmatrix}.$$

This checks out with the solution using the alternative methods.

We next turn to example two. Here we have

$$\mathbf{b} = \begin{pmatrix} 3 \\ 2 \\ 4 \end{pmatrix}, A = \begin{pmatrix} 4 & 3 & 2 \\ -3 & 5 & -2 \\ -6 & 10 & -4 \end{pmatrix}.$$

First we find the determinant: $\det(A) = 4(-20+20) + 3(-12-20) - 6(-6-10) = 0 - 96 + 96 = 0$. Since the determinant is zero, the matrix is singular, and we can't invert it. So we can't directly use this method. We can check to see how

[14] We are partial to the R programming language (`http://www.r-project.org/`), which is powerful, free, and has many free packages that can make it more user-friendly for people who have no programming background.

[15] Guess what? You should check these.

many solutions there are, though, by computing the rank of the matrix. First, we note that row three is a multiple of row two, and b_3 is the same multiple of b_2, so we delete row three, leaving

$$\begin{pmatrix} 4 & 3 & 2 \\ -3 & 5 & -2 \end{pmatrix}.$$

Next, we note that the columns of this new matrix are linearly dependent, as they must be since there are three of them in a 2-D space. Removing the second one yields the matrix[16]

$$E = \begin{pmatrix} 4 & 2 \\ -3 & -2 \end{pmatrix}.$$

We can't remove the column though, as we could a row, because each equation has a RHS, and these must remain balanced. But we can eliminate the third row from $\mathbf{b}$ (since it corresponded to the eliminated third equation) and subtract the eliminated second column multiplied by y to get the vector[17]

$$\mathbf{c} = \begin{pmatrix} 3 \\ 2 \end{pmatrix} - y \begin{pmatrix} 3 \\ 5 \end{pmatrix} = \begin{pmatrix} 3 - 3y \\ 2 - 5y \end{pmatrix}.$$

Now we find the inverse of E. First, its determinant is $\det(E) = -8+6 = -2$, so the matrix is nonsingular with rank 2. Its inverse is

$$\begin{pmatrix} 1 & 1 \\ -\frac{3}{2} & -2 \end{pmatrix}.$$

We can can apply E^{-1} to $\mathbf{c}$ to yield a 2-D vector consisting of the solutions for x and z in terms of y. This yields the solution

$$E^{-1}\mathbf{c} = \begin{pmatrix} 1 & 1 \\ -\frac{3}{2} & -2 \end{pmatrix} \begin{pmatrix} 3 - 3y \\ 2 - 5y \end{pmatrix} = \begin{pmatrix} 5 - 8y \\ -\frac{17}{2} + \frac{29y}{2} \end{pmatrix}.$$

Again, the solution matches that obtained from our earlier methods.

Finally, we turn to our third example and form the required matrix and vector:

$$\mathbf{b} = \begin{pmatrix} 5 \\ 3 \\ 5 \end{pmatrix}, A = \begin{pmatrix} 3 & 6 & -6 \\ 1 & 2 & -2 \\ 1 & 1 & -1 \end{pmatrix}.$$

The determinant of A is $\det(A) = 3(-2+2) - 1(-6+6) + 1(-12+12) = 0 + 0 + 0 = 0$, so the matrix is singular, and we can't invert it. However, now

[16] We could have removed any of the three; however, since the second corresponds to y, and y is the variable we chose earlier to remain unsolved, we removed the second column to be consistent.

[17] To elaborate a bit more, first we delete the third row of both A and $\mathbf{b}$. This we can always do. If we revert back to equation form, we see we have two equations and three unknowns. So, we then move all the ys over to the RHS. Converting back to matrix form simultaneously gives us E and $\mathbf{c}$.

we can't remove an equation, as the RHS of equation two is not one-third of the RHS of equation one, while the LHS of equation two is one-third of the LHS of equation one. So we're stuck, and there is no solution because the equations are contradictory.

13.2.4 Cramer's Rule

The final method we explore is one that works only when there an equal number of equations and unknowns (i.e., a square matrix A) and A is nonsingular. So we add another example along with example one to provide more practice.[18]

The method is called **Cramer's rule** after the eighteenth-century Swiss mathematician Gerald Cramer. It states that we can solve for $\mathbf{x}$ using the formula $x_i = \frac{|B_i|}{|A|}$, where the matrix B_i is formed by replacing the ith column of A (the column corresponding to variable x_i) with $\mathbf{b}$.[19] We start with a new example:

$$\mathbf{b} = \begin{pmatrix} 0 \\ 1 \\ 0 \end{pmatrix}, A = \begin{pmatrix} 1 & 2 & 0 \\ -1 & 1 & 1 \\ 1 & 2 & 3 \end{pmatrix}.$$

First we take the determinant of A, both to check to make sure that we can apply Cramer's rule and to determine the denominator of each x_i. $|A| = 1(3-2) + 1(6-0) + 1(2-0) = 9$ is the required determinant, and it is clearly non-zero. So we may proceed.

Next we form the B_i by replacing each of the three columns by $\mathbf{b}$. These are

$$B_1 = \begin{pmatrix} 0 & 2 & 0 \\ 1 & 1 & 1 \\ 0 & 2 & 3 \end{pmatrix}, B_2 = \begin{pmatrix} 1 & 0 & 0 \\ -1 & 1 & 1 \\ 1 & 0 & 3 \end{pmatrix}, B_3 = \begin{pmatrix} 1 & 2 & 0 \\ -1 & 1 & 1 \\ 1 & 2 & 0 \end{pmatrix}.$$

As the penultimate step, we compute the determinant of these three matrices: $|B_1| = 0(3-2) - 1(6-0) + 0(2-0) = -6$, $|B_2| = 1(3-0) + 1(0-0) + 1(0-0) = 3$, and $|B_3| = 1(0-2) + 1(0-0) + 1(2-0) = 0$.

To finish, we now apply Cramer's rule to get

$$\begin{aligned} x_1 &= -\frac{6}{9} = -\frac{2}{3}, \\ x_2 &= \frac{3}{9} = \frac{1}{3}, \\ x_3 &= \frac{0}{9} = 0. \end{aligned}$$

[18] At this point, the reader may be feeling that the variety of solution methods we offer is a bit excessive. We offer two responses. One, in our experience, different students find different methods more intuitive. By offering a variety of methods to solve linear equations, we improve the chances that any particular student will find one that works for her. Two, we intend this book as a reference in addition to a class text, and to that end we prefer to err on the side of overkill.

[19] Cramer's rule follows logically from some properties of the determinant, but that is not important for our purposes.

We can check these values by plugging them back into the three equations represented by A and $\mathbf{b}$, and verifying:

$$\begin{aligned}
0 &= (-\frac{2}{3}\cdot 1) + (\frac{1}{3}\cdot 2) + (0\cdot 0) = -\frac{2}{3} + \frac{2}{3},\\
1 &= (-\frac{2}{3}\cdot -1) + (\frac{1}{3}\cdot 1) + (0\cdot 1) = \frac{2}{3} + \frac{1}{3},\\
0 &= (-\frac{2}{3}\cdot 1) + (\frac{1}{3}\cdot 2) + (0\cdot 3) = -\frac{2}{3} + \frac{2}{3}.
\end{aligned}$$

Before concluding this section, we also tackle example one yet another time. Recall that

$$\mathbf{b} = \begin{pmatrix} 9 \\ -6 \\ 2 \end{pmatrix}, A = \begin{pmatrix} 2 & -1 & 3 \\ 1 & 4 & -5 \\ 1 & -1 & 1 \end{pmatrix}.$$

Also recall that $\det(A) = 2(4-5) - 1(-1+3) + 1(5-12) = -2-2-7 = -11$. Now we need the B_i:

$$B_1 = \begin{pmatrix} 9 & -1 & 3 \\ -6 & 4 & -5 \\ 2 & -1 & 1 \end{pmatrix}, B_2 = \begin{pmatrix} 2 & 9 & 3 \\ 1 & -6 & -5 \\ 1 & 2 & 1 \end{pmatrix}, B_3 = \begin{pmatrix} 2 & -1 & 9 \\ 1 & 4 & -6 \\ 1 & -1 & 2 \end{pmatrix}.$$

We must calculate the determinants of these, which are $|B_1| = 9(4-5) + 6(-1+3) + 2(5-12) = -11$, $|B_2| = 2(-6+10) - 1(9-6) + 1(-45+18) = -22$, and $|B_3| = 2(8-6) - 1(-2+9) + 1(6-36) = -33$. To finish, we now apply Cramer's rule to get

$$\begin{aligned}
x_1 &= \frac{-11}{-11} = 1,\\
x_2 &= \frac{-22}{-11} = 2,\\
x_3 &= \frac{-33}{-11} = 3.
\end{aligned}$$

This matches the solution from our previous methods.

Those interested in seeing Cramer's rule put to use in the context of Richardson's arms race model can consult Huckfeldt, Kohfeld, and Likens (1982, pp. 87–89).

13.3 WHY SHOULD I CARE?

We have left this question to the end in this chapter because much of the chapter builds on itself. Before tackling more advanced topics in the next chapter, though, we explain where you will most likely use this material in statistics and formal theory.

13.3.1 Use in Statistics

We start with statistics and its relationship to linear independence, which we mentioned briefly at the beginning of the chapter. Now we can do better. After changing notation a bit, the equation we've been solving over and over in this chapter is this: $\mathbf{y} = X\boldsymbol{\beta}$, which is a linear equation. Add a constant, $\boldsymbol{\alpha}$, and an error term, $\boldsymbol{\epsilon}$, to the RHS, and we have the standard linear model used in all sorts of regression analysis, which has dominated quantitative studies in the social sciences for decades, from OLS to what is known as the general linear model (GLM):[20] $\mathbf{y} = \boldsymbol{\alpha} + X\boldsymbol{\beta} + \boldsymbol{\epsilon}$. Thus, all the discussion we've been having about the properties of such systems apply to almost all statistical studies in the social sciences!

There's a lot to get into here, and you'll cover it in your statistics classes in proper detail. But there's one thing worth discussing here, and this relates to the notion of linear dependence. The matrix X is the data matrix: it holds information on the values of all the independent variables for each data point. Each column is an independent variable, and each row is a data point. The vector $\mathbf{y}$ contains data on the dependent variable. The solution for which we'd want to solve, $\boldsymbol{\beta}$, tells you how changes in the values of the independent variables are, on average, translated into changes in the values of the dependent variable, given a sample of data.

As we saw above, if the number of rows is less than the number of columns in X, then one does not get a unique solution for $\boldsymbol{\beta}$. Thus, to be able to get anywhere, one should have at a minimum as many data points as independent variables. Usually, though, there will be many more rows than columns. We don't have to worry so much about a contradiction leading to no solution, as we're only looking for the closest solution anyway (that's what the $\boldsymbol{\epsilon}$ is for). However, if it turns out that the column rank is less than the number of columns, so that one or more of the columns are linear combinations of each other, then we actually don't have a set of *independent* variables! This is known as multicollinearity, and it leads one to be unable to identify the coefficient vector $\boldsymbol{\beta}$. That is, one cannot determine a solution to the regression equation.

The connection between singular matrices and linear dependence implies that we can tell easily when this is present, even when not directly comparing the data. If the matrix $X^T X$ has a determinant of zero, so that it is singular and thus noninvertible, then multicollinearity is present and the coefficient vector $\boldsymbol{\beta}$ is not identified. So if one is estimating a regression and one's software reports that the matrix $X^T X$ is singular, that suggests that it is time to return to the theoretical drawing board and rethink the independence of the "independent" variables in one's theory.

One can go further with this argument. The assumption that the matrix X is full rank (i.e., has a rank equal to k when X has k columns) is central to the proof that the OLS estimator is BLUE. In other words, OLS provides the best estimate of an unbiased linear relationship between a set of independent variables and a

[20] See Gill (2001), among others, for an introduction to the GLM.

dependent one, and one need not look further for other estimation schemes that are also linear and unbiased. A full proof and explication of this must wait until your statistics classes, but we can provide the flavor of the argument. First, recall from the previous chapter that the OLS estimator is $\hat{\boldsymbol{\beta}} = (X^T X)^{-1} X^T \mathbf{y}$ and that the linear model is $\mathbf{y} = X\boldsymbol{\beta} + \boldsymbol{\epsilon}$. We'll make three assumptions. The first is that the matrix X is full rank. This ensures we have a set of independent variables, as we've noted. Second, we assume that the expectation of $\mathbf{y}$ conditional on X is equal to $X\boldsymbol{\beta}$.[21] This means that the expected value of the error term, $\boldsymbol{\epsilon}$, is zero, which ensures that expected values of the dependent variable follow from the independent variables and not a biased error. Third, we assume that all errors are independently and identically distributed, so that correlations between errors or differences in the size of errors do not determine variation in the dependent variable. In formal notation, this means that there is a constant variance, $\sigma^2 > 0$, such that if $\boldsymbol{\epsilon} = \mathbf{y} - E[\mathbf{y}|X]$, then $var[\boldsymbol{\epsilon}|X] = \sigma^2 I$, where I is the identity matrix of the appropriate size and $var[\ ]$ is the variance of the random variable within the square brackets.[22]

These assumptions guarantee that the estimator, $\hat{\boldsymbol{\beta}}$, is linear and unbiased. The question is whether it's the best linear and unbiased estimator, often shortened to, is OLS BLUE? Here "best" means closest to the true value of $\boldsymbol{\beta}$ that we are attempting to estimate. To prove that OLS is BLUE, we need to write down a general linear and unbiased estimator and show that it is further from the true value. We'll sketch a proof of this here, mirroring the proof by White and Cho (2012).

First, note that all linear estimators have the form $\tilde{\boldsymbol{\beta}} = B\mathbf{y}$ for some matrix B. This is because matrices define linear transformations. What is B? Well, it turns out that our first two assumptions are satisfied, implying that $\tilde{\boldsymbol{\beta}}$ is conditionally unbiased, if and only if $BX = I$. To see why this is so, note that the expected value of $\mathbf{y}$ is $X\boldsymbol{\beta}$ by the second assumption. This means that the expected value of $\tilde{\boldsymbol{\beta}} = B\mathbf{y}$ is $BX\boldsymbol{\beta}$. But this only equals $\boldsymbol{\beta}$ if $BX = I$.

Now let the matrix $A = (X^T X)^{-1} X^T$, i.e., the matrices that multiply $\mathbf{y}$ in the OLS estimator. A full-rank X implies that this matrix exists and is unique. Following our first two assumptions, then, both $\tilde{\boldsymbol{\beta}}$ and the OLS estimator $\hat{\boldsymbol{\beta}}$ are linear and unbiased as long as $BX = I$, which will be assumed.

Next, note that $BA^T = AB^T = (X^T X)^{-1} X^T B^T = (X^T X)^{-1} I = AA^T$.[23] Finally, we can use our third assumption to produce a measure of the "distance" of an estimate of $\boldsymbol{\beta}$ (e.g., $\tilde{\boldsymbol{\beta}}$) from its true value, $cov(\tilde{\boldsymbol{\beta}}|X)$, where $cov[\ ]$ produces the covariance matrix of the random variables in the brackets. The covariance matrix generalizes the covariance you saw in Chapter 11 to multiple dimensions. You'll learn a great deal more about covariance matrices in your statistics classes.[24] As far as this proof goes, we just need to

[21] We covered expectations of random variables in Chapters 10 and 11.

[22] See Chapters 10 and 11 for a formal definition of variance.

[23] The first equality results from the fact that $(BA^T)^T = (A^T)^T B^T = AB^T$. The third equality results from the fact that $BX = I$, which implies $X^T B^T = (BX)^T = I^T = I$.

[24] The diagonal elements of these matrices encode the variance of each component of the

know that the covariance is equal to $E[(\tilde{\boldsymbol{\beta}} - \boldsymbol{\beta})(\tilde{\boldsymbol{\beta}} - \boldsymbol{\beta})^T|X]$, which is equal to $E[B\boldsymbol{\epsilon}\boldsymbol{\epsilon}^T B^T|X] = \sigma^2 BB^T$ by virtue of assumption three. The same manipulations with $\hat{\boldsymbol{\beta}}$ yield $cov(\hat{\boldsymbol{\beta}}|X) = \sigma^2 AA^T$.

We're almost done now. All that's left is to show that $cov(\tilde{\boldsymbol{\beta}}|X) - cov(\hat{\boldsymbol{\beta}}|X)$ produces a positive semidefinite matrix, which is the analogue to one number being greater than the second whenever the difference between the two is positive. This difference is $\sigma^2(BB^T - AA^T)$. Since $\sigma^2 > 0$, we need to show that $(BB^T - AA^T)$ is positive semidefinite. We manipulate matrices: $AA^T = AB^T = BA^T$ by our above manipulations, so $(BB^T - AA^T) = (BB^T - AB^T) - BA^T + AA^T$.[25] Grouping terms produces $(BB^T - AA^T) = (B - A)(B - A)^T$. Since all matrices of the form given by the RHS of this last equation are positive semidefinite (note the connection to $x^2 \geq 0$ for all real numbers x), our proof is complete, and OLS is BLUE.

13.3.2 Use in Formal Theory

The primary use of the content of these last two chapters in formal theory is more straightforward. Whenever there is more than one actor in a model, and sometimes when there is more than one action one could take, then the set of variables for which one needs to solve is a vector, and one must solve a system of equations to produce that vector. This system is not always linear, but when it is, you can use any of the methods here to solve for it. Actually, even when the system is not linear, substitution and elimination are useful. You have to be more creative, using some of the techniques discussed in Chapter 2 to solve for or eliminate variables.

Perhaps the most likely place you'll see this used is when trying to derive an equilibrium from a pair of what are called "best response functions" in game theory. You saw these in Chapter 3, but to reiterate, a best response function tells one player how to act optimally,[26] *given a particular action by the other player*. So, for example, if there are two parties engaged in electoral competition, each trying to win the election by appealing via a policy platform to the majority of the electorate, but each also caring about the platform they choose, not wanting it to be too far from their ideal policy, then the platform of each depends on the platform of the other party. If party B, say, decides to locate toward the extreme right, then a left-leaning party A fares differently if it attempts to seize the middle, more likely winning the election but with a worse policy outcome, than if it offers a far left platform, putting the election more in doubt but leading to a more satisfying outcome should it win. In game theory, one postulates utility functions that represent preferences, as we saw in Chapter 3, and there will be an optimum action for every action of one's opponent, which

relevant vector-valued random variable, and the off-diagonal elements encode the covariance of two different elements.

[25] First we substituted in for $AA^T = AB^T$, and then we subtracted BA^T and added AA^T, which we can do because they're identical.

[26] That is, decide which of the choices available to her will maximize her expected utility.

we derive by maximizing this utility function, as we saw in Chapter 8. We saw this maximization in action under uncertainty in Chapter 11, in a similar electoral location scenario, only with one actor.

What we're adding here is the presence of two players acting at the same time in choosing platforms. If we constrain policies to be between 0 and 1 on a line, we might get that the optimum strategy for party A is to choose a platform one-third as large as party B's, and the optimum strategy for party B is to choose a platform equal to two-thirds plus one-third of party A's platform. We can write this formally as

$$\begin{aligned} p_A(p_B) &= \frac{p_B}{3}, \\ p_B(p_A) &= \frac{2}{3} + \frac{p_A}{3}. \end{aligned}$$

Here p_A and p_B are the platforms of the parties. We want to know when both equations are simultaneously satisfied, as then both players will be playing optimally.[27] These are two linear equations in two unknowns, and we can use any of the methods provided in this chapter to yield the solution: $p_B = \frac{3}{4}, p_A = \frac{1}{4}$. We call this the *equilibrium* of the game in that, because each player is responding optimally to the other's action at this point, there is no reason to unilaterally alter one's action if one chooses the equilibrium platform. The payoff to this is in the intuition it yields: we see that the players in equilibrium don't go to the middle ($\frac{1}{2}$), but instead split the difference between trying to win in the middle and trying to get what they want at the end. The techniques in this chapter enable this intuition.

13.4 EXERCISES

1. From the previous chapter, let $\mathbf{a} = \begin{pmatrix} 10 \\ 2 \\ 5 \\ 2 \end{pmatrix}$, $\mathbf{b} = \begin{pmatrix} 4 \\ 15 \\ 6 \\ 8 \end{pmatrix}$, $\mathbf{e} = (14, 17, 11, 10)^T$, and $\mathbf{f} = (20, 4, 10, 4)^T$. Answer each of the following questions, saying that it's not possible if there is a calculation you cannot perform.

   a) Write the most general vector that is a linear combination of **a** and **b**.

   b) Are all four of the vectors given in the problem linearly independent? If not, choose independent ones (this is not a unique choice) and write the others in terms of them.

   c) What dimensional space do the four four-dimensional vectors span?

[27] That is, each player will choose among his or her available options that choice that provides the highest expected payoff *given what each expects the other player to choose to do.*

2. Solve the following systems of equations using substitution or elimination, or both:

   a)

$$\begin{aligned} x - 3y + 6z &= -1, \\ 2x - 5y + 10z &= 0, \\ 3x - 8y + 17z &= 1. \end{aligned}$$

   b)

$$\begin{aligned} 4x + 2y - 3z &= 1, \\ 6x + 3y - 5z &= 0, \\ x + y + 2z &= 9. \end{aligned}$$

   c)

$$\begin{aligned} -4x + 6y + 4z &= 4, \\ 2x - y + z &= 1. \end{aligned}$$

3. Use elimination to determine for what values of the parameter k the system given by

$$\begin{aligned} x_1 + x_2 &= 1, \\ x_1 - kx_2 &= 1, \end{aligned}$$

   has no solution, one solution, and more than one solution.

4. From the previous chapter, let $D = \begin{pmatrix} 1 & 2 \\ 3 & 4 \end{pmatrix}$, $E = \begin{pmatrix} 3 & 1 \\ 6 & 2 \end{pmatrix}$, $\mathbf{g} = \begin{pmatrix} 2 \\ 3 \\ 1 \end{pmatrix}$, and $\mathbf{h} = (1, 2, 3)$. Calculate each of the following, indicating that it's not possible if there is a calculation you cannot perform.

   a) Rank of D.
   b) Rank of E.
   c) Solve $D\mathbf{x} = \mathbf{g}$ for $\mathbf{x}$ using matrix inversion.
   d) Solve $\mathbf{x}E = \mathbf{h}$ for $\mathbf{x}$ using matrix inversion.

5. Use Cramer's rule to solve the following systems of equations:

   a)

$$\begin{aligned} x - y + 2z &= 4, \\ -x + 2y + 3z &= 1, \\ x + y - z &= 5. \end{aligned}$$

b)

$$\begin{aligned} 3x - 2y - z &= 4, \\ -3x + 2y - 5z &= 2, \\ 2x + 2y - 2z &= 5. \end{aligned}$$

6. Why is Cramer's rule useful?

7. What does it mean if you have a singular matrix? Provide a different and/or more complete answer than for the same question in the previous chapter.

13.5 APPENDIX

The Khan Academy has some videos that illustrate ways to solve systems of equations that you may find helpful. They include videos on the use of matrices (`http://tinyurl.com/KAmatrix`), the substitution method (`http://tinyurl.com/KAsubmthd`), and the elimination method (`http://tinyurl.com/KAelimmthd`).

Chapter Fourteen

Eigenvalues and Markov Chains

Unlike the previous two chapters, this one provides neither key tools for computation nor core insights into linear algebra that are widely useful for *all* students. Rather, it introduces a handful of more advanced topics that are coming into increasing use in both statistics and formal theory in political science, and contains material one needs before tackling these more advanced topics. Because of this, we suggest that only readers tackling a full semester math course, as opposed to a math camp, should take on this chapter now, with the rest returning later should these topics arise.

The first section of this chapter covers eigenvalues, their associated eigenvectors, and a useful form of matrix decomposition that these make available. In statistics, eigenvalues are used to understand dynamical systems, like time series. The German word *eigen* translates to English as *characteristic*, and for a matrix that represents a linear mapping, the eigenvalues tell you about the characteristics of the mapping. Some dynamical systems represented by time series are explosive, which means that once they get going, they go off the rails, so to speak.[1] The values of a variable continue to increase without limit, typically until something in the system "breaks."

Other dynamical systems may be comparatively stable in that they stay within prescribed boundaries. In this case, the values of a variable never get too big. They may settle stably on some steady state, or they may shrink. The magnitude of the eigenvalues—particularly the largest one—tells you in which of these worlds your system is. In other words, is your system explosively unstable, or will it settle down? As a bonus, the computation of eigenvalues also allows the computation of eigenvectors, which are a set of basis vectors that have some nice properties, as we will see. In particular, they allow one to decompose a matrix in a way that allows one to raise it to a power and readily take its determinant.

Eigenvalues are also useful when studying interdependence of actors, as modeled by network relationships. Network analysis is an increasingly used technique in the social sciences (e.g., Huckfeldt and Sprague, 1995; Centola and Macy, 2007; Siegel, 2009; Jackson, 2010). It focuses on mapping out the relationships between actors, under the presumption that these relationships help

[1]To offer a loose sense, explosive behavior in a system can occur when changes in one or more variables in a system feed back on themselves or each other, producing ever-increasing and/or ever-decreasing values in one or more of the variables in the system as the system moves forward in time.

to dictate, and are in part dictated by, social, political, and economic outcomes. For example, one's voting behavior is a consequence of the voting behavior of others with whom one interacts (e.g., Huckfeldt and Sprague, 1995). In situations in which behavior is affected by one's network of connections, one often wants to understand the relative importance of the actors in driving behavior. One measure of this importance is the *centrality* of an actor. There are many measures of centrality, and an often used one is known as *eigenvector centrality*. To compute it, one uses the techniques discussed in this chapter, and we briefly discuss its computation and meaning at the end of the first section.

The second section of this chapter provides an introduction to the topic of Markov processes. This is a rapidly growing topic in both statistics and formal theory, and we would guess even students who skip this chapter at first will be back here at some point. Stochastic processes, of which Markov processes or Markov chains are perhaps the most notable, describe dynamical systems with some element of randomness. In other words, starting from any state of the system at any time, there is some chance that the system moves into other states. This is a very general framework that has wide utility. In statistics, it is commonly used as the foundation of Markov chain Monte Carlo methods, often abbreviated as MCMC. These are methods designed to produce a probability distribution and are vital to the use of Bayesian statistics, as we describe below when we discuss ergodicity.

In formal theory, Markov chains arise in some game theoretic models that utilize random events, and they have an associated equilibrium concept: Markov perfect equilibrium (e.g., Slantchev, 2003). But they are used more frequently in models of bounded rationality, in which utility optimization is replaced with different models of behavior (e.g., Bendor et al., 2011). In these models there is typically some probability of taking any action that is dependent not only on parameters in the model but also on the previous actions taken. Stochastic processes are often the natural modeling framework in this case, and Markov chains are the most common type of these chosen.

14.1 EIGENVALUES, EIGENVECTORS, AND MATRIX DECOMPOSITION

In the following three subsections we present eigenvalues, their associated eigenvectors, and a technique to decompose matrices based on them. Following the pattern in this part of the book, we first provide the tools necessary to compute each, and then discuss related intuition.

14.1.1 Eigenvalues

14.1.1.1 Computation

Eigenvalues may be computed for any square matrix. There are two steps necessary to calculate them. First, one finds what is called the **characteristic**

equation. This is a polynomial equation with order n, where n is the number of rows and columns of the matrix. Second, one solves for the roots of the characteristic equation. These roots are the eigenvalues of the matrix. We'll take each step in turn.

An **eigenvalue** of a matrix is the solution to the equation $A\mathbf{x} = \lambda\mathbf{x}$, which we can call the **eigenvalue equation**. The $\mathbf{x}$ in this equation is called an eigenvector, and we'll get to it shortly. The λ is an eigenvalue and there will be n of them, though not all may be real-valued (some may be complex numbers).[2]

To find the eigenvalues, we make use of the properties we learned in the previous two chapters. We first walk through the steps, then work an example. To begin, left-multiply both sides by the identity matrix, I. That doesn't change anything, because it's the identity, and now we have $A\mathbf{x} = \lambda I\mathbf{x}$, where we've moved the eigenvalue scalar to the outside. Next, subtract the RHS from both sides to get $A\mathbf{x} - \lambda I\mathbf{x} = 0$. Third, combine terms on the LHS: $(A - \lambda I)\mathbf{x} = 0$. Fourth, recall that for a non-zero $\mathbf{x}$, this equation states that the columns of the matrix $(A - \lambda I)$ are linearly dependent, since you can add a linear combination of them to get the zero vector on the RHS.[3] This means that the matrix $(A - \lambda I)$ has less than full rank, and is singular. Fifth and finally, recall that a singular matrix has determinant zero. Thus the eigenvalue equation implies that $|A - \lambda I| = 0$, which is the characteristic equation.

The LHS of this last expression, $|A - \lambda I|$, provides the **characteristic polynomial**.[4] Finding this is the first step. Since λI gets subtracted from A, there will be a λ in every row and column of $A - \lambda I$, and so no matter how one computes the determinant, it will end up producing a polynomial in λ of order n if A is an $n \times n$ matrix.

Setting this polynomial equal to zero as the characteristic equation requires produces a polynomial equation that we must solve for λ. To solve it, the techniques of Chapter 2, such as factoring, will come in handy. Even so, for all but small matrices of dimension two or three most social scientists will turn to computers to compute eigenvalues.[5]

Let's try a couple of examples of small matrices to see how this works. We begin with a two-by-two matrix. First we find $A - \lambda I$, then we take its deter-

---

[2] The Greek letter λ is typically used for eigenvalues.

[3] Here the columns of the matrix are the vectors in question, and the elements of $\mathbf{x}$ are the coefficients on these vectors, because each column vector ends up multiplied by one of the elements of $\mathbf{x}$. Since these elements aren't all zero (because $\mathbf{x}$ is non-zero) but the linear combination of the columns of the matrix adds to zero, the columns are linearly dependent.

[4] Because the solutions to the characteristic polynomial are the eigenvalues, you will sometimes see **characteristic root** used to signify an eigenvalue. They are the same thing.

[5] Many software options are available, ranging from open source freeware such as FreeMat (`http://freemat.sourceforge.net/`) and Maxima (`http://maxima.sourceforge.net/`) to online resources such as Wolfram Alpha (`http://www.wolframalpha.com/`) and commercial software such as Maple (http://www.maplesoft.com/products/maple/), MatLab (http://www.mathworks.com/academia/), and Mathematica (`http://www.wolfram.com/solutions/highered/`).

minant, and then we write it as a polynomial.

$$
\begin{aligned}
A &= \begin{bmatrix} 4 & 1 \\ 2 & 5 \end{bmatrix} \\
A - \lambda I &= \begin{bmatrix} 4-\lambda & 1 \\ 2 & 5-\lambda \end{bmatrix} \\
|A - \lambda I| &= (4-\lambda)(5-\lambda) - 2(1) \\
&= \lambda^2 - 9\lambda + 18.
\end{aligned}
$$

This shows that the characteristic polynomial for the matrix A is $\lambda^2 - 9\lambda + 18$. Now we set this equal to zero and solve for all values of λ that solve this equation. We can solve this equation by factoring the characteristic polynomial:

$$
\begin{aligned}
\lambda^2 - 9\lambda + 18 &= 0 \Rightarrow \\
(\lambda - 6)(\lambda - 3) &= 0 \Rightarrow \\
\lambda = 6 \text{ and } \lambda = 3.
\end{aligned}
$$

The eigenvalues of A are thus 6 and 3.

Now consider a more difficult example. We perform the same steps on this three-by-three matrix, but now it is easier to factor as we go.

$$
\begin{aligned}
A &= \begin{bmatrix} 1 & 0 & 1 \\ 0 & 1 & 1 \\ 1 & 1 & 0 \end{bmatrix}. \\
A - \lambda I &= \begin{bmatrix} 1-\lambda & 0 & 1 \\ 0 & 1-\lambda & 1 \\ 1 & 1 & -\lambda \end{bmatrix}. \\
|A - \lambda I| &= (1-\lambda)(-\lambda(1-\lambda) - 1) - 0(0-1) + 1(0 - (1-\lambda)) \\
&= (1-\lambda)[\lambda^2 - \lambda - 1 - 1] \\
&= (1-\lambda)[\lambda^2 - \lambda - 2] \\
&= (1-\lambda)(\lambda - 2)(\lambda + 1).
\end{aligned}
$$

This shows that the characteristic polynomial for the matrix A is $(1-\lambda)(\lambda - 2)(\lambda + 1)$. We set this equal to zero and solve for all values of λ that solve this equation. Since it's already factored, the solutions are straightforward. They are $\lambda = 1, \lambda = 2, \lambda = -1$. These are the eigenvalues of A.

14.1.1.2 Intuition

While there is a lot one could say about eigenvalues (Stewart, 1973), for our purposes the intuition behind both them and their use is fairly straightforward. Recall again the eigenvalue equation, $A\mathbf{x} = \lambda\mathbf{x}$. If we view A as a linear mapping—that is, a linear function that takes the vector $\mathbf{x}$ into some other vector $\mathbf{y}$—then A has a special effect on certain vectors. Specifically, if $\mathbf{x}$ is a vector that, when multiplied by A, ends up as a scalar multiple of itself—that is, it stays in the same direction but gets shorter or longer—then the eigenvalue tells us how much shorter or longer it gets. From this is derived the descriptor "characteristic": for certain vectors (the characteristic ones, discussed next), eigenvalues tell you how they evolve with repeated applications of A.

If the vector $\mathbf{x}$ represents the state of some system at a given time, then the eigenvalues tell us how the system evolves over time along certain vectors. We are typically interested in the largest of the eigenvalues, as this is the one that determines the biggest a vector can get over time. If the largest eigenvalue is greater than one, then the vector gets bigger and bigger without bound with each application of A. This is known as explosive growth, and it leads to all sorts of problems in analyzing systems, both in terms of formal models and in terms of time series. One can readily see why this might be. Consider a series of data over time (called longitudinal data) describing war initiation, bilateral international trade, or political participation. Explosive growth would mean each of these would grow ever larger without end, clearly not a likely scenario. At some point, something would have to give, and the entire system would change.

In contrast, stable systems have a largest eigenvalue of 1 or less. A perfectly stable system, one that maintains a constant value over time, has a largest eigenvalue of exactly 1. A system with a largest eigenvalue that is less than 1 is shrinking (i.e., getting smaller and smaller). The canonical physical example of a stable system is a damped oscillator, such as any conventional pendulum or spring that experiences friction. These systems are more manageable, either because of external forces or because of negative feedbacks that keep them in line.

Before considering an example of a dynamical system in political science, we need one additional definition. A **difference equation** is one in which past values (at discrete times) of the variable of interest are included in the equation. So the equation $y_t = \beta y_{t-1}$ is a difference equation as it states that y is a function of its past value. You will sometimes encounter discussion of the "order" of a difference equation. This is a first-order difference equation as it includes only one past value. By contrast, $y_t = \beta_1 y_{t-1} + \beta_2 y_{t-2}$ is a second-order difference equation because it includes both the first and the second lagged values of y. Thus, the order of a difference equation indicates how many past values of the variable of interest are included in the equation.

Now for the example. Richardson (1960) proposes a difference equation model of a possible arms race between two countries, A and B. He proposes a deceptively simple-looking model with only two variables (the level of arms expenditures in A at time t, represented by X_t, and the level of arms expenditures in B at time t, represented by Y_t) and three constants for each country $i = A, B$ (the level of grievances each country has toward the other, represented by γ_i; the strength of the incentive to respond to one's opponent by accumulating more arms, represented by κ_i, and the strength of the disincentive to continue to acquire arms as one's own arms increase, represented by α_i.). He suggests that the expected level of armaments in each country can be modeled via the following two equations:

$$\Delta X_t = -\alpha_A X_t + \kappa_A Y_t + \gamma_A.$$

$$\Delta Y_t = -\alpha_B Y_t + \kappa_B X_t + \gamma_B.$$

The symbol Δ represents the difference operator. The dependent variable for each equation, then, is the change in the level of arms expenditures in each country for a given period of time, say one year. Note that each of the equations is endogenous to the other: the value of one depends on the value of the other. This is what makes them a system of equations. Further, the values of the variables change over time, as dictated by the equations. These equations capture Richardson's intuition about the process. For example, the negative sign in front of the present value of each country's arms expenditures indicates that Richardson does not expect countries to increase their expenditures via a bureaucratic budgeting process that leads to more arms the more one has, but rather one that slows down runaway expenditures.

What is theoretically exciting about models like this is that they permit one to explore the implications of one's theoretical intuition in a very precise way to evaluate what sort of outcomes will "fall out" of the model as one "puts it in motion" (i.e., evaluates how the system changes over time). In Richardson's case, he wanted to know what might lead to an arms race in which two countries would continue to build weapons stockpiles well beyond what they would require to defend themselves from attack by the other. His study rather presciently anticipated the nuclear arms race between the US and the USSR (and, to a lesser extent, China). Returning to our discussion of eigenvalues, Richardson wanted to establish what range of parameter values (i.e., what range of α_i, κ_i, and γ_i for $i = A, B$) would be associated with an explosive versus stable versus shrinking system. A stable system is one which two countries maintain relatively constant levels of arms expenditures, whereas an explosive system is an arms race, and a shrinking one represents a period of disarmament. Work by Huckfeldt, Kohfeld, and Likens (1982, pp. 41–60) walks the reader through Richardson's use of matrix algebra to answer this question.

Other examples that may be of interest can be found in Blalock (1969, pp. 100–140), which describes a simultaneous equation model of social interaction; Francisco (1995), which offers a simultaneous equation model of coercion and protest; and Gottman et al. (2005), which presents a model of marital dynamics, the study of which will permit the reader to participate in parlor games debating the probability of divorce among various couples. Those interested in further study of these models will want to consult Goldberg (1958), which is widely regarded as the classic statement of the use of difference equation models in the social sciences.

14.1.2 Eigenvectors

14.1.2.1 Computation

We have worked with the eigenvalue equation, which could be, and often is, called the eigenvector equation. An **eigenvector** is the vector that makes the eigenvalue equation true for a particular eigenvalue. That is, there will be one eigenvector for each eigenvalue, and you can find each of them by solving $A\mathbf{x} = \lambda\mathbf{x}$ for each eigenvalue. Doing so in general involves solving a system of linear equations, as one can see by substituting $\mathbf{b} = \lambda\mathbf{x}$ on the RHS. There is one tricky part, though. Since $\mathbf{x}$ is on both sides of the equation, multiplying the entire vector by any scalar leaves the equation still true. Consequently, the eigenvectors are determined only up to a multiplicative constant. In other words, if $\begin{pmatrix}1\\0\\3\end{pmatrix}$ is an eigenvector for some eigenvalue, $2\begin{pmatrix}1\\0\\3\end{pmatrix} = \begin{pmatrix}2\\0\\6\end{pmatrix}$ is an eigenvector for the same eigenvalue.

The good news is that this actually makes our lives easier, not harder. Because only the relative values of the elements of the eigenvectors matter, we can freely assign a value to one of these elements. To make our lives easy, we use 1 for this value. Then, all we need to do is solve the equation $A\mathbf{x} = \lambda\mathbf{x}$, where one of the elements of $\mathbf{x}$ is replaced by 1 and the rest of the elements are free variables (unknowns) for which we must solve. Thus we will have one fewer unknown to solve for than the dimension of the matrix and vector. Usually we can assign the 1 to the first element, which will generally work as long as the first element would not otherwise have been zero. However, if it doesn't work, we can always assign the 1 to a different element of $\mathbf{x}$; one of them will work. We can use any of the methods from the previous chapter to solve the resulting equation. We exhibit this procedure for the matrices and eigenvalues from the previous subsection to make this clearer.

Our first example is

$$A = \begin{bmatrix} 4 & 1 \\ 2 & 5 \end{bmatrix}.$$

Recall that this has eigenvalues 6 and 3. Each of these has an associated eigenvector, so we'll find one at a time, starting with 6. This equation is $A\mathbf{x} = 6\mathbf{x}$, where $\mathbf{x} = \begin{pmatrix}1\\c\end{pmatrix}$ and c is some scalar for which we will solve. Since there are two equations and only one unknown, substitution works well, so we multiply out the matrix equation to get the two equations, $4 + c = 6$ and $2 + 5c = 6c$. It is quick work to see that these equations are the same, and that they produce the same $c = 2$.[6] This is no accident: because the vector's elements are defined only

[6] If you are unable to do the algebra in your head, please do it on a piece of paper to see that this is true.

up to some multiplicative constant, not all the equations you must solve will be unique. There will always be one of them that is a linear combination of the others, so don't be concerned about that. Thus, the eigenvector of A associated with the eigenvalue 6 is $\begin{pmatrix} 1 \\ 2 \end{pmatrix}$.

We can repeat the procedure with the eigenvalue 3 to finish the job. Now our two (identical) equations are $4 + c = 3$ and $2 + 5c = 3c$, both of which reduce to $c = -1$. Thus, the eigenvector of A associated with the eigenvalue 3 is $\begin{pmatrix} 1 \\ -1 \end{pmatrix}$.

Now for our second example:

$$A = \begin{bmatrix} 1 & 0 & 1 \\ 0 & 1 & 1 \\ 1 & 1 & 0 \end{bmatrix}.$$

Recall that this has eigenvalues $1, 2, -1$. While we could solve for the eigenvectors a number of ways, because you can write one of the equations as a linear combination of the others we try substitution again.[7] We let $\mathbf{x} = \begin{pmatrix} 1 \\ c \\ d \end{pmatrix}$ and multiply out the matrix by the first eigenvalue, for $\lambda = 1$. This yields three equations in two unknowns:

$$\begin{aligned} 1 + d &= 1, \\ c + d &= c, \\ 1 + c &= d. \end{aligned}$$

The first equation implies $d = 0$. Plugging $d = 0$ into the second equation produces $c = c$, implying that this equation is the unnecessary one. Plugging $d = 0$ into the third equation implies $c = -1$. Thus the eigenvector corresponding to the eigenvalue 1 is $\begin{pmatrix} 1 \\ -1 \\ 0 \end{pmatrix}$.

We now repeat this for the other two eigenvalues. For $\lambda = 2$ we get the same three equations, but with 2 multiplying each RHS:

$$\begin{aligned} 1 + d &= 2, \\ c + d &= 2c, \\ 1 + c &= 2d. \end{aligned}$$

[7]One could use matrix inversion via the method presented in the previous chapter, but it's simpler to do it this way. We could also use elimination, but the equations we get were chosen to be simple algebraically, so it would not help much.

The first equation implies $d = 1$. Plugging $d = 1$ into the second equation produces $c + 1 = 2c$, or $c = 1$. Plugging $d = 1$ into the third equation implies $c + 1 = 2$, or again $c = 1$. There is no new information there, but no contradiction either, which is a good way to check our algebra. Thus the eigenvector corresponding to the eigenvalue 2 is $\begin{pmatrix} 1 \\ 1 \\ 1 \end{pmatrix}$.

Finally, for $\lambda = -1$, the three equations are

$$\begin{aligned} 1 + d &= -1, \\ c + d &= -c, \\ 1 + c &= -d. \end{aligned}$$

The first equation implies $d = -2$. The second produces, after plugging in $d = -2$, $c - 2 = -c$, or $c = 1$. The third again doesn't provide new information, but is satisfied by $c = 1, d = -2$ (this yields $2 = 2$), so it checks out. Thus the eigenvector corresponding to the eigenvalue -1 is $\begin{pmatrix} 1 \\ 1 \\ -2 \end{pmatrix}$.

14.1.2.2 Intuition

As noted in the previous subsection, eigenvectors are vectors that, when multiplied by A, end up as scalar multiples of themselves, so the transformed vectors point in the same direction as they did originally, only potentially changing in length. They therefore indicate what may be thought of as invariant vectors under the linear transformation represented by A. This is easiest to see geometrically, as when A is something like a reflection across the x-axis. In this case, any vector parallel to the x-axis is unchanged in direction by the reflection, and any vector parallel to the y-axis has its sign flipped, so it is an eigenvector with eigenvalue -1.

However, while eigenvectors are extremely important in fields like physics, for the reasons stated above, eigenvalues tend to be used more in political science. There are three reasonably common situations in which eigenvectors do play a role, though. One is as the constituent parts in matrix decomposition and principal components analysis (e.g., Rummel, 1970). We discuss this in the next subsection. The second is when assessing the degree of centrality of an actor in a network (e.g., Hafner-Burton, Kahler, and Montgomery, 2009). We discuss this in the subsection following. The third is when you interpret $\mathbf{x}$ as the state of the system and A as the transition matrix between states, as one does in stochastic processes. We consider this more in the next section.

14.1.3 Matrix Decomposition

There are times when one wants to go beyond the matrix operations we've presented, particularly in dynamical systems, or when one has already calculated several objects of interest and wants a shortcut to other calculations. For example, let's say one wanted to observe how a dynamical system evolved over time with repeated applications of A. The first application gets you $A\mathbf{x}$, the second $AA\mathbf{x}$, and so on. We might write all the As multiplied together as A^z for some power z, but that could take a very long time to compute. Is there an easier way?

A **matrix decomposition** is a way of factoring a matrix into more than one matrix such that the product of these new matrices equals the original one. The point of doing so is that in some cases it is easier to work with the decomposed matrix.

There are a large number of potentially useful decompositions out there. For example, one can factor a square matrix so that $A = LU$, where L and U are lower and upper triangular matrices, respectively. This can be useful for some numerical calculations and for more advanced topics in statistics. However, most of these are sufficiently uncommon so that we choose not to take up space going through them all.[8]

There is one decomposition that proves useful in multiple ways, though, that we will cover here, and that is called **spectral decomposition** or **eigenvector decomposition**. It can be accomplished when you have a square $n \times n$ matrix in which all n eigenvalues are *distinct*. In other words, no two eigenvalues can be the same.[9] When this is the case, one can decompose one's matrix in the following way: $A = QDQ^{-1}$, where D is a diagonal matrix with all the eigenvalues along the diagonal, and Q is a matrix with columns composed of the corresponding eigenvectors.

This technique is useful in several ways. First, it solves the problem of raising the matrix to a power. One can write $A^z = (QDQ^{-1})(QDQ^{-1})\ldots(QDQ^{-1})$, where we've replaced each of the A with its spectral decomposition. But every adjacent $QQ^{-1} = I$, and so we have $A^z = (QDID\ldots DQ^{-1}) = QD^zQ^{-1}$. Raising a matrix that can be so decomposed to a power thus involves raising a diagonal matrix to a power, which one does by raising each eigenvalue along the diagonal to that power. You can even define matrix roots in this way, letting z be a fraction like $\frac{1}{2}$ in the expression $A^z = QD^zQ^{-1}$.

Second, once you have the eigenvalues and vectors of a matrix, it allows for easier calculation of things like the determinant. To see this, recall that $\det(A) = \det(QD^zQ^{-1}) = \det(Q)\det(DQ^{-1}) = \det(Q)\det(D)\det(Q^{-1})$. Each determinant is just a scalar, so they commute. Further, we know that $\det(Q^{-1}) =$

---

[8] An Internet search on "matrix decomposition" will produce a great deal of information, if you are interested.

[9] The Jordan decomposition generalizes this to the case of repeated eigenvalues.

$\frac{1}{\det(Q)}$. So we have

$$\det(A) = \det(Q)\det(D)\det(Q^{-1}) = \det(Q)\det(Q^{-1})\det(D) = \det(D).$$

Once we have this, we can use the fact that the determinant of a diagonal matrix is just the product of the elements on the diagonal, so $\det(A) = \det(D) = \prod_{i=1}^{n} \lambda_i$. This can be very useful for large matrices.[10]

Finally, this decomposition is closely related to the method of principal components analysis, which is related to the commonly used factor analysis. The goal of each is to explain variation in observed variables via variation in a smaller number of factors, often called latent variables. It turns out then when A is both symmetric and positive semidefinite (see Chapters 12 and 13, respectively), then the set of eigenvectors of a matrix is orthogonal (see Chapter 12), and all eigenvalues are real-valued and nonnegative. Since covariance and correlation matrices are positive semidefinite matrices, one can use the eigenvector decomposition on either matrix to determine the principal components of that matrix. These components, which are the eigenvectors of the matrix, help evaluate the extent to which the variables are dependent on one another. The variance explained by these components is given by the associated eigenvalues. The largest eigenvalues thus correspond to the components that explain the most covariance in the data. Another usage occurs in (Bayesian) vector autoregression modeling (e.g., Brandt and Freeman, 2006).

14.1.4 Networks and Eigenvector Centrality

As we have seen, matrices represent relationships between objects. When capturing a system of linear equations, they represent relationships between variables; when capturing dynamical systems, they represent relationships between the values of variables at different moments in time. Their facility in capturing relationships is a consequence of their structure: with two dimensions, they can specify how elements in one dimension (e.g., the values of variables at a given time) relate to elements in the other dimension (e.g., the values of different variables or variables at different times).

One can extend this logic to capture other relationships as well. In these relationships, researchers' concern is not with the way in which the values of variables are related but rather with the presence or absence of a connection between objects. For example, abundant research (e.g., Huckfeldt and Sprague, 1995; Ryan, 2011; Klofstad, Sokhey, and McClurg, 2012; Sinclair, 2012) indicates the importance of social connections in voting behavior. Thus, when modeling voting behavior either empirically or theoretically, we might want to keep track of the set of other individuals to whom each person is connected, since this set will help determine when and how one votes. The same is true if one believes that one's likelihood of engaging in dissent or terrorism (e.g., McAdam, 1986;

[10] It is also the case that the trace of A is equal to the sum of the eigenvalues, but this fact is used less often.

Siegel, 2011) depends on others' choices, or if a country's set of trading, alliance, or conflict partners affects with whom the country trades or fights (e.g., Ward, Siverson, and Cao, 2007; Ward, Ahlquist, and Rozenas, 2013). We call the full set of relationships between actors a **network**; network analysis consists of the set of methods designed to analyze networks.

Matrices are well suited to capture networks. For example, let each of the indices i and j correspond to an actor (e.g., an individual or a country). Then one can define the matrix A such that each element a_{ij} is equal to 1 if actors i and j are connected in some fashion (e.g., are friends, family, co-workers, or trading partners, have or are presently engaged in conflict, etc.) and equal to 0 if not.[11] A matrix A formed in this way is known as an **adjacency matrix**.

There's a lot one can do with an adjacency matrix, far more than we can cover here.[12] We focus on one question: how can we measure the relative importance of an actor in the network? This is an important measure, as it describes the power each actor possesses that can be attributed to his, her, or its network connections, and it has a name: the **centrality** of the actor.

The concept of importance is a bit nebulous, though, and fittingly, there are many measures of centrality. One of the more straightforward is called **degree centrality**, because it counts the number of connections to an actor.[13] This measure expresses the direct influence of an actor as it counts all its connections, but it fails to incorporate information about the larger network. For example, there might be an actor with a lot of direct influence over a small subset of actors in the network, but because this subset is not well connected to the rest of the network, the actor's absolute influence is small. Therefore we turn to other measures.

A commonly used measure (e.g., Hafner-Burton, Kahler, and Montgomery, 2009; Fowler, 2006; Victor and Ringe, 2009) that does a better job incorporating this information is known as **eigenvector centrality** (Bonacich, 1972) owing to its close relationship to the eigenvector. Loosely speaking, it incorporates additional information by considering not only an actor's number of connections but also *each connection's* number of connections. Connections to better-connected actors count more than those to less well-connected actors.[14]

Because all actors' eigenvector centrality scores are interrelated, computation of all of them occurs at once. For each actor, one adds the eigenvector centrality

[11]One can also capture the strength of relationships by using a number other than 1, with larger numbers corresponding to stronger relationships, but that is beyond the scope of this brief discussion.

[12]Network analysis is a broad interdisciplinary field with a growing footprint in political science and across the social sciences. Interested readers may want to start with a general text (e.g., Wasserman and Faust, 1994).

[13]In network analysis, actors are called *nodes*, and connections or links between them are called *edges*. Edges can be undirected, in which case they are symmetric (if I am connected to you, you are connected to me), or directed, in which case they are asymmetric (I can be connected to you even when you are not connected to me). Our discussion is appropriate either for undirected edges or for directed edges in which we focus on edges specifying the application of influence out from a node.

[14]Some Internet search engines work via a similar principle (e.g., Google PageRank).

scores for all other actors to which that actor is connected. This amounts to multiplying a vector of eigenvector centrality scores by the adjacently matrix in order to produce (a multiple) of that same vector. The equation that captures this is identical to the eigenvalue equation discussed above, whence the name of this measure arises. Each element of the eigenvector is the eigenvector centrality measure for the corresponding actor. Though there may be many eigenvectors for the adjacency matrix, it turns out that requiring all centrality measures be positive forces us to consider only the eigenvector associated with the largest eigenvalue. Thus, after computing the eigenvalues of a matrix, to find the eigenvector centrality measures for all actors in the network one need only compute the eigenvector associated with the largest eigenvalue. Once one computes these, they can be used as a descriptive measure of the network (e.g., is there a lot of variance in centrality, implying variation in power?) or as independent variables in testing the effect of network position on other variables.

14.1.5 Why Should I Care?

In regression diagnostics one can appeal to the characteristic roots to evaluate the strength of relationships among variables. Similarly, eigenvalues and eigenvectors are central to principal component analysis and factor analysis (e.g., Kim and Mueller, 1978). They tend to be of particular interest to those studying time series (e.g., Reinsel and Ahn, 1992).

When studying difference and differential equations, as well as other dynamical systems, the characteristic roots provide information about the stability of solutions: whether the solution to a system of equations is explosive or settles down to some equilibrium (e.g., Francisco, 2010). For example, imagine that you are studying the conflictual behavior of two countries or political parties (e.g., the India-Pakistan nuclear arms competition or the propensity of two candidates for office to spend money on negative ads in a tightly contested election). One can specify a dynamical model to study this behavior theoretically. A given range of values of the model's parameters will be associated with interactions that will lead to an ever-growing spiral of expenditures, while another range of parameter values will be associated with a steady-state equilibrium where the two parties will match one another's expenditures but neither raise nor lower them. Another potential outcome is oscillation, where the expenditures of each side rise and fall in cycles. Finally, other ranges of parameter values will generate mutual reduction in expenditures.

Once one has specified a set of equations to describe the relations among the relevant independent variables and the dependent variables one can use a matrix representation to solve the equations and determine the eigenvalues. These will tell one whether the solution will produce a steady-state equilibrium oscillation, an explosive competitive spiral, or a reduction spiral. Thus, eigenvalues (characteristic roots) are of central importance in some statistical and formal models. Further, as we have seen, they aid us in calculations, allowing us to easily raise matrices to powers or find determinants. Finally, eigenvectors are of use in characterizing the degree of importance (centrality) of actors within a network.

14.2 MARKOV CHAINS AND STOCHASTIC PROCESSES

The theory of stochastic processes is rich, and one section in one chapter cannot remotely do it justice. However, as it is an increasingly important theory in the social sciences, it is worth sketching out the basics here.[15] As usual, we forgo formalism and focus on the broad strokes and practical applications. First we define stochastic processes in general and Markov chains in particular. Then we discuss the important notions of ergodicity and the steady state and introduce notation and terminology necessary for using Markov chains. Finally, we show how to do some elementary calculations with them, including calculation of the steady state.

14.2.1 Basic Definitions

A **stochastic process** is a set of random variables that dictate how the state of a system evolves. To understand what this means, it is necessary to define a few terms first. We begin with the state of a system, which we've referenced informally several times. A **state**, formally, is a representation of some system at some time. There is enormous flexibility in this definition. A state can represent the set of approval ratings for all potential candidates, the conflict status between any two countries or subnational groups, each country's GDP or level of foreign direct investment, the set of parties taking part in a coalition government, stock prices in a stock market, the weather, or anything else that can be conceptualized as a set of values. Further, the state need not represent only values known with certainty, such as war or peace. A state might tell you the probability of being at war, for example. The **state space**, roughly speaking, may be thought of as the set of all possible states of the system. If this set is discrete—that is, countable (or even finite)—then it is a discrete state space; otherwise the state space is continuous.

The stochastic part of the term stochastic process refers to the fact that the way each element in the state changes over time typically has a random component to it. This is why the stochastic process can be conceptualized as a set of random variables. For our purposes, though, we'll stick to a simpler conceptualization and refer to a stochastic process as a state space and a set of (stochastic) rules that dictate (probabilistically) how each state in the state space evolves over time, conditional on a set of parameters and the prior history of the system. In a sense, then, a stochastic process provides probabilities that any particular state will transition to any other state, conditional on the pattern

[15] There exist no shortage of texts and other courses to which the interested reader can turn later.

of states in which the system had been previously. We call the probabilities that these transitions will occur **transition probabilities**.

Stochastic processes are as general as the states whose evolution they describe. Approval ratings can go up or down based on candidates' speeches, economic factors, other candidates' actions and prior approval ratings, militarized conflicts and existential threats, and shifts in support for social policies the candidates favor. International conflicts can start or end based on economics, land use, shifts in power or demography, or any one of a multitude of other things. The same is true for all our examples of states. In general, transition probabilities can depend on the full history of states' evolution and vary with time as well.

This is all well and good, but in practice, analyzing *models* of stochastic processes in their full generality is not likely to be a fruitful endeavor. So, typically we add simplifying assumptions. We've already seen that the state space can be discrete or continuous; discrete state spaces are far more manageable, and often are realistic simplifications. For example, while it may be theoretically possible to define conflict as, say, the level of economic damage done per second down to fractions of a penny, often we just want a binary indicator of conflict. The states in this case might be "conflict" and "no conflict." That drastically limits the state space of a model of conflict, making it much more tractable. This is why you will usually see discrete state spaces in political science models.

Other common assumptions apply to the transition probabilities themselves. We consider three widely used ones. The first is known as **stationarity**. If a stochastic process is stationary, its transition probabilities do not change in time, other than possibly via their dependence on prior history. In other words, if the chance of staying in some state A (e.g., no conflict) is given at some time conditional on having observed the state A in all prior periods (i.e., no prior conflict), then, as long as the system stays in that state (i.e., no conflict), the chance of staying in that state stays the same (i.e., the chance of transitioning to conflict does not vary over time). This assumption is useful whenever the underlying causal mechanisms that are driving transitions do not change.

The second common assumption you will see in the social sciences is known as the **Markov assumption** or **Markov property**. Processes that satisfy this property are known as **Markov processes**. The Markov property removes the possibility of dependence of the transition probabilities on past history, save for the present state. In other words, the process is **memoryless**: transitioning between states can depend on where the system is *now*, but not on where it's *been*.

Markov processes are useful simplifications of reality, and so are useful models, when a detailed past history is unlikely to be strongly determinative of future behavior. This may seem overly constraining to be useful, but recall that states can encode lots of different types of information. So you may think that a Markov process is a poor model of the outbreak of civil conflict because we know empirically that the likelihood of conflict depends on the history of past conflict. However, one can include state variables (i.e., elements in the state) corresponding to (1) the number of past conflicts and (2) the time since the last conflict. In this manner the transition probabilities can depend on these two proxies for the entire history of conflict. In such a scenario, Markov processes can be useful in describing a system in which conflict depends on past conflict, but not on the exact timing of past conflicts. In contrast, they would not be useful for describing a system that required knowledge of the exact, detailed history of prior conflict to predict future conflict.

Markov processes are part of a class of finite memory processes, with which you're already familiar. Take weather forecasts, for example. Tomorrow's forecast takes the weather of the past few days into account to predict what the weather will be like in the future, but the fact that there was a hurricane in Florida last year does not have a direct influence on the weather forecast for Tallahassee, Florida, for tomorrow. Other examples of finite memory processes are board games like Monopoly. The position of each piece on the board is stochastic because it depends on the roll of a die or of multiple dice. But to predict the position of any particular piece at the end of the next turn we need only know the current position.[16]

Card games, however, require more memory, as the distribution of available cards varies as cards are drawn from the deck. This is the whole idea behind card counting. If we took a deck of cards and drew cards from the top, we would have a stochastic process in that we cannot absolutely predict what card will be drawn next. However, each card can be drawn only once, and therefore we can make better predictions by remembering which cards have been drawn so far. In fact, a person with perfect memory could make progressively better predictions about what card will be drawn next, and could perfectly predict what the last card will be! The more decks used at once, the more memory is required. An infinite memory process would allow transitions to depend on the entire history of the process, no matter how long it got.

The third and final assumption we consider is the assumption of a **discrete time** process. A process that satisfies this assumption does not evolve continuously but rather at discrete time intervals, such as days, years, or turns. Discrete-time processes tend to be the most commonly used in the social sciences, both because many actions occur at discrete intervals, such as elections in the United States, and because data are by necessity observed at discrete intervals.

You will typically see the phrase **Markov chain** used to refer to a stochastic process that has a discrete (usually finite, too) state space and that satisfies the Markov property. Sometimes you will see in addition "discrete-time," "continuous-time," or "stationary" preceding this term. Somewhat confusingly, you will also encounter discrete-time, stationary Markov chains referred to as simply Markov chains. All the stochastic processes we consider in this book are

[16]Note that we might also know where the piece was during the previous turn, but this does not help us make a better prediction about where it will be at the end of the next turn.

discrete-time, stationary Markov chains. Having said that, we (hopefully less confusingly) refer to them simply as Markov chains.

14.2.2 Ergodicity, Limiting Distributions, and Other Definitions

In a Markov process, or any other stochastic process, states evolve according to defined transition probabilities. There are two broad questions one is usually interested in regarding such processes. First, one might want to know about the long-run behavior of the process. That is, if states evolve according to transition probabilities for a very long time, what does the system end up looking like? Second, one might want to know about the short-run behavior of the process. For example, do states subject to this process change rapidly or slowly? Is there a consistent change driving the states to some particular state, or is there oscillation between states?

Short-run questions address what are known as **pathwise properties** of the stochastic process. This term arises from the probabilistic nature of stochastic processes. Transitions between states occur with some level of randomness; thus, starting from any particular state, the set of states through which the system evolves over time might be different every time. Think of the board game example mentioned above. Everyone starts the game at some spot, such as the Go square in Monopoly. After that, though, different things can happen based on consecutive rolls of the dice. One time playing you might take short hops around the board, landing on a property of each color. Another time you might traverse the board quickly, landing on fewer properties.

Because of this randomness, we must think not of a single evolution of the system, as we would if the system were *deterministic*—that is, if each state in sequence was uniquely determined by previous states and the transition rules. Rather, we must turn to the intuition gained from our study of probability and refer instead to **sample paths**. Each sample path corresponds to one particular realization of the stochastic process. Because stochastic processes typically evolve over time, we also often refer to a sample path as a **history**. To tie to basic concepts of probability, one can loosely think of a sample path or history as an *event* comprising a particular sequence of states. The set of all possible sample paths is then the **sample space**.

A **pathwise property** is one that applies to *every* sample path. Consequently, pathwise properties are extremely useful. They tell you not only where the system might eventually end up, given enough time, but also the manner it which it might get there. This is important, as many systems may never settle at anything like equilibrium. As one might expect, though, pathwise properties are difficult to discern and are not particularly common in the social sciences. Their presence often requires transition probabilities that remain consistent with changes in state.

Bendor et al. (2011) provide an example of a model that explores pathwise properties of a system. The authors deduce in a (stochastic) formal model that if one individual is more likely than another to vote for a particular party, and has interests more in line with that party to start, then that individual remains more likely to vote for that party in the future. This is a pathwise property because it holds for all time and for all sample paths, but note that it is a fairly weak statement and requires several assumptions to support it.

Although pathwise properties are preferred for their generality and utility, they are not usually the focus of analysis for a couple of reasons. One, as mentioned, is feasibility. It is often difficult to obtain them. Another is comparability: we are often interested in comparing results obtained via stochastic processes to those obtained via other means. In statistics, Markov chains are often used to derive posterior distributions that are the endpoints of a process, allowing, among other things, comparison to asymptotic properties of other estimators. In formal theory, dominant game theoretic approaches provide the equilibria of games, which are points at which the system does not change; the steady states of a Markov chain serve a similar role. Consequently, we typically focus on the long run when analyzing stochastic processes.

Long-run questions address the likelihood that particular states recur or are common in the long run, after the system has had a long time to evolve and has perhaps settled down. There are several properties relevant here; we'll tackle the simpler ones first. Let's start by considering the fate of an individual state of the system.

In any system, a state may either be **recurrent** or **transient**. The former means that as the system evolves, the state may occur again and again. For instance, conflict may repeatedly blight international interactions. The state of conflict would then be a recurrent state. The latter means that as the system evolves, at some point the state no longer occurs henceforth, and so it is transient. The state may reoccur for a while, but eventually the stochastic process leaves that state and never returns. Normatively, we'd all like conflict to be a transient state of the global system.

Recurrent states may also be subdivided. A recurrent state may be **absorbing** or **nonabsorbing**. An absorbing state is one in which the process never leaves once it reaches it. In a two-state system, for example, if one of the states is transient (e.g., conflict) then the other (no conflict) must be absorbing. But it is also possible for multiple states to be absorbing. Which one the system ends up in might depend on the history of the process. Continuing our conflict example, starting from a state of conflict might ensure conflict, while starting from a state of no conflict might ensure no conflict. Or perhaps the reverse might be true.

This brings up another important long-run concept: the **initial distribution**. The initial distribution specifies the probability distribution over states at the beginning of a Markov process, i.e., at time zero. For example, a two-state process might have an initial distribution that specifies that the system starts in a state of no conflict with certainty, or the system might start with a two-thirds chance of being in a state of conflict and a one-third chance of being in a state of no conflict. In other words, the initial distribution over the states tells us where the process starts. Even in a Markov process, where the system is going can depend on where it has been, so for this reason the initial distribution can be important.

The most important long-run property in which we are interested, however, is the existence and structure of the **steady state** or **limiting distribution** of a stochastic process. These are more subtle than transient or absorbing states, which have intuitive interpretations, but we'll gloss over the fine distinctions here. For our purposes, a steady state or limiting distribution of a system is the distribution of states that one would observe if one were to look at the "end" of a history of infinite length. In other words, it is the distribution of states that one observes in the limit as time approaches infinity, conditional on the initial distribution.

It is important to note that this is a distribution over sample paths (or histories), not over actions within a path. For example, it might be that one sample path of a model of conflict ends up in a permanent state of conflict, while another ends up in a permanent state of no conflict. What the limiting distribution tells us is the probability that each of these (and other) paths occurs; consequently, it tells us the probability, given a particular initial distribution over the states, that we might be in each state at some long time in the future. More formally, the limiting distribution is a distribution over sample paths of the stochastic process. Sometimes the system may go in one way and sometimes in another, but, if a limiting distribution exists, then one can assign a likelihood that any particular sample path will occur given a specific initial distribution over states.[17] Then, given some initial distribution, the sample path you end up in is drawn from the limiting distribution in the same manner that random variables are drawn from distributions, as discussed in Chapters 10 and 11.

As mentioned above, limiting distributions will allow us to compare to equilibria in game theory and asymptotic behavior in statistics. However, it is not the case that limiting distributions exist for all stochastic processes. The question of when they do exist is a complicated one, but focusing as we will on discrete-time, stationary Markov chains makes it much simpler. Again, though, we must first provide a few definitions. All these apply to discrete-time, stationary Markov chains, though extensions exist in some cases for more general stochastic processes.

A state is **accessible** from another state if it is possible to get to the first state from the second in a finite number of steps with positive probability. Two states **communicate** if each is accessible from the other. Accessibility and communication thus address the ability to move between states, either directly via a single transition or indirectly via interim states and multiple transitions.

[17] There is a great deal of nuance in the concept of a limiting distribution, and it may take a while for it to make sense. There are also numerous other ways of expressing the concept, such as steady-state distribution, stationary distribution, and limiting stationary distribution. Sometimes these different words are imbued with different shades of meaning as well. We'll focus on the most straightforward concept, and provide some examples to compute in the next subsection.

A **distinguished state** is reachable with strictly positive probability from all other states.

The **period** of a state is the fastest frequency with which the system may return to that state. In some processes it is possible to stay in a state rather than transitioning to a new one. Such a state in such a process has a period of 1. In other processes there are states that the process must always leave immediately. The number of transitions, or steps, it takes to get back to that state again determines its period, which will be greater than 1. A Markov chain in which all states have periods of 1 is called **aperiodic**.

Finally, we present the most important definition. A Markov chain is **ergodic** if it (1) has a unique limiting distribution and (2) converges to that distribution from all initial distributions. Ergodicity is an extremely powerful concept. It tells us not only that a limiting distribution exists, implying that the system actually moves over time toward something concrete, comprehensible, and comparable, but also that *where the system ends up is unrelated to where the system starts*. When modeling an ergodic system, therefore, one does not need to vary or even think too hard about the initial distribution, as long as one runs the system out long enough—that is, as long as one takes enough steps, and applies the transition probabilities enough times. The ergodic theorem plays a foundational role in the Gibbs sampler, which is commonly used in estimation of Bayesian statistical models: it establishes that the Monte Carlo Markov chains produced by the sampler cover the full space, which is a central assumption of those statistical models (Gelman et al., 2003).

So, what makes a Markov chain ergodic? There is no single answer to that question, but we'll provide two ways by which one can check whether a Markov chain is ergodic. The first: If all states communicate and the process is aperiodic, then the process is ergodic.[18] The second: A finite Markov chain is ergodic if and only if the process is aperiodic and there is at least one distinguished state. The good news is that neither of these is that difficult to check, as we will see in the next subsection, so we can productively apply the concept of ergodicity.

Before getting to this and calculating steady states, it is worth discussing what happens when a process is *not* ergodic. A nonergodic process violates some aspect of the definition of ergodicity. It might have a unique limiting distribution but not converge to it from all initial distributions, or it might not have a unique limiting distribution at all. We'll see some examples in the next subsection. While this means concrete results are harder to come by and comparison with extant results using other methods is more difficult, nonergodic systems are both common and important, and they underlie a couple of very important concepts in the social sciences.

Two of these concepts are **path dependence** and **history dependence**.[19] The hallmark of ergodicity is that the path a process takes has limited im-

[18] If the process is not aperiodic but all states do communicate, then there is a unique steady state of the Markov chain, but the process doesn't have to converge to this state from all initial conditions.

[19] Note that the term path dependence is also used to describe a broadly similar but less

portance with respect to the eventual limiting distribution. Path and history dependence run counter to this. Almost all stochastic processes betray some level of what we might colloquially refer to as path dependence, as the state at any time can depend on the state at the previous time. But that's a weak statement, and not typically what is formally called path dependence. Rather, a truly path-dependent process has the endpoint of the process depend on the path the process took to get there. Such processes are often characterized by positive feedback, lock-in, or both. The first might occur when the benefit to taking some action increases in the number or proportion of people taking some action, as in some social processes, such as protest or turnout (Siegel, 2009). The second might occur when the benefits from switching to another option decline with the number of people supporting your option. The examples typically given here are economic, such as the use of the QWERTY keyboard over the DVORAK one, or VHS versus Betamax (Arthur, 1994). Both, however, are of the flavor that a few chance early events might lead to a cascade in which almost everybody congregates on one option. These sorts of models are called cascade or tipping point models, and are usually path dependent in this way.

It is also possible that the actual path doesn't matter, but where the process started does. In other words, the initial distribution determines the endpoint of the process, but apart from that, random events during the process don't alter the end point. For example, whether or not a system starts in a state of conflict might determine the likelihood that a system ends in a state of conflict, but whether or not there are brief lulls of no conflict in the middle may have no effect on the eventual outcome. This sort of sensitivity to initial conditions is typically called history dependence.[20] It is a form of path dependence, since part of the path does matter, but it is weaker than that concept, since only part of the path matters. Some have said that history and path dependence are common in political processes (e.g., Pierson, 2000); stochastic processes can help us better understand what this means (see Page, 2006 for more on path dependence).[21]

14.2.3 Computation

Given the importance of limiting distributions in Markov chains in the social sciences, we now turn to a brief primer on how to know whether they exist, and how to find them if they do. We'll limit ourselves to discrete-time stationary Markov chains; in this case the framework we've built up in this part of the book is sufficient for analysis, and we need only present a bit of notation before getting going.

precise (non-mathematically defined) process by many political scientists and sociologists (e.g., Mahoney, 2000; Pierson, 2000).

[20] Though the equation of the terms "history" and "sample path" perhaps makes this term a poor choice.

[21] A similar interesting approach known as sequential analysis uses rigorous statistical methods to study history. See Abbott (1995) for an overview, and Stovel (2001) for an interesting application.

Let's begin with a state. Since we're assuming a discrete stochastic process, we can represent the present state of the system by a vector. A vector that corresponds to a state in this way is known as a **state vector**. In general, this vector may be infinite dimensional, but we'll stick to finite dimensional states to make our lives easier.[22] Each component of the state vector represents a different state in which the system could be. So, for example, the elements of a two-dimensional state vector might correspond to "conflict" and "no conflict," the elements of a three-dimensional state vector might correspond to "yes," "no," and "don't know," and the elements of a four-dimensional state vector might correspond to "civil war and democracy," "civil war and autocracy," "peace and democracy," and "peace and autocracy."[23]

In this formulation, the value of each component of the state vector represents the *probability* that the system is in that state at that time. We can use either a row or a column vector for the state vector. Since doing so will alter slightly the definition of the transition matrix, which we'll get to shortly, we'll stick with a column vector. Thus the state vector $\begin{pmatrix} .5 \\ .2 \\ .3 \end{pmatrix}$ corresponds to a system with three states, and says that there is a 0.5 chance of being in the first state, a 0.2 chance of being in the second state, and a 0.3 chance of being in the third state at some time. Note that the sum of the elements in any state vector must add to 1, since the system must be *somewhere*. Since the state changes with time and encodes probabilities, we label states $\mathbf{p}_t$ here for the state at time t. The initial distribution occurs at $t = 0$, and so is $\mathbf{p}_0$.

Each state vector represents a probability distribution over states of the system that is conditional, in a Markov chain, on the state of the system at the previous time step. Accordingly, we often call the initial distribution the unconditional distribution, and the later state vectors conditional distributions. The limiting distribution, should it exist, is then the state vector in the limit as $t \to \infty$. We'll label this vector $\boldsymbol{\pi}$ to distinguish it from intermediate states.

State vectors evolve according to transition probabilities. Since the distributions are given by a vector, you might guess that the transition probabilities can be expressed as a matrix; that guess would be correct. We call this matrix the **transition matrix** and label it M. In this framework, the transition rule is very simple: $\mathbf{p}_t = M\mathbf{p}_{t-1}$. In words, you apply the transition matrix to a state vector to transition to the next state vector. Each element m_{ij} of the transition matrix M has the following interpretation: m_{ij} is the probability that the system transitions from state j to state i. Because each state in the vector must go *somewhere*, each column in the matrix must add to 1.[24]

[22]One can extend the terminology that follows to infinite or uncountable spaces, such as the real numbers, by using the language of continuous distributions and functions. However, ergodicity becomes much more complex in this case so we do not address such systems here.

[23]We can do this because the state space is discrete.

[24]If we were to represent the state vector as a row vector instead, then the transition rule would be $\mathbf{p}_t = \mathbf{p}_{t-1}M$ instead, the elements m_{ij} would be the probability of transitioning

Let's work an example. First we define an initial distribution and a transition matrix:

$$\mathbf{p}_0 = \begin{pmatrix} 1 \\ 0 \end{pmatrix}, M = \begin{pmatrix} .5 & .5 \\ .5 & .5 \end{pmatrix}.$$

In words, this example specifies that there are two states—we'll call them A and B—and that the system starts in state A with certainty. We could have instead specified starting in state B at first with $\begin{pmatrix} 0 \\ 1 \end{pmatrix}$, or set up an initial distribution in which each state was equally likely, with $\begin{pmatrix} .5 \\ .5 \end{pmatrix}$. Or anything else in which the elements add to 1. The transition matrix specifies that state A has a one-half chance of staying in A and a one-half chance of staying in B, and the same for B.

A pathwise property in this setting would relate to how $\mathbf{p}_{t-1}$ becomes $\mathbf{p}_t$. We'll focus instead on the steady state. The steady-state distribution is the long-run distribution in which the system has settled down. Settling down here means that state vectors no longer transition to new ones, which in turn means that the steady-state vector remains constant under repeated applications of the transition matrix. Thus, it is the state vector for which $\boldsymbol{\pi} = M\boldsymbol{\pi}$ is true.

This assumes that the limiting distribution exists. To check this, we look at the transition matrix. Under one theorem, we need all states to communicate and the system to be aperiodic. Well, from our verbal description of the transition matrix, we see that state A can go to B in one step, and the same for B to A. Thus all states communicate. Also, there is a one-half chance that both A and B will stay in the states A and B, respectively, so the system is aperiodic. Therefore, we know the process is ergodic, and a steady state exists.

Let's find it. We need to solve the equation $\boldsymbol{\pi} = M\boldsymbol{\pi}$. This, however, looks just like an eigenvalue equation with eigenvalue 1. This is no accident; recall that eigenvalues tell you the rate of growth of the system, and the steady state implies a completely stable system with no growth or shrinking. Consequently, the exact same method we used to find eigenvectors will work here to find the steady state, with only one minor change: because the elements in a state vector must add to 1, we need to make sure this is true for our steady-state vector.

We start by letting $\boldsymbol{\pi} = \begin{pmatrix} 1 \\ c \end{pmatrix}$. Multiplying by the matrix M yields two identical equations: $.5 + .5c = 1$ and $.5 + .5c = c$. Both yield a solution of $c = 1$ and a vector of $\begin{pmatrix} 1 \\ 1 \end{pmatrix}$. This adds to 2, not 1, so we'll need to divide both elements by 2 to yield the steady state[25] $\boldsymbol{\pi} = \begin{pmatrix} .5 \\ .5 \end{pmatrix}$. Thus, in the steady state there is an equal chance of being in state A or state B. This is true *regardless* of

from state i to state j, and each row of the transition matrix would add to 1, rather than each column.

[25]Note that this procedure is different from *normalizing* a vector. We are not looking for the length of the vector to be one, but rather for the sum of its elements to be one. As such

where the system starts; we never used the initial conditions in the derivation of the steady state. One could check this by repeatedly applying M to any state vector. Doing so would move it closer and closer to $\begin{pmatrix} .5 \\ .5 \end{pmatrix}$. Of course, the actual *path* one takes to that point depends on the initial distribution (i.e., where one starts). If one had started at $\begin{pmatrix} .5 \\ .5 \end{pmatrix}$, then the state vector would never change at all!

What about situations in which there is not ergodicity? Let's see one example:

$$\mathbf{p}_0 = \begin{pmatrix} 1 \\ 0 \end{pmatrix}, M = \begin{pmatrix} 0 & 1 \\ 1 & 0 \end{pmatrix}.$$

Multiplying M by $\mathbf{p}_0$ takes the system to $\begin{pmatrix} 0 \\ 1 \end{pmatrix}$, and we see that all this transition matrix does is switch you back and forth between states A and B. This is because, while all states communicate—in fact, they're forced to switch every time step—all states in the system have periods of 2, not 1. Thus the system is not aperiodic, and our theorem says that, while there is a unique limiting distribution, the system does not converge to it from all initial conditions. The unique steady state is $\begin{pmatrix} .5 \\ .5 \end{pmatrix}$ (we encourage you to check this), but the system only converges to this when it starts at this. Otherwise it converges to nothing at all.

Let's work one more example:

$$\mathbf{p}_0 = \begin{pmatrix} 1 \\ 0 \end{pmatrix}, M = \begin{pmatrix} 1 & 0 \\ 0 & 1 \end{pmatrix}.$$

Here you may note that the transition matrix is the identity matrix, which means it maps all vectors to themselves. Thus, the state vector never changes. If you start in state A, you stay there, and the same for state B, or for any probability distribution over them. While this system is aperiodic—you can't leave a state, after all—all states do not communicate and there is no unique limiting distribution. Any initial distribution one selects yields a different steady state. This system is entirely history dependent.

14.2.4 Why Should I Care?

A revolution in statistics started about two decades ago, and Markov chains figure prominently in the revolution. The debate is between frequentists (i.e., those who subscribe to statistical inference as developed in the early twentieth century) and Bayesians (e.g., Kruschke, 2011). A prominent method for

---

we don't divide through by the square root of the length of the vector (which would be $\sqrt{2}$ in this case) as we would if normalizing the vector, but rather divide through by the sum of the elements in the vector (which is 2 in this case).

estimation within the Bayesian paradigm involves what are called Monte Carlo Markov chains (MCMCs), and these techniques are becoming increasingly popular among political methodologists (e.g., Gill, 1999; Jackman, 2009). We expect this approach to estimation to grow in popularity, in no small part because it makes it possible to estimate new models that people simply could not reasonably estimate before (e.g., time series models of binary data). The elementary treatment of Markov chains offered here provides you with the first stepping stone to learning how those models work.

Markov chains are also used in formal theory. In game theory they contribute to the solution concept of Markov perfect equilibrium, in which players respond not only to each other's actions, but also to the state of the system. The state of the system must relate to the actors' payoffs, and actors can only condition strategies on the present state (whence arises the "Markov" in the name). For example, Slantchev (2003, p. 623) models war as a Markov chain with the absorbing states "victory" and "loss."

Markov chains are also commonly used as part of behavioral models, such as those based on reinforcement learning, aspiration-based search, or hill climbing. In behavioral models, Markov chains specify how actors alter their actions over time and allow the derivation of comparative statics: how the behavior of the model varies with the model's parameters (we discuss comparative statics more in Chapter 17). Behavioral models containing Markov chains have been used to model, for example, electoral dynamics (e.g., Kollman, Miller, and Page, 1998; De Marchi, 1999; Laver, 2005; Bendor et al., 2011), government formation (Golder, Golder, and Siegel, 2012), and state repression (Siegel, 2011).

14.3 EXERCISES

1. Find the eigenvalues and corresponding eigenvectors for $A = \begin{pmatrix} 3 & 1 \\ 1 & 3 \end{pmatrix}$. Use them to diagonalize A.

2. Find the eigenvalues and corresponding eigenvectors for $A = \begin{pmatrix} 2 & 1 & 0 \\ 1 & 2 & 0 \\ 0 & 1 & 2 \end{pmatrix}$.

   Use them to diagonalize A.

3. Consider the transition matrix $M = \begin{pmatrix} .25 & .5 \\ .75 & .5 \end{pmatrix}$. Is it ergodic? If so, find the steady-state distribution $\boldsymbol{\pi}$ over the two states. If not, explain why not and discuss the consequences for path or history dependence.

4. Consider the transition matrix $M = \begin{pmatrix} \frac{9}{10} & \frac{1}{3} & 0 \\ \frac{1}{10} & \frac{1}{6} & \frac{1}{10} \\ 0 & \frac{1}{2} & \frac{9}{10} \end{pmatrix}$. Is it ergodic? If so find the steady-state distribution $\boldsymbol{\pi}$ over the two states. If not, explain why not and discuss the consequences for path or history dependence.

5. Consider the transition matrix $M = \begin{pmatrix} 1 & \frac{1}{3} & 0 \\ 0 & \frac{1}{6} & 0 \\ 0 & \frac{1}{2} & 1 \end{pmatrix}$. Is it ergodic? If so find the steady-state distribution $\boldsymbol{\pi}$ over the two states. If not, explain why not and discuss the consequences for path or history dependence.

Part V

Multivariate Calculus and Optimization

Chapter Fifteen

Multivariate Calculus

The majority of this book has thus far dealt with functions of a single variable, and operations such as differentiation on these functions. This is sufficient for many applications; however, we do not want to limit our use of calculus in the social sciences to functions of a single variable. For example, we might believe multiple independent variables act in concert to determine the value of a dependent variable, or a political actor might have to make multiple choices at once. In these situations, we must use multiple variables to model the real world. This introduces some complexity, as we see in this final part of the book.

Despite our prior focus on a single variable, though, we have already slipped some multivariate analysis in via the back door. In Chapter 3 we defined functions of multiple variables as functions with multiple arguments, and talked a little about graphing them and about their meaning, including their use in interaction terms in statistics. We go into more depth on interaction terms in Section 2 of this chapter. In Chapter 11 we briefly discussed joint distributions. These distributions are nothing other than functions of two or more variables with particular properties. Further, the entirety of Part IV of the book, covering linear algebra, was really a story of multivariate analysis that was confined to linear operators (matrices). After all, vectors may be loosely thought of as collections of variables. We see that the techniques developed in our study of linear algebra will prove useful in studying more general operators. Finally, we even gave a hint of this chapter's content in Chapter 5 in our brief discussion of the partial derivative there.

The first section of this chapter introduces functions of several variables, providing some common notation. It briefly ties to the material in Parts I, III, and IV of this book as well, to make clear how many concepts we've already discussed generalize to the case of more than one variable.

The second section provides an overview of the major concepts with which you will need to be familiar in multivariate calculus. This includes the partial derivative and its connection to multiple regression, the total derivative and chain rule in multiple dimensions, the Jacobian and derivatives of multidimensional functions, the Hessian matrix and higher-order derivatives, and finally, very briefly, integration in more than one dimension. We do our best to provide intuition along with some formalism in this section.

The third section uses some of the second section's formalism to introduce multivariate analogues to the concepts of concavity and convexity covered in Chapter 8. Finally, we should note that in this chapter, and in this part of the book in general, we are leaving out steps in differentiation when these steps were covered in Part II of the book. However, in all cases you would be well served to attempt the derivative yourself and check it against the answer provided. To help, we've at times not simplified the solution in order to make it more clear where, for example, a product rule might have been used.

15.1 FUNCTIONS OF SEVERAL VARIABLES

In Chapter 3 we introduced a function as a mapping between two sets A and B. We wrote this as $f(x) : A \rightarrow B$ and described the function as taking a value x from the set A, applying some function to x, such as raising it to a power or taking its logarithm, and then finding the result of this in the set B. Working in more than one dimension does not change this in the slightest. Just replace x with the vector $\mathbf{x}$, which is a collection of variables $x_1, x_2, \ldots, x_m$, and the exact same sentences are true.

Now, this isn't quite fair, as before the sets A and B were sort of implicitly one-dimensional, and now they are explicitly m-dimensional. But conceptually that's not a big deal. We're still taking an element of one set and operating on it in some fashion, with the output an element of some other set. Because of this, almost all of the basic properties of functions we discussed in Chapters 3 and 4 carry over pretty directly to more than one dimension, though the notation does get a bit messier. To see this, we use vectors of real numbers, since this is what we primarily encounter.

Let the set A be an m-dimensional vector in the real numbers, which we denote as $A \in \mathbb{R}^m$, and the set B be an n-dimensional vector in the real numbers, which we denote as $B \in \mathbb{R}^n$. Then we can define the function $\mathbf{f}(\mathbf{x}) : \mathbb{R}^m \rightarrow \mathbb{R}^n$.[1] This is the most common function of several variables that we use. Sometimes we see f written as a function of a vector, e.g., $\mathbf{f}(\mathbf{x})$, and sometimes as a function of multiple variables, e.g., $\mathbf{f}(x, y, z)$ or $\mathbf{f}(x_1, x_2, x_3, x_4)$. These mean the same thing: that the function takes some number of variables as arguments, whether elements of one vector or independently defined variables, and returns some other set of values.

It is important to note that in this general statement of the function it is a *vector*, not a scalar. In other words, the range of the function[2] can also be multidimensional, in addition to the domain of the function.[3] When f is only a scalar, we can write $f(\mathbf{x}) : \mathbb{R}^m \rightarrow \mathbb{R}$ instead. This may well be confusing at this

---

[1]Note that we don't need to do this. We could have easily left A and B as they were and defined them as subsets of $\mathbb{R}^m$ and $\mathbb{R}^n$, respectively. We chose not to do so to make things more concrete, and because these particular sets A and B are those most commonly used in applications.

[2]Recall that the range of a function is the set of values it can take over the domain of the function (Chapter 3).

[3]Recall that the domain of a function is the set of values it can accept as arguments (Chapter 3).

point, but we assure you it is only a notational issue and it will resolve quickly. To get there faster, let's look at some examples of functions.

We begin with some scalar functions:

$$\begin{aligned} f(x,y,z) &= 3xy - y^2z + 2, \\ f(x,y) &= x^y - xy, \\ f(w,x,y,z) &= w(xz + wy) - 2wz(e^{x+y}) - 15z, \\ f(\mathbf{x}) &= x_1x_3 - x_2x_5. \end{aligned}$$

Note that in every case, all that's happening is that the various variables of which f is a function are being combined algebraically into a single scalar. This is true regardless of how many variables there are or whether they're expressed individually or in the form of a vector.

In the case of a function that is a vector, i.e., one that has multiple components, we must define each component separately. However, other than having to keep track of this, defining such a vector-valued function is no different from defining a scalar function. For example, we could define $\mathbf{f}(x,y,z)$ as

$$\mathbf{f}(x,y,z) = \begin{pmatrix} 3xy - y^2z + 2 \\ x^y - xy \\ z(xz + y) - 2y(e^{x+y}) - 15z \\ xyz - 1 \end{pmatrix}.$$

Other than having to keep track of various components (e.g., $f_4(x,y,z) = xyz - 1$), most of the common manipulations and properties of functions we discussed earlier still hold. For example, in Chapter 4 we talked about the notion of *continuity*, which meant intuitively that there were no "jumps" in the function. Well, we can talk about continuous functions in more than one dimension as well. The same requirements apply (see Chapter 4); we just need them to hold for every variable of which $\mathbf{f}(\mathbf{x})$ is a function, for every component of $\mathbf{f}(\mathbf{x})$. While this might end up being a lot of checking—for the previous example of three arguments and four components of $\mathbf{f}(\mathbf{x})$, we'd need to check for continuity twelve ways—it is straightforward conceptually.

Other important properties hold as well. A function can be monotonic in each of several variables, and can have a limit in each of several variables. A multidimensional function is *linear* if $f_i(\mathbf{x} + \mathbf{y}) = f_i(\mathbf{x}) + f_i(\mathbf{y})$ and $f_i(r\mathbf{x}) = rf_i(\mathbf{x})$ for each component f_i of $\mathbf{f}$, each scalar r, and each vector $\mathbf{x}$ and $\mathbf{y}$. We saw in the previous part of the book that one can represent linear functions as matrices, so that, if $\mathbf{f}$ is linear, we can write $\mathbf{f}(\mathbf{x})$ as $A\mathbf{x}$ for some A. Here, if $\mathbf{f}(\mathbf{x}) : \mathbb{R}^m \to \mathbb{R}^n$, then A is an $n \times m$ matrix. Note here that the difference between an *affine* function and a linear function becomes important. An affine function can be represented as $A\mathbf{x} + \mathbf{b}$, and does not have the same properties.

Quadratic forms, mentioned briefly in Chapter 13, can also stand in for functions—quadratic ones, not surprisingly. Recall that a quadratic form looks like $Q = \mathbf{x}^T A\mathbf{x}$, where A is a symmetric matrix. We address these below once we get into calculus.

Utility functions, which are a centerpiece of rational choice theory, naturally map to functions of more than one variable as well. A common example is the utility often used in formal models of legislative behavior (e.g., Austen-Smith and Banks, 1988) in which both policy, labeled x, and something else such as cabinet share, office perquisites, or other particularistic good, labeled g_i for each relevant party, are both valued. Then each party's utility function might be written as $u_i(x, g_i) = g_i - (x - z_i)^2$, where z_i is the ideal policy of party i.[4]

Finally, probability distributions also naturally accommodate more than one variable. If a probability distribution of one variable details the relative likelihood of any particular realization of that variable, then a probability distribution of more than one variable details the relative likelihood of any particular simultaneous realization of *all* the variables. We illustrated this for the case of two variables in Chapter 11, and adding additional variables does not change anything.

15.1.1 Why Should I Care?

As we have noted, functions of more than one variable are essential to model situations in which there are multiple independent variables thought to be important, or in which actors have multiple choices. They also help us understand the way in which multiple variables interact to affect some outcome measure. A **level set** is a good example of this. In a level set, you hold one particular variable constant as you vary the other ones. The most common level set you'll see is a constant utility curve. This curve tells you the region over which one's utility is unchanging. In other words, it provides the regions within which one is indifferent between different choices. As such, these are more typically known as **indifference curves**.

Consider the example of legislative utility given in this section. For simplicity, we let $z_i = 0$ and consider one legislator's utility only. Then $u(x, g) = g - x^2$. When g is fixed at 0, $u = -x^2$. In this case, the legislator is indifferent between any points on a line the same distance away from zero, so her indifference curves are sets of the form $\pm d$ for $d \geq 0$. Substantively, this means that a legislator in this model doesn't care if a policy is three to the right or three to the left of zero, and that she prefers both to anything four away from zero. When you allow g to vary, you can sketch out indifference curves in two dimensions. To see how this works, let $u = -1$. Then the legislator is indifferent between the points $g = 0, x = \pm 1$, $g = 1, x = \pm\sqrt{2}$, $g = 2, x = \pm\sqrt{3}$, and so on. In general, the further one goes in the dimension of particularistic goods, the further one can also go in policy away from zero to achieve the same level of utility. This gives us a nice way of conceptualizing trade-offs occasioned by choices, and is particularly useful in social choice theory when trying to understand what policies groups of people might prefer. Let's say there are three people voting on some topic. If two of them are indifferent between policies A and B because both policies lie

---

[4]This type of utility is known as quasi-linear utility, as it is linear in the good g_i.

on the same indifference curves for each person, but one person strictly prefers policy A to policy B, then policy A is something on which the group may be able to agree. Analysis involving level sets can be extended to concepts beyond utility as well, for example, isoquant analysis in which the quantity of something is held constant.[5]

15.2 CALCULUS IN SEVERAL DIMENSIONS

We've seen that many of the concepts introduced in the context of one variable generalize to multiple variables. Much of what we learned in the second part of the book, covering calculus in one dimension, generalizes as well. However, there is one big sticking point here. Calculus may allow us to deal with continuous change in a consistent manner, but once there is more than one variable, the question arises of *what* change. Take, for example, the function $f(x, y, z) = 3xy - y^2z + 2$. When we talk of change in f, to what are we referring? With $f(x) = 3x^2$, it's clear: we want to know how f changes as x changes. To this end we can calculate first derivatives and higher-order derivatives. But with this $f(x, y, z)$, are we talking about changes in x, y, z, or some combination? And what's a higher-order derivative in this case? It turns out the answers to these questions are straightforward conceptually but messy analytically, in much the same way linear algebra is. In this section we provide some new framework for dealing with multivariate calculus, and we give some reasons for why you'd want to use it.

15.2.1 Partial Derivatives

The simplest answer to the question of what to do with a derivative when there is more than one variable is to ignore all the other variables and consider one at a time. This isn't facile: sometimes one is interested in taking the derivative of a multivariate function with respect to only one of its variables. For example, consider the function $y = f(x, z) = x \cdot z$. We can take the derivative of y with respect to x, or we can take the derivative of y with respect to z. Each of those is known as a partial derivative as it explains only a part of the variation in y. In fact, the **partial derivative** represents the instantaneous rate of change in y due to x (or z) while holding the other variable, z (or x), *constant.*

This is an important concept and it warrants unpacking. The partial derivative describes how a function changes with a variable while all other variables are held constant. In other words, it provides an answer to the question of how the function changes with some variable *with all else equal* (or *ceteris paribus*). Because this is a question we have frequent need of answering, we end up frequently needing the partial derivative. We discuss this more at the end of the subsection, but both statistics and game theory centrally require understand-

[5]In general, one adds the prefix *iso* to indicate constancy.

ing how isolated change in one variable affects variation in another, and this is where the partial derivative is essential.

It is so essential, in fact, that it's worth some new notation to describe it. One can use what looks like an italicized and backward "6," ∂, to indicate a partial derivative. So for the function $y = f(x, z) = xz$, the partial derivative of y with respect to x can be written $\frac{\partial y}{\partial x}$. Sometimes you will also see this notation shortened to $\partial_x = \frac{\partial}{\partial x}$. You may also even see f_x, which is an even shorter way than $\partial_x f$ to specify the partial derivative of f with respect to x. They all mean the same thing. This partial derivative operator has the same properties as does the one-dimensional derivative operator we discussed in Chapter 6. Namely, it is also a linear operator, follows the chain and product (and quotient) rules, and yields the same derivatives of functions as does its one-dimensional counterpart.

How does one calculate a partial derivative? Happily, it is not difficult, once you can calculate a derivative, of course: take the derivative, treating all other variables as constants. Thus, ∂_x and $\frac{\partial}{\partial x}$ both mean "treat every variable other than x as a constant, and take the derivative with respect to x." The partial derivative with respect to x represents the instantaneous change in y as a function of x when we hold all other variables constant (i.e., set their rate of change to zero). So to take the partial derivative, we calculate the derivative with respect to one variable and treat all variables other than the one of interest as constants. Let's do some examples.

Consider the function above, $y = f(x, z) = xz$. The partial derivative of y with respect to x is $\frac{\partial y}{\partial x} = z$, since the partial derivative of zx with respect to x is no different from the derivative of any other cx, where c is a constant.[6] Similarly, the partial derivative of y with respect to z is $\frac{\partial y}{\partial z} = x$. In both cases the variable we are not interested in is a constant, so we are able to use the fact that the partial derivative is a linear operator to compute it.

Let's have a look at an example that contains a multivariate function with three arguments: $y = f(x_1, x_2, x_3) = 6 + 3x_1 + \frac{5}{2}x_2 - x_3^2$. What is the partial derivative of y with respect to x_2? Take the derivative of the function treating both x_1 and x_3 as constants. Doing so yields $\frac{\partial y}{\partial x_2} = 0 + 0 + \frac{5}{2} - 0 = \frac{5}{2}$. Similarly, $\frac{\partial y}{\partial x_1} = 0 + 3 + 0 - 0 = 3$, and $\frac{\partial y}{\partial x_3} = 0 + 0 + 0 - 2x_3 = -2x_3$.

This next example is of particular interest in statistical modeling: $y = f(x_1, x_2, x_3) = \beta_0 + \beta_1 x_1 + \beta_2 x_2 + \beta_3 x3 + \beta_4 x_1 x_3$. This is an affine function with an interaction term (the term with coefficient β_4) that is typically called a linear interaction model. The presence of the interaction term gives nuance to the partial derivatives with respect to x_1 and x_3: $\frac{\partial y}{\partial x_1} = \beta_1 + \beta_4 x_3$ and $\frac{\partial y}{\partial x_3} = \beta_3 + \beta_4 x_1$. What does this mean? Well, consider first that $\frac{\partial y}{\partial x_2} = \beta_2$. This means that for each one-unit variation in x_2, y changes by β_2, all other variables held constant. In other words, the dependence of y, the dependent variable, on the independent variable x_2 is independent of all other variables. This allows us to interpret the coefficient β_2 cleanly.

[6]If you don't recall why the derivative of cx is c, please revisit Chapters 5 and 6.

However, since $\frac{\partial y}{\partial x_1} = \beta_1 + \beta_4 x_3$, we do not have quite the same interpretation available for β_1. In fact, β_1 provides only the isolated effect of x_1 on y when $x_3 = 0$! For all other values of x_3 we need also to consider the product of the coefficient β_4 and x_3. In other words, the dependence of the dependent variable on the independent variable x_1 is conditional on the value of x_3, and vice versa. Consequently, we cannot isolate the dependence of either of these variables on y; we must always consider them in concert (Friedrich, 1982; Braumoeller, 2004; Brambor, Clark, and Golder, 2006).

Finally, to remind you of some of the rules of calculus from Chapter 6, we provide an example that is a bit more difficult: $f(x, y, z) = x^y \ln(z) - y^3 x^2 z + 2yz - x + 1$. Let's take each partial derivative in turn: $\frac{\partial f}{\partial x} = yx^{y-1}\ln(z) - 2y^3 xz - 1$, $\frac{\partial f}{\partial y} = x^y(\ln(x))\ln(z) - 3y^2x^2z + 2z$, and $\frac{\partial f}{\partial z} = \frac{x^y}{z} - y^3x^2 + 2y$. Note again that any term, no matter how complex, that does not contain the variable with respect to which we're differentiating yields a partial derivative of zero.

15.2.1.1 *Why Should I Care?*

The application of partial differentiation that most people will see most often is in multiple regression. The example of the interaction term above highlighted the degree to which partial derivatives underlie the technique of multiple regression in statistics. Specifically, they provide a mechanism by which one can isolate changes in a dependent variable arising from variation in an independent variable. The overwhelming majority of quantitative empirical papers in the social sciences use some variant of multiple regression, and every time results from one of these regressions is interpreted, partial derivatives underlie the interpretation (though few textbooks raise this point).

Partial derivatives, more generally, also underlie discussions of *marginal effects.* A marginal effect is an isolated change in one variable due to variation in another, and is used to understand how small changes in variables alter outcomes (e.g., Cameron and Trivedi, 2005, pp. 122–24, 317). We've already talked about this in the context of statistics, but it is common in formal theory, too, even down to a basic level. For example, a common result in economics and models of political economy is that one continues to take an action until the marginal cost equals the marginal benefit of that action. In the next chapter, where we address optimization, we encounter more of this. Of course, the entire topic of multivariate optimization, which is also central to statistics and game theory, relies on partial differentiation, as does comparative statics, discussed in Chapter 17.

Variations of the concept of marginal change exist as well. Another common term you'll encounter is **elasticity**. An elasticity is like a marginal change, but it is a change relative to the value of the variable, and is designed to remove dependence on units. One can write it as $\frac{\partial \ln(y)}{\partial \ln(x)} = \frac{\frac{\partial y}{y}}{\frac{\partial x}{x}} = \frac{x}{y}\frac{\partial y}{\partial x}$.

15.2.2 Gradients and Total Derivatives

You might at this point be wondering about the term "partial." If each partial derivative is only part of the story, providing the isolated effect of only one variable, is there something that tells a more complete story? The answer, as you may have guessed from the existence of this subsection, is yes.

Actually, there are two answers. There are two ways we can join together information about partial derivatives. One is the same way we joined together information about multiple variables in the previous part of the book: by putting them in vectors. Recall that the partial derivative is a linear operator. We can therefore put all possible (first-order) partial derivatives together into a vector to make a vector-valued operator. This is called the **gradient vector**, and it is denoted $\boldsymbol{\nabla}$.

The gradient vector is an operator defined as $\boldsymbol{\nabla} = \begin{pmatrix} \partial_{x_1} \\ \partial_{x_2} \\ \vdots \\ \partial_{x_m} \end{pmatrix}$. That is, each element of the vector is the partial derivative with respect to one of the available variables.[7] We write $\boldsymbol{\nabla} f$ for the gradient of the scalar function f. We've already computed some examples of the gradient in the previous subsection without really knowing it. For example, we computed the partial derivatives of the function $f(x_1, x_2, x_3) = 6 + 3x_1 + \frac{5}{2}x_2 - x_3^2$. The gradient does nothing more than put these together into a column vector: $\boldsymbol{\nabla} f = \begin{pmatrix} 3 \\ \frac{5}{2} \\ -2x_3 \end{pmatrix}$. So there's nothing special about computing a gradient.

Because the gradient is a vector, so is $\boldsymbol{\nabla} f$, and like all vectors it points somewhere. In this case, where it points has an important substantive meaning: $\boldsymbol{\nabla} f$ points in the direction in which the function f increases most rapidly, and its magnitude is the rate of this increase. An example will help clarify. Let $f(x_1, x_2, x_3) = x_1$. Then $\boldsymbol{\nabla} f = \begin{pmatrix} 1 \\ 0 \\ 0 \end{pmatrix}$, and the gradient of f points in the direction of the x-axis, indicating that the function increases most rapidly in that direction. In fact, this function increases only in that direction; however, the concept extends to more complicated functions that increase in some other direction not aligned with one of the coordinate axes.

You might also want to know how fast the function increases in some other direction. To do this you can take the dot product of the gradient of the function with a vector in the direction of interest.[8] In general, this looks like $\boldsymbol{\nabla} f(\mathbf{x}) \cdot \mathbf{v}$

[7] The gradient is usually written as a column vector. Sometimes you will also see the notation $\mathbf{D}$ for a gradient, and D_{x_i} for a partial derivative with respect to x_i. Often $\mathbf{D}$ refers to a row vector instead.

[8] Recall from Chapter 12 that the dot product addresses the projection of one vector onto another.

for some vector $\mathbf{v}$. This tells you the rate at which f changes as one moves from $\mathbf{x}$ in the direction of $\mathbf{v}$. To return to our previous example, if $\mathbf{x} = \begin{pmatrix} 1 \\ 1 \\ 1 \end{pmatrix}$ and $\mathbf{v} = \begin{pmatrix} 1 \\ 1 \\ 1 \end{pmatrix}$, then the rate of change of function f, starting at $\begin{pmatrix} 1 \\ 1 \\ 1 \end{pmatrix}$ and moving further in that same direction, is

$$\nabla f(\mathbf{x}) \cdot \mathbf{v} = \begin{pmatrix} 3 \\ \frac{5}{2} \\ -2 \end{pmatrix} \cdot \begin{pmatrix} 1 \\ 1 \\ 1 \end{pmatrix} = 3 + \frac{5}{2} - 2 = \frac{7}{2}.$$

The gradient is one way to incorporate information about the isolated effects of multiple variables, but it is not the only way. Another way takes advantage of the fact that each of the variables with respect to which we are differentiating may *itself* be a function of some other variable. So, for example, x_1, x_2, and x_3 might all be functions of some variable t. What if we want to know how f changes with t? The partial derivative with respect to any x_i doesn't tell you this, as it isolates only the effect of that x_i. Nor does the partial derivative with respect to t. This tells you the *direct* effect of t on f, but not the *indirect* effect of t on f, mediated through the various x_i. For example, if $f(x, y, t) = xy + t$, then $\partial_t f(x, y, t) = 1$. Even though x might depend on t, so that when t changes, x changes too, which in turn changes f, the partial derivative ignores this indirect effect because it treats x as a constant.

To incorporate both direct and indirect effects of t on f, we need to compute not only partial derivatives but also the **total derivative** (or full derivative). There is no new notation for this; we use the same $\frac{d}{dt}$ that we used back when there was only one dimension to worry about.[9]

The formula for the total derivative, below, may seem complex at first glance, but it is closely related to the chain rule. This is for good reason: the chain rule may be thought of as a mechanism allowing us to change variables (from x to u in the notation of Chapter 6), and this is really all that's going on here (with t in the role of x and each x_i in the role of u). Here's the formula for the total derivative of the function $f(x_1, x_2, \ldots, x_m, t)$ with respect to t, making no assumption about the functional dependence of any of the x_i:

$$\frac{df}{dt} = \frac{\partial f}{\partial t}\frac{dt}{dt} + \frac{\partial f}{\partial x_1}\frac{dx_1}{dt} + \ldots + \frac{\partial f}{\partial x_n}\frac{dx_n}{dt}.$$

Refer back to the definition of the chain rule in Chapter 6, if you've forgotten. The total derivative is the RHS of the chain rule equation, summed over all the possible variables, and with partial derivatives instead of a one-dimensional derivative on the first term in each product, to account for having more than

[9] In fact, we were implicitly using the total derivative in that case.

one variable. In plain English, all we're doing with a total derivative is adding all the ways in which changing t leads to a change in f. The first term is the direct effect: since $\frac{dt}{dt} = 1$, that term is $\frac{\partial f}{\partial t}$, which is the isolated effect of t on f. The rest of the terms constitute the indirect effects of t as it acts through all the x_i. If in any case one of the x_i does not depend on t, then the corresponding $\frac{dx_i}{dt} = 0$, and that particular indirect effect vanishes. Otherwise, the term in the sum is non-zero and contributes to the indirect effect. As $\frac{dt}{dt} = 1$, you sometimes will see the total derivative written as

$$\frac{df}{dt} = \frac{\partial f}{\partial t} + \sum_{i=1}^{m} \frac{\partial f}{\partial x_i}\frac{dx_i}{dt}.$$

Let's pause for a moment to observe that this is pretty powerful stuff! We elaborate on this point below, where we discuss why you should care.

Now for an example. We again use $f(x_1, x_2, x_3) = 6+3x_1+\frac{5}{2}x_2-x_3^2$, but now we let $x_1 = t^5, x_2 = t-t^2$, and $x_3 = e^t$. We've already computed all three partial derivatives $\frac{\partial f}{\partial x_i}$, and $\frac{\partial f}{\partial t} = 0$, since there is no direct dependence of the function on t. So we're left with the three derivatives, $\frac{dx_i}{dt}$: $\frac{dx_1}{dt} = 5t^4$, $\frac{dx_2}{dt} = 1 - 2t$, and $\frac{dx_3}{dt} = e^t$, so, plugging this all into the equation for the total derivative, we have $\frac{df}{dt} = 0 + (3)(5t^4) + (\frac{5}{2})(1-2t) - 2x_3(e^t)$. We want to substitute in for x_3 to put the whole thing in terms of t. This gives $\frac{df}{dt} = 15t^4 + \frac{5}{2} - 5t - 2e^{2t}$. Note that one could have gotten the same answer by plugging in $x_1(t), x_2(t)$, and $x_3(t)$ into f, producing a function of one variable (t), and then differentiating that with respect to t. You can (and should) check that this is the case. This is in general true, but the total derivative is particularly useful for situations in which the function is very complex or you have already calculated the partial derivatives for another reason.

We framed this in terms of dependence on t, but the idea of a total derivative is more general. One way to express this generality is through the use of the **total differential**, dF. The total differential is the total derivative without the dt:

$$df = \sum_{i=1}^{m} \frac{\partial f}{\partial x_i} dx_i.$$

We referred to differentials in a footnote in Chapter 7, as one of their uses is in calculating integrals.

15.2.2.1 Why Should I Care?

Total derivatives have as many uses as partial derivatives, and in particular are essential for optimization and comparative statics. Since we discuss these two topics in the next two chapters, we do not elaborate here, but instead take this opportunity to offer another way of looking at both partial and total derivatives.

You can think of a partial derivative as a mathematical analogue to the experimentalists' effort to evaluate the impact of a given cause on a given effect, all other things held constant. Consider the structure of a classic experiment. One wants to know whether a given x has a causal impact on a given y, as suggested by some theory. In an ideal experiment one exposes the object of study to the treatment x, while holding all other known relevant factors constant. The goal is to determine whether the expected change in y occurs, as compared to a control object that is not exposed to the treatment. A partial derivative, in comparison, is the instantaneous rate of change in y given a small change in x while holding the other variables constant.

We use this similarity, among other places, in interpreting observational work. For example, both differences and derivatives are often used to interpret the impact of a given variable while holding other variables constant after estimating a logit or other nonlinear model (see, e.g., the discussion in Long, 1997, pp. 69–76).

This same logic can be used to help us understand the behavior of multidimensional functions near a point, which can be useful in some game theoretic applications. The gradient of a function can be used to approximate a function of more than one variable with a linear equation in the same manner as we saw in Chapter 8 for a function of a single argument with a Taylor series. This was a way of approximating a function with a series of derivatives of higher and higher order. We noted after introducing the Taylor series that it provides a good approximation of a function near any point. This is most commonly used in making linear approximations of functions near some point a: $f(x) \approx f(a)+f'(a)(x-a)$. This approximation is useful whenever the function does not change too wildly near the point. We can extend this to functions of more than one variable by writing $f \approx f(\mathbf{a}) + \nabla f(\mathbf{a})(\mathbf{x} - \mathbf{a})$, where $\nabla f(\mathbf{a})$ is the gradient evaluated at the vector $\mathbf{a}$. Again, we're looking at isolated changes, except we are now doing so across several dimensions at once.

But what if, upon performing an experiment, the experimentalist comes to believe that some other factor actually is affecting the handful of variables she is exploring, and that this other factor, through these variables, influences the outcome? This is where the total derivative becomes useful, as it allows her to understand the indirect effect of this new factor on the outcome of the experiment, via her measured isolated effects of each of the variables. Analogues to this exist in statistics and formal theory as well. Spurious causation occurs when some third variable affects both the dependent and independent variables; the total derivative allows one to model and understand this. And, as we show below, understanding the effect of a parameter on an equilibrium or steady state of a formal model requires understanding both the direct and indirect effects of that parameter, and so requires the use of the total derivative.

15.2.3 Derivatives of Multidimensional Functions: The Jacobian

So far we've been dealing with scalar functions, which take a vector of arguments and produces as output a scalar. Much of what we've said in this section thus far, though, applies to the vector-valued functions that we introduced in the previous section. Recall that these are defined as $\mathbf{f}(\mathbf{x}) : \mathbb{R}^m \to \mathbb{R}^n$. They thus take a vector of m arguments, each of which we label x_k, and produce an n-dimensional vector in the range of the function, with components f_i. We mentioned in the previous section that if one wanted to show continuity when there were three variables as arguments and four components of $\mathbf{f}$, then one would need to check twelve ways. The same is true when computing first derivatives: to compute the first derivative we must compute all twelve partial derivatives corresponding to each combination of x_k and f_i.

In general, if there are n components of $\mathbf{f}$ and m components of $\mathbf{x}$, then there are $n \times m$ partial derivatives to take to get a full accounting of the first derivative of the function. We saw in the previous part of this book how to keep track of all these—use a matrix. So let's do that. Define the **Jacobian matrix**[10] J (or $\mathbf{Df}$) of the function $\mathbf{f}$ as the the matrix in which element (i, k) is the partial derivative of component f_i with respect to variable x_k. In symbols, $j_{ik} = \frac{\partial f_i}{\partial x_k}$. The easiest way to remember this is to note that each row of J is the transpose of the gradient vector for the corresponding component of the vector $\mathbf{f}$.

Let's do the example from the previous section that we just mentioned. Recall that the function $\mathbf{f}$ has components

$$\begin{aligned}
f_1(x, y, z) &= 3xy - y^2 z + 2, \\
f_2(x, y, z) &= x^y - xy, \\
f_3(x, y, z) &= z(xz + y) - 2y(e^{x+y}) - 15z, \\
f_4(x, y, z) &= xyz - 1.
\end{aligned}$$

Then the Jacobian matrix J is

$$J = \begin{pmatrix} 3y & 3x - 2yz & -y^2 \\ yx^{y-1} - y & \ln(x)x^y - x & 0 \\ z^2 - 2y(e^{x+y}) & z - 2(e^{x+y}) - 2y(e^{x+y}) & 2xz + y - 15 \\ yz & xz & xy \end{pmatrix}.$$

While time-consuming to compute, there is nothing conceptually difficult about the computation of the Jacobian matrix beyond knowing how to compute a partial derivative. Nor does the Jacobian matrix have a particularly novel meaning. It is a generalization of the gradient vector that describes change along multiple vectors at once, like a kind of stretching. But an easier intuition can be gained by again recalling the Taylor series from Chapter 8.

The Jacobian matrix allows us to make the same linear approximation for a vector-valued function $\mathbf{f}$ with a vector-valued argument $\mathbf{x}$ that the Taylor series allows for a scalar-valued function with a single variable argument. We can write such a linear approximation as $\mathbf{f} \approx \mathbf{f}(\mathbf{a}) + J(\mathbf{a})(\mathbf{x} - \mathbf{a})$, where $J(\mathbf{a})$ is the Jacobian matrix evaluated at the vector $\mathbf{a}$. Thus the Jacobian serves to describe linear change across several dimensions of the function's image.

[10] The Jacobian matrix was named after the German mathematician Carl Gustav Jacob Jacobi.

15.2.3.1 Why Should I Care?

There are several applications of the Jacobian matrix, but the one you will most likely encounter is its use in the implicit function theorem in more than one dimension, which lets us compute comparative statics in more than one dimension. We see how this works in Chapter 17. In that application you'll see both a few varieties of the Jacobian matrix, J, as well as its determinant, $|J|$. This determinant is seen often enough so that it is sometimes referred to (a bit confusingly) simply as the Jacobian.

15.2.4 Second-Order Derivatives: The Hessian

Thus far we have dealt with only first derivatives, but one can compute higher-order ones as well. This, as you might imagine, gets really complicated really quickly, and we mostly avoid the topic here. However, we have significant occasion to use a multidimensional analogue to the second derivative, known as the **Hessian matrix**, so we describe that here and show you how to find it.[11]

Let's begin by asking what a second derivative would look like in a multidimensional setting. There are two possible derivatives we might consider. First, we could just repeat the same partial derivative we just took, and get something like $\frac{\partial^2}{\partial x_j^2}$. This is no different from the one-dimensional second derivative $\frac{d^2}{dx^2}$, and is computed in the same way: take the partial derivative of the same variable twice. Thus, if $f = x^2y$, then $\frac{\partial f}{\partial x} = 2xy$ and $\frac{\partial^2 f}{\partial x^2} = 2y$.

Second, one could first take the partial derivative with respect to one variable, and then with respect to a *different* variable. We write this as $\frac{\partial^2}{\partial x_j \partial x_i}$ and call it a **cross-partial derivative** or a **mixed-partial derivative**. For instance, let's use the previous example, $\frac{\partial^2 f}{\partial x \partial y} = 2x$. We could have gotten this in one of two ways. The way this cross-partial derivative was written, we first take the partial derivative of f with respect to x, yielding $2xy$, and then the partial derivative of *this* with respect to y, yielding the answer $2x$. Or we could have gotten this by computing $\frac{\partial^2 f}{\partial y \partial x}$, which means first take the partial derivative with respect to y, yielding x^2, and then take the partial derivative of this with respect to x, yielding $2x$. These two are the same, implying that the order of differentiation doesn't matter.

This is a general result, $\frac{\partial^2}{\partial x_j \partial x_i} = \frac{\partial^2}{\partial x_i \partial x_j}$, which makes our lives easier in that there are fewer second-order derivatives to compute. Specifically, there are $\frac{m(m+1)}{2}$ different second-order partial derivatives to compute whenever there are m variables. We can also simplify our notation, either using ∂_{xx} and ∂_{xy}, or just f_{xx} and f_{xy}, to refer to second-order partial derivatives.

[11]The Hessian matrix is sometimes referred to as simply the Hessian. This is not necessarily confusing except insofar as the Jacobian refers to the determinant of the Jacobian matrix, and not the matrix itself. We apologize for the confusion, but we did not make this stuff up. The Hessian is named for the German mathematician Ludwig Otto Hesse.

As we did in the Jacobian matrix, we can put our second-order partial derivatives together into a matrix, called the Hessian matrix. Element (i, j) of the Hessian matrix is $h_{ij} = \frac{\partial^2}{\partial x_i \partial x_j}$. Elements on the diagonal have $i = j$, and are $\frac{\partial^2}{\partial x_i \partial x_i} = \frac{\partial^2}{\partial x_i^2}$. Elements off the diagonal are the cross-partial derivatives. Because it doesn't matter what order you take these in, the Hessian is a symmetric matrix: $h_{ij} = h_{ji}$.

Let's do an example, using the more difficult example from the first subsection: $f(x, y, z) = x^y \ln(z) - y^3 x^2 z + 2yz - x + 1$. We've already found the first-order partial derivatives, which we reproduce here: $\frac{\partial f}{\partial x} = yx^{y-1}\ln(z) - 2y^3 xz - 1$, $\frac{\partial f}{\partial y} = x^y(\ln(x))\ln(z) - 3y^2x^2z + 2z$, and $\frac{\partial f}{\partial z} = \frac{x^y}{z} - y^3x^2 + 2y$. Now let's find the six unique second-order partial derivatives, and place them in our matrix. This yields $H =$

$$\begin{pmatrix} y(y-1)x^{y-2}\ln(z) - 2y^3 z & (x^{y-1} + yx^{y-1}\ln(x))\ln(z) - 6y^2xz & \frac{yx^{y-1}}{z} - 2y^3x \\ (x^{y-1} + yx^{y-1}\ln(x))\ln(z) - 6y^2xz & x^y(\ln(x))^2\ln(z) - 6yx^2z & \frac{x^y(\ln(x))}{z} - 3y^2x^2 + 2 \\ \frac{yx^{y-1}}{z} - 2y^3x & \frac{x^y(\ln(x))}{z} - 3y^2x^2 + 2 & \frac{-x^y}{z^2} \end{pmatrix}.$$

Because of the symmetry of the Hessian matrix, we find it a bit more intuitive to remember than the Jacobian matrix. However, if you disagree, the latter can give you the former, as the Hessian of f is the Jacobian of the gradient of f, $\boldsymbol{\nabla} f$. Of perhaps more importance: since the Hessian matrix is symmetric, if it is also positive definite (negative definite), then, as noted in Chapter 14, its eigenvectors are orthogonal and its eigenvalues are positive (negative) real numbers. Since the product of the eigenvalues in this case is the determinant of the matrix, we can use the determinant of the matrix to discern whether the Hessian is positive or negative definite, or neither. We make use of this fact in the next chapter.

We can also use the Hessian matrix, along with our knowledge of quadratic forms, to approximate a scalar function of more than one variable more precisely than a linear approximation could do. Harking back to our discussion of a Taylor series, we can write $f \approx f(\mathbf{a}) + \boldsymbol{\nabla} f(\mathbf{a})(\mathbf{x} - \mathbf{a}) + \frac{1}{2}(\mathbf{x} - \mathbf{a})^T H(\mathbf{a})(\mathbf{x} - \mathbf{a})$, where $\boldsymbol{\nabla} f(\mathbf{a})$ and $H(\mathbf{a})$ are the gradient vector and the Hessian matrix evaluated at the vector $\mathbf{a}$. Comparison with the $f''(a)$ term in a one-dimensional Taylor series helps us understand what the Hessian is intuitively: like the second derivative, it provides information on the curvature of the function, only across more than one dimension.

Finally, we should note quickly that we have confined the discussion of higher-order partial derivatives to scalar functions and second-order derivatives. Similar concepts could be defined for vector-valued functions or even higher-order derivatives, but then we would be dealing with *tensors*, rather than just matrices. Tensors are similar to and more general than matrices, but can have higher orders. For example, for a vector-valued function with n components and m arguments we would need a third-order tensor of dimension $n \times m \times m$ to describe the equivalent of a Hessian. Dealing with such objects is beyond the scope of this book.

15.2.4.1 Why Should I Care?

Like the Jacobian matrix, the Hessian matrix has many uses, but the one you'll primarily see is as an aid in optimization, where it helps to determine whether one has a maximum, a minimum, or a saddle point. In this way the Hessian serves the same purpose as the second derivative did in one-dimensional optimization, which makes sense, since it also describes the curvature of the function. We show how this works in both the next section and the next chapter.

15.2.5 Integration

For those readers who felt that integration gave them the most trouble when they first learned calculus, you will be happy to know that there is far less additional material to cover in extending the integral to multiple variables than there was when extending the derivative.[12] All we need to do is understand how to integrate a function of multiple variables in the same manner as we integrated a function of one variable. And we have even already covered this in the case of a continuous probability distribution. As given in Chapter 11, the joint cumulative distribution function (CDF) of two random variables X and Y is $Pr(X \leq x \cap Y \leq y) = F(x,y) = \int_{-\infty}^{x} \int_{-\infty}^{y} f(x',y')dx'dy'$. This is known as a double integral, but we are not limited to two integrals. We could, for instance, have the triple integral $\int_0^1 \int_0^1 \int_0^1 f(x,y,z)dxdydz$.

In computing these multiple integrals, just as in the case of partial derivatives, we can largely ignore the other variables in integrating over one of them. That is to say, we can treat all the other variables as constants when integrating over one of them, and then iteratively integrate over one variable after another.[13] In the CDF example just given, if we were to integrate over x', we could treat y' in $f(x',y')$ as a constant.

Further, for almost all our purposes we can choose to do the integrals in whatever order we want.[14] So, we could find the CDF either by first integrating over x' and then over y', or first over y' and then over x'. In the first case we could write $F(x,y) = \int_{-\infty}^{y} \left(\int_{-\infty}^{x} f(x',y')dx'\right) dy'$, while in the second we could write $F(x,y) = \int_{-\infty}^{x} \left(\int_{-\infty}^{y} f(x',y')dy'\right) dx'$ to make clear the order of integration. They are both the same.

The only exception you are likely to run across, and so the only real complication to watch out for, is when one of the variables appears in the bounds of

[12]This is a function not of the scope of mathematics but rather of our interests in it as social scientists. There are line, surface, and volume integrals, along with rules connecting them, among other things, but these have been of little utility to most if not nearly all political scientists.

[13]This is not in general true. However, for the vast majority of the situations in which we might find ourselves as social scientists (possessed of finite integrals), this will be true, and we treat it as such in this subsection. For more on this, see Fubini's theorem and Tonelli's theorem. Also note that we consider only definite multiple integrals here.

[14]See the previous footnote.

one of the definite integrals.[15] Let's say the variable y appears in the bounds of a definite integral, and that we are also integrating over y. The definite integral over y must eliminate y from the expression.[16] In this case, one must perform all integrals that have y in their bounds *before* integrating over y. The reason is that, since we are integrating over y as well, the y from the bounds must also be included in that integral, and this can't happen until the integrals for which y is in the bounds are resolved.

This is less confusing to see in an example than it is to say. So, for instance, in the double definite integral $\int_{-\infty}^{\infty}\int_{-\infty}^{y} f(x,y)dxdy$, one would need to do the integral over x first, because there is a y in the upper bound of the integral over x. In symbols, $\int_{-\infty}^{\infty}\left(\int_{-\infty}^{y} f(x,y)dx\right)dy$. We can't integrate y first, because then we'd be left with a y in the result, and one can never end up with the variable over which one is integrating in the result of a definite integral.

Let's do a few examples to make this clearer. We start with $\int_0^1\int_0^1\int_0^1 xyzdxdydz$. We can do this in any order, since there are no variables in any of the bounds, so let's start with x and work our way out. We treat y and z as constants in that integral, yielding $\int_0^1\int_0^1(\int_0^1 xyzdx)dydz = \int_0^1\int_0^1(\frac{1}{2}yz)dydz$. Then we do y and z in the same way: $\int_0^1(\int_0^1 \frac{1}{2}yzdy)dz = \int_0^1(\frac{1}{4}z)dz = \frac{1}{8}$. And that's the answer.[17]

Here's an example of integration of a uniform distribution over a rectangle. Let's check that it integrates to one over all space, as it must. The integral is $\int_c^d\int_a^b \frac{1}{(b-a)(d-c)}dxdy$, since if we assume that draws on both dimensions are independent then the PDF is the product of the PDF of a uniform distribution from a to b and one from c to d. Let's do x first after pulling out the constant: $\frac{1}{(b-a)(d-c)}\int_c^d\left(\int_a^b dx\right)dy = \frac{1}{(b-a)(d-c)}\int_c^d(b-a)\,dy$. Now we do the y integral, after again pulling out the constant: $\frac{b-a}{(b-a)(d-c)}\int_c^d dy = \frac{b-a}{(b-a)(d-c)}(d-c) = 1$. So everything checks out.

Finally, let's do one where the bounds of the integral come into play more: $\int_0^1\int_0^y x^2y^3dxdy$. We want to integrate over x first because there is a y in the bounds of the integral over x. So we do: $\int_0^1\left(\int_0^y x^2dx\right)y^3dy = \int_0^1\left(\frac{1}{3}(y^3-0)\right)y^3dy$ $= \int_0^1 \frac{1}{3}y^6dy$. Now the integral is straightforward: $\frac{1}{3}\int_0^1 y^6dy = \frac{1}{3}\frac{1}{7} = \frac{1}{21}$. Note that if we hadn't integrated over x first, we'd be left with a y in the answer, which we can't have when doing definite integrals.

---

[15] Recall that the difference between a definite and an indefinite integral is that the former has bounds.

[16] Recall the fundamental theorem of calculus: $\int_a^b f(x)dx = F(b) - F(a)$. The RHS is evaluated at b and a, and no x remains. This is true as well with more than one variable in a definite integral. The variable over which one integrates in a definite integral cannot appear after taking that integral.

[17] If you don't remember how to take the integral of x from 0 to 1, look back at Chapter 7. Okay, here it is: $\int_0^1 xdx = \frac{1}{2}x^2|_0^1 = \frac{1}{2}(1-0) = \frac{1}{2}$.

15.2.5.1 Why Should I Care?

Truth be told, you will most likely not spend a great deal of time doing multiple integrals. However, they do come up when assessing joint probabilities, as noted in Chapter 11, and it is important to know what you're looking at when presented with a double or triple integral. Also, there may be circumstances in which one has to take an expected utility over two random variables and their joint probability distribution function in a game theoretic context. In this case you'd need to take the double integral $\int_{-\infty}^{\infty}\int_{-\infty}^{\infty} u(x,y)f(x,y)dxdy$.

15.3 CONCAVITY AND CONVEXITY REDUX

Back in Chapter 8 we discussed the properties of concavity and convexity in one dimension. We close out the core of this chapter by providing analogous definitions for more than one variable. This section is brief, as not only the formalism but also the conceptual framework underlying concavity and convexity in the one-dimensional setting carry over directly to the multidimensional setting.

A function is concave if, for $\lambda \in [0,1]$, $f(\lambda \mathbf{x_1} + (1-\lambda)\mathbf{x_2}) \geq \lambda f(\mathbf{x_1}) + (1-\lambda)f(\mathbf{x_2})$, convex if $f(\lambda \mathbf{x_1} + (1-\lambda)\mathbf{x_2}) \leq \lambda f(\mathbf{x_1}) + (1-\lambda)f(\mathbf{x_2})$, and strictly concave or convex if the inequalities are strict. If you flip back to Chapter 8, you will note that these are the same definitions we used there for these properties, except now we're using vector arguments rather than scalar ones. Also note that, from these definitions, if f is concave, $-f$ is convex.

If this was all we could do there would be little point in repeating these definitions, it would seem. However, we can use some of the formalism we built up in the previous section to provide some more intuition. A function f is concave if and only if $f(\mathbf{x_2}) - f(\mathbf{x_1}) \leq \nabla f(\mathbf{x_1}) \cdot (\mathbf{x_2} - \mathbf{x_1})$. This is roughly equivalent to saying that the gradient (the tangent) of the function is more steeply sloped than the secant to the function, implying that the rate of increase is decreasing. Check back at the accompanying figure to the definition of concavity in Chapter 8 to see why this is so. If $f(\mathbf{x_2}) - f(\mathbf{x_1}) \geq \nabla f(\mathbf{x_1}) \cdot (\mathbf{x_2} - \mathbf{x_1})$, then the function is convex, since the tangent is less steeply sloped than the secant and the rate of increase of the function is increasing.

This is one way to use multivariate calculus to help us check for concavity and convexity, but it not necessarily the simplest one. You may recall from Chapter 8 that concavity and convexity are closely tied to the curvature of a function, and that the curvature can be assessed by looking at the second derivative. This remains true with more than one variable, except now we must use the Hessian matrix, the analogue of the second derivative.

If the Hessian, H, is negative semidefinite (has eigenvalues that are 0 or negative) at all points x, then the function f is concave. If H is positive semidefinite (has eigenvalues that are 0 or positive) at all points $\mathbf{x}$, then f is convex. If H is either negative or positive definite at all points $\mathbf{x}$ (has eigenvalues that are all negative or all positive), then f is either strictly concave or strictly convex, respectively. These conditions should be recognizable from Chapter 8:

H negative semidefinite is just like $f'' \leq 0$, and H positive semidefinite is just like $f'' \geq 0$. As it is often easier to discern definiteness via the determinant of H, particularly with computational aid, this can be an easier way to determine concavity or convexity of a function than using the previous definitions. We can also use the fact that the sum of concave (convex) functions multiplied by positive constants (e.g., $af(\mathbf{x})+bg(\mathbf{x})$, $a, b > 0$) is also concave (convex) to build up more complex concave (convex) functions from less complex functions whose Hessians are easier to manage.

15.3.1 Why Should I Care?

Concave and convex functions are important in multiple dimensions for the exact same reasons they are in one dimension: they describe important properties of functions such as acceleration, increasing or decreasing returns, and risk preferences in utility. Since usually more than one variable is of importance, the multidimensional version of these properties is often what we encounter.

There is another related concept frequently used by economists and in game theory that you might see, known as quasi-concavity (or quasi-convexity). This has a similar definition to concavity: a function is quasi-concave if $f(\mathbf{x_1}) \geq f(\mathbf{x_2}) \Rightarrow f(\lambda\mathbf{x_1} + (1-\lambda)\mathbf{x_2}) \geq f(\mathbf{x_2})$ or, equivalently, if $f(\lambda\mathbf{x_1} + (1-\lambda)\mathbf{x_2}) \geq \min\{f(\mathbf{x_1}), f(\mathbf{x_2})\}$, both for $\lambda \in [0,1]$. Quasi-convex functions are defined similarly except that we flip the inequality from $\geq$ to $\leq$ in each case and change the minimum to a maximum. Analogously to concavity, we can also say f is quasi-concave if and only if $f(\mathbf{x_2}) \geq f(\mathbf{x_1}) \Rightarrow \nabla f(\mathbf{x_1}) \cdot (\mathbf{x_2} - \mathbf{x_1}) \geq 0$.

Concavity implies quasi-concavity but not the reverse, so that concavity is a stronger condition than quasi-concavity. This is why it is of use in some formal theory contexts: some important results on the existence of optima and equilibria hold under the weaker condition of quasi-concavity (see, e.g., McCarty and Meirowitz, 2007, p. 108), and it is always good to use a weaker assumption if one can, as this allows the model to apply to a wider range of substantive settings.

15.4 WHY SHOULD I CARE?

We have already talked about a significant number of reasons why one should care about working with multiple variables, mostly centered on the need to model more complex settings than can be captured with a single variable. And once one accepts the need for such models, one requires a modified set of tools to accomplish tasks such as differentiation, integration, and optimization, which are required in applications in both statistics and game theory. These tools also allow us to engage in comparative statics to draw more and better conclusions from our models.

We address these applications more in the next two chapters, but before doing so we briefly discuss a topic that has gotten sporadic play in political science over the years: difference and differential equations used in dynamical models. Differential and difference equation models are useful for studying the change in a variable over time. If one can develop some expectations about the functional form of a variable with respect to its own past values and the values of other variables, then one can use the tools that have been developed to study these models to develop expectations about the qualitative trajectory of the variable of interest and about the impact of each variable in the equation on the variable of interest. Doing so with a single equation is not terribly difficult, and the formal use of math may not yield insight beyond what one would have obtained by thinking carefully. However, when one expands one's dynamical model to include two or more equations, then the relationships are not at all clear, and the mathematics makes it possible to ensure that the implications that one deduces have a solid foundation. Many of us think that there are non-trivial feedback relationships in political processes, and these models make it possible to first specify expected relationships and then analyze them to deduce implications.

The topic of differential and difference equations is the subject of entire courses, and they have been used in social science research for decades (e.g., see Goldberg, 1958). We discuss difference equations briefly in Chapter 14 in the context of modeling dynamical processes, and we provide a brief complementary sketch of differential and difference equations here. As we note there, dynamical models such as systems of differential and difference equations are used to formally study growth over time. Population growth is a popular topic, and biologists have used dynamical models to study the growth of species and the spread of disease. Political scientists might use dynamical models to study the growth of budgets, of support for a policy, of power accumulation among countries, or of a dissident organization. There are two types of dynamical models of concern to us here: difference equation models and differential equation models. The first study growth over discrete units of time, the second study growth over continuous time. Only the latter use the tools of multivariate calculus, but both address the same topic.

Let's begin with a discussion of difference equation models, which we have seen make it possible to write down explicit functions that capture intuitions we might have about the growth of something over time. Of particular interest is whether the process one is modeling will (1) tend to settle down into some equilibrium, (2) explode into ever-increasing values, or (3) decline to zero. The methods one uses to evaluate a given difference equation to determine this are well worked out, and are, if you worked through Chapter 14, already somewhat familiar to you, at least for the difference equation $y_t = \beta y_{t-1}$. This equation describes a Markov process, with the transition rule given by β, and we learned (when y can only take a finite number of values) when we can and how to compute a steady state for such processes in the previous chapter. Adding multiple linear (though still first-order) equations, so that there would be many state variables like y and a matrix of transition rules, might greatly complicate the analysis but would not change its character.[18]

When we turn our attention to differential equation models we need to use calculus as well as algebra as we are studying continuous growth rather than discrete growth. Sometimes we use a one-dimensional derivative, usually when all variables are only functions of one other variable, such as time, and sometimes we use a partial derivative. The study of equations involving the former is the study of ordinary differential equations, and the study of the latter is the study of partial differential equations.

Differential equations can also model the change in a variable as a function of its own past values, except that the measurement of the variable is continuous rather than discrete. The order of the equation indicates how many past values are used, and in this case, as with difference equations analysis, focuses primarily on the qualitative dynamics of the equation (i.e., does its growth explode, collapse, or settle down to some equilibrium?). For example, $\frac{dy}{dt} = \beta y$ is an analogue of the difference equation given earlier. We can guess the solution to this is $y = e^{\beta t}$, which we can verify by plugging it into the ordinary differential equation. This tells us that the equation models exponential growth, and the system, devoid of any other influences, will experience explosive growth as long as $\beta > 0$. However, as with difference equations, going further is beyond the scope of this book.

15.5 EXERCISES

1. The following variables derive from Fair (2009), who argues that the two-party vote share in US presidential elections can be modeled from a handful of variables, including:
   $y =$ Two Party Vote Share, an integer variable for the Democratic candidate's vote share that ranges from 0 to 100.
   $x =$ Macroeconomic Growth Rate, a continuous measure of the rate of growth of GNP, which ranges from $-\infty$ to ∞.
   $w =$ Good News Quarters, an integer count variable of the number of quarters in which macroeconomic growth was positive during the sitting president's term, which ranges from 0 to 15.
   $z =$ Incumbent, a binary measure coded 1 if the Democratic Party currently holds the presidency.

[18] Doing this for higher-order equations requires complexity beyond the scope of this book. That said, the methods of solution to such problems have elements recognizable from our discussion of eigenvalues in Chapter 14. In particular, for a single difference equation representable in homogeneous form (i.e., it equals zero rather than some other function), one method of solution involves associating with the equation a characteristic polynomial and exploring this polynomial's roots to determine whether the system oscillates, converges, or diverges. For difference equation models with more than one equation, graphs (known as state-space and phase plots) of the behavior of two variables relative to one another are often of great interest.

a) First consider the model $y = f(x, w, z) = \beta_0 + \beta_1 x + \beta_2 w + \beta_3 z$. Find the partial derivatives $\frac{\partial f}{\partial x}$, $\frac{\partial f}{\partial w}$, and $\frac{\partial f}{\partial z}$, and interpret each.

b) Now consider the model with an interaction term, $y = f(x, w, z) = \beta_0 + \beta_1 x + \beta_2 w + \beta_3 z + \beta_4 xw$. Again, find the same three partial derivatives. How do they differ from those in the previous model? Interpret each.

2. The following variables derive from Hafner-Burton (2005), who studies governments' human rights violations as a function of preferential trade agreements, among other things.
z = Political Terror Score, an ordinal variable that measures the extent to which a government violates its obligations to respect people's human rights.
x_1 = Preferential Trade Agreements, a binary measure coded 1 if the government has signed at least one trade treaty that lowers tariffs imposed by an OECD country *if* that government respect labor rights.
x_2 = Human Rights Agreements, an ordinal count of whether the government has ratified two key human rights treaties.
x_3 = Democracy, a 21-point integer-level variable from the Polity IV data where pure autocracy has a value of -10 and pure democracy a value of 10.

a) First consider the model $z = f(x_1, x_2, x_3) = \beta_0 + \beta_1 x_1 + \beta_2 x_2 + x_3^{\beta_3}$. Find the partial derivatives $\frac{\partial f}{\partial x_1}$, $\frac{\partial f}{\partial x_2}$, and $\frac{\partial f}{\partial x_3}$, and interpret each.

b) Now consider a revised version of the model, $z = f(x, w, z) = \beta_0 + \beta_1 x_1 + \beta_2 x_1^2 + \beta_3 x_3$. Again, find the same three partial derivatives. How do they differ from those in the previous model? Interpret each.

3. A promising young graduate student has decided, after taking every core seminar, that our theories of politics are far too simplistic. Undaunted by complexity, the student proposes that turnout is governed by the model $y = \beta_1 x_2^2 x_1^{\beta_2 x_2^2 - \beta_3 x_3}$, where y is some measure of turnout, x_1 a measure of free time, x_2 a measure of education, and x_3 a measure of taste for burgers. Find the marginal effects of all three of x_1, x_2, x_3 on y.

4. Let $z = 3xy + y^2$ and $f = 3xyz$. Find both $\frac{\partial f}{\partial x}$ and $\frac{df}{dx}$.

5. Consider the function $f = 3xy + \frac{3}{2}y^2 + \frac{5}{2}x^2$. Find ∇f. Calculate the Hessian matrix. Is f concave, convex, or neither?

6. Find ∇f, where $f = 2x^2 y + zx - y^z$. Compute the Hessian matrix. Is f concave, convex, or neither?

7. Find $\int_0^1 \int_0^x x^5 (y-1)^3 dy dx$.

Chapter Sixteen

Multivariate Optimization

Multivariate optimization is needed in statistics when trying to find the vector of coefficients that minimize least squared error or maximize a likelihood function, and in game theory (and decision theory) when trying to make the best choice among several options. Thus, it is central to quantitative analysis in the social sciences. Because of this, it was our primary justification for delving into multivariate calculus in the previous chapter. The purpose of this chapter is to provide techniques to optimize functions of multiple variables.

We consider two types of optimization here: unconstrained and constrained. **Unconstrained optimization**, the topic of the first section, is intuitively and practically the same as what we did in Chapter 8. We locate critical points using the gradient of a function, the analogue to the one-dimensional first derivative. And we check these critical points to see if they're minima, maxima, or neither using the Hessian matrix, the analogue of the one-dimensional second derivative. That's it. There's no new intuition here that wasn't either covered in Chapter 8 when introducing the optimization procedure, or in Chapter 15 when showing how the gradient and the Hessian correspond to the first and second derivatives. Consequently we focus on procedure rather than intuition, and on examples that illustrate how to find the extrema of functions of more than one variable. However, if it's been a while since you've worked through the relevant sections of either Chapter 8 or Chapter 15 and this intuition is fuzzy to you, it may be worth revisiting those chapters.

Constrained optimization is a new topic, and a messier one. It deals with the common scenario, particularly in game theory, of having constraints on the values that variables can take, which makes optimization more difficult. A common example of such a constraint is a budget constraint. You may want (i.e., it would be the optimum of your utility function over all values of the variable) to buy a really fancy car, but because you can't afford it you must buy a less fancy car. This decision is straightforward, but when there is more than choice variable and maybe more than one constraint, it becomes much more difficult to see what one should do just by looking at the problem, and the technique of constrained optimization is useful.

In fact, it turns out that it is unusual for social scientists to face optimization problems that have no constraints on them. For example, one might be interested in studying politicians under the assumption that they desire to maximize the number of votes they will receive in an upcoming election. We might write down a function composed of the concepts that we believe might influence the number of votes one gets and then optimize the function. If we were to do so, should we look for the maximum given *any values of those concepts*? This might seem reasonable until we ask ourselves whether we should include one's campaign budget as one of those concepts. If so, is any value from $-\infty$ to ∞ a reasonable value to consider when optimizing one's vote function? Surely not! A better model would find the maximum of the function over the range from \$0 to the maximum number of dollars one has in one's campaign budget. That is, we would want to maximize the vote function subject to the constraint placed on it by the size of one's campaign war chest. In other words, we would only want to study the function over the range of values between \$0 and the maximum number of dollars one could spend. Since the level of spending might alter the impact of other concepts—e.g., it might be hard to convey support from local figures without advertising this fact—the constraint on spending alters the impact of other concepts both directly and indirectly.

We cover two types of constrained optimization in this chapter. The easier type to perform is an optimization problem subject to what are known as **equality constraints**. These are constraints that *must* be precisely satisfied. This kind of thing is common in fields like physics, where, for example, a ball might be constrained to the surface of a sphere. However, usually problems in social science give you more leeway. You can spend up to a certain amount of money, or can save some. You can provide particularistic benefits to your constituency up to the amount you have available, but no more. You can't remove existing benefits from individuals; that is, all disbursements must be non-negative. And so on. These sorts of constraints are known as **inequality constraints** because they can be expressed as an inequality: e.g., the amount spent must be less than or equal to the budget available. Sometimes you can convert an inequality constraint to an equality constraint. For instance, if one's utility function is increasing in all variables (representing nonsatiable preferences), then one will always spend all of one's budget if given the opportunity to do so, in which case the budget constraint is equivalent to the requirement that the total amount spent must equal the budget available. But usually you can't convert inequality constraints in this way, which is a shame because they're harder to manage.

Nevertheless, we present procedures to optimize functions of more than one variable subject to both equality and inequality constraints. These are the focus of Sections 2 and 3. Because we believe the basic intuition behind optimization will prove sufficient for most readers, we largely eschew discussing the underlying intuition behind the methods we use. We again focus instead on technique and examples in Sections 2 and 3.

16.1 UNCONSTRAINED OPTIMIZATION

As we noted above, the procedure to find optima of functions of more than one variable is the same as the procedure to find optima of functions of one variable. The only differences are what we use for the first and second derivative. With the gradient in place of the first derivative, and the Hessian matrix in place of the second derivative, we can reprise our summary of a method of unconstrained optimization from Chapter 8. We write down the method first, discuss it briefly, and then provide a couple of examples in another subsection. More examples can be found in the exercises.

16.1.1 Method

To find the extrema and saddle points of any given (at least) twice-differentiable function $f(\mathbf{x})$, follow these steps:

1. Find $\nabla f(\mathbf{x})$.

2. Set $\nabla f(\mathbf{x}^*) = \mathbf{0}$ and solve for all $\mathbf{x}^*$. These are stationary points of the function.

3. Find the Hessian matrix H for $f(\mathbf{x})$.

4. For each stationary point $\mathbf{x}^*$, substitute $\mathbf{x}^*$ into H.

   - If H is negative definite, $f(\mathbf{x})$ has a local maximum at $\mathbf{x}^*$.
   - If H is positive definite, $f(\mathbf{x})$ has a local minimum at $\mathbf{x}^*$.
   - If H is neither negative nor positive definite but $|H| \neq 0$, then $f(\mathbf{x})$ has a saddle point at $\mathbf{x}^*$.
   - If $|H| = 0$ then the second-derivative test is inconclusive.

5. Compare the local maxima and minima. The largest local maximum is the global maximum, and the smallest local minimum is the global minimum.

We discussed the correspondence of the gradient and the first derivative, and the Hessian matrix and the second derivative, in the last chapter, and talked at length about optimization in Chapter 8, so there is little to say here by way of new intuition. In fact, the intuition is more or less the same.

There are a few simplifications of this method, and a few complications, that bear discussing before moving to examples. As we don't want to be computing higher-order derivatives at stationary points, we call any point at which the Hessian matrix is indeterminate (i.e., neither positive nor negative definite), but $|H| \neq 0$, a saddle point, and leave it at that. We also leave alone inconclusive tests. Further, we do not compare the values of the function at local extrema to those at the boundary of the function's domain here. In general, the boundary is a lot more complicated in multiple dimensions than a pair of points, and we address this as part of our study of constrained optimization. Consequently, this method applies only to situations in which the domain over which one is optimizing is *unconstrained* (i.e., situations in which we are trying to find the extrema over all possible values of the arguments of the functions).

Any simplification this buys us, though, is outweighed by the complexity of steps one through four. Computation of the gradient is not too bad unless the number of variables is very large, in which case you do not want to be optimizing by hand anyway! But step two requires you to set this gradient vector equal to the zero vector and solve for all stationary points. This equality implies that each component of the gradient must equal zero at a stationary point, which produces a system of m equations in m unknowns if there are m arguments of the function. This can be difficult if m is large, and particularly difficult if the system one obtains is not linear. Solving this system if it is linear requires the techniques we developed in Chapter 13, so flip back there if it's been a while since you reviewed or used them. We stick to $m = 2$ or $m = 3$ in our examples, as well as functions that produce linear systems of equations. We assume that you'll turn to computational methods for larger m or complicated nonlinear systems.[1]

Step three requires computing the Hessian, which we've seen in the previous chapter can be time consuming for complex functions. Step 4 requires plugging each stationary point into the Hessian we've just computed, and then checking if the Hessian at that point is positive definite, negative definite, or indeterminate. The most convenient way to do this is often via the use of eigenvalues, which we discussed in Chapter 14. If all the eigenvalues of the Hessian at a point are positive, then the matrix is positive definite there (and we have a minimum). If all are negative, then the matrix is negative definite there (and we have a maximum). And if some are positive and some are negative, then the Hessian is indeterminate (and we have a saddle point). Connecting to our discussion of concavity and convexity in the previous chapter, a minimum occurs where the function is locally convex, a maximum where it is locally concave, and a saddle point where local concavity and local convexity depend on the direction in which one approaches the stationary point. If the determinate is zero, though, the Hessian has at least one zero eigenvalue, and the test is inconclusive for the same reason a second derivative of zero was inconclusive in the one-dimensional case.

Step five is not difficult. Let's use the steps and try a couple of examples to see how this all works.

16.1.2 Examples

Let's start with a two-dimensional example, and then work a three-dimensional example. More can be found in the exercises. In each case we follow the five steps in order. Let's begin with $f(x, y) = \frac{3}{2}x^2 - 2xy - 5x + 2y^2 - 2y$.

1. To find $\nabla f(\mathbf{x})$, we need to compute two partial derivatives, one for x and one for y. These are $\partial_x f(x, y) = 3x - 2y - 5$ and $\partial_y f(x, y) = -2x + 4y - 2$.

[1] A number of software options are available, ranging from open source freeware such as FreeMat (`http://freemat.sourceforge.net/`) and Maxima (`http://maxima.sourceforge.net/`) to online resources such as Wolfram Alpha (`http://www.wolframalpha.com/`), and commercial software like Maple (`http://www.maplesoft.com/products/maple/`), Matlab (`http://www.mathworks.com/academia/`), and Mathematica (`http://www.wolfram.com/solutions/highered/`).

2. We next set $\nabla f(\mathbf{x}^*) = \mathbf{0}$ and solve for all $\mathbf{x}^*$. We have two equations in two unknowns:

$$3x - 2y - 5 = 0,$$
$$-2x + 4y - 2 = 0.$$

Let's use substitution. Solving the second equation for x yields: $x = 2y-1$. Plugging this into the first equation yields $3(2y-1) - 2y - 5 = 0$, which simplifies to $4y - 8 = 0$ or $y = 2$. Plugging $y = 2$ back into the equation for x yields $x = 3$. So our only stationary point is $x = 3, y = 2$.

3. To find the Hessian matrix H for $f(\mathbf{x})$ we need f_{xx}, f_{yy}, and f_{xy}. $f_{xx} = 3, f_{yy} = 4$, and $f_{xy} = -2$. (Note that either order of differentiation produces the same number for the cross-partial derivative.) So the Hessian is $H = \begin{pmatrix} 3 & -2 \\ -2 & 4 \end{pmatrix}$.

4. There is only one stationary point, and the Hessian matrix is independent of the function's arguments anyway (as it will be whenever the system of equations arising from step two is linear), so we need to check the eigenvalues of the matrix. The eigenvalue equation is $|H - \lambda I| = 0$, which implies $(3-\lambda)(4-\lambda) - 4 = 0$, or $\lambda^2 - 7\lambda + 8 = 0$. Using the quadratic equation, we get $\lambda = \frac{7}{2} \pm \frac{1}{2}\sqrt{49-32} = \frac{7\pm\sqrt{17}}{2}$. Since $\sqrt{17} < 7$, both of these eigenvalues are positive, implying that the Hessian matrix is positive definite. Thus, the stationary point $x = 3, y = 2$ is a local minimum.

5. There is only one local minimum, so $x = 3, y = 2$ is the global minimum. The function shoots off to infinity in multiple directions, so it has no global maximum.

Now for the more difficult example: $f(x, y, z) = -3x^2 - 2xy + xz - \frac{1}{2}y^2 - yz - 4z^2 + 5x + 7y + 25z$

1. To find $\nabla f(\mathbf{x})$, we need to compute three partial derivatives, for x, y, and z. These are $\partial_x f(x, y, z) = -6x - 2y + z + 5$, $\partial_y f(x, y, z) = -2x - y - z + 7$, and $\partial_z f(x, y, z) = x - y - 8z + 25$.

2. We next set $\nabla f(\mathbf{x}^*) = \mathbf{0}$ and solve for all $\mathbf{x}^*$. We have three equations in three unknowns:

$$-6x - 2y + z + 5 = 0,$$
$$-2x - y - z + 7 = 0,$$
$$x - y - 8z + 25 = 0.$$

Let's use Cramer's rule (from Chapter 13). The matrix and vector corresponding to this system of equations are $A = \begin{pmatrix} -6 & -2 & 1 \\ -2 & -1 & -1 \\ 1 & -1 & -8 \end{pmatrix}$, $\mathbf{b} =$

$\begin{pmatrix} -5 \\ -7 \\ -25 \end{pmatrix}$. We also need the three matrices, B_1, B_2, B_3, in which the relevant column of A is replaced by $\mathbf{b}$. These are $B_1 = \begin{pmatrix} -5 & -2 & 1 \\ -7 & -1 & -1 \\ -25 & -1 & -8 \end{pmatrix}$, $B_2 = \begin{pmatrix} -6 & -5 & 1 \\ -2 & -7 & -1 \\ 1 & -25 & -8 \end{pmatrix}$, $B_3 = \begin{pmatrix} -6 & -2 & -5 \\ -2 & -1 & -7 \\ 1 & -1 & -25 \end{pmatrix}$. Cramer's rule requires that we take the determinants of each of these four matrices. You should verify our answers, going back to Chapter 12 if you need a reminder as to how to take a determinant, but these are $|A| = -6(7) + 2(17) + 1(3) = -5$, $|B_1| = -5(7)+7(17)-25(3) = 9$, $|B_2| = -6(31)+2(65)+1(12) = -44$, and $|B_3| = -6(18) + 2(45) + 1(9) = -9$. Then $x = \frac{|B_1|}{|A|} = -\frac{9}{5}, y = \frac{|B_2|}{|A|} = \frac{44}{5}$, and $z = \frac{|B_3|}{|A|} = \frac{9}{5}$. So our only stationary point is $x = -\frac{9}{5}, y = \frac{44}{5}, z = \frac{9}{5}$.

3. To find the Hessian matrix H for $f(\mathbf{x})$, we need f_{xx}, f_{yy}, f_{zz}, and f_{xy}, f_{xz}, f_{yz}. $f_{xx} = -6, f_{yy} = -1, f_{zz} = -8$, and $f_{xy} = -2, f_{xz} = 1, f_{yz} = -1$. (Note that either order of differentiation produces the same numbers for the cross-partial derivatives.) So the Hessian is $H = \begin{pmatrix} -6 & -2 & 1 \\ -2 & -1 & -1 \\ 1 & -1 & -8 \end{pmatrix}$.

4. There is only one stationary point, and the Hessian matrix is independent of the function's arguments anyway (as it will be whenever the system of equations arising from step two is linear), so we need to check the eigenvalues of the matrix. The eigenvalue equation is $|H - \lambda I| = 0$ which implies $(-6-\lambda)((-1-\lambda)(-8-\lambda)-1)+2(-2(-8-\lambda)+1)+1(2-1(-1-\lambda)) = 0$, or $-\lambda^3 - 15\lambda^2 - 56\lambda - 5 = 0$. This is a cubic equation, and further, one that does not factor easily. However, we can tell immediately that $\lambda = 0$ is not a solution by plugging in 0, so the second derivative test will not be inconclusive. Further, since all we care about is whether or the eigenvalues are all negative or all positive, rather than the specific eigenvalues, we can learn all we need to know by graphing the characteristic polynomial. The cubic polynomial will have no more than three real roots, and if they are all the same sign the second derivative test will indicate a maximum or a minimum. The graph is shown in Figure 16.1.

   As we can see, there are three real roots of the characteristic polynomial, and all three are negative. Thus the Hessian matrix is negative definite, and the stationary point we have found is a local maximum.

5. There is only one local maximum, so $x = -\frac{9}{5}, y = \frac{44}{5}, z = \frac{9}{5}$ is the global maximum. The function shoots off to negative infinity in multiple directions, so it has no global minimum.

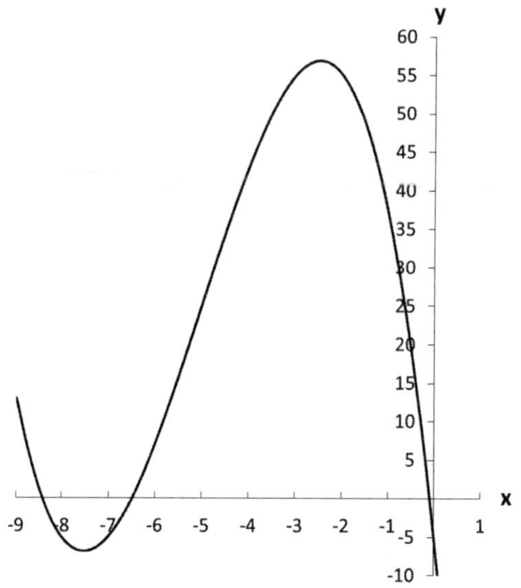

Figure 16.1: Graph of Cubic Characteristic Polynomial

16.1.3 Why Should I Care?

Optimization lies at the heart of quantitative political science, as we've noted in numerous places in this text. Whenever one has more than one independent variable and is minimizing the sum of squared errors in OLS or is maximizing a likelihood function, one is performing a multidimensional optimization problem. Similarly, whenever an actor's decisions in a game theoretic model involve more than one choice variable, one must perform a multidimensional optimization problem.

These optimization problems are *unconstrained* when the variables of interest need not take a particular set of values (e.g., need not be all positive) and there do not exist trade-offs between the variables that are not directly encoded in the function being maximized or minimized (e.g., a budget constraint). This is the case in the typical example of linear regression, since the parameters of these linear models (i.e., the coefficients of the independent variables in the regression) are free to take any values. Thus, there is a multitude of examples of unconstrained optimization in empirical political science. However, in most cases the large dimensionality of the problem requires the analyst to rely on computational means to perform the optimization. Consequently, quantitative examples of this procedure are not terribly enlightening.

While game theoretic models typically employ only a small number of choice dimensions, there is often some constraint involved. In the next two sections we consider an example that uses constrained optimization. Here we note that a lack of constraint implies that each choice variable can take any real value. A common example of this is multidimensional policy choice. We have stated above that one can model policy preference in one dimension according to the utility $u(x) = -(x-z)^2$, where z is an actor's ideal policy. This generalizes in a straightforward fashion to multiple dimensions. For example, in two dimensions: $u(x_1, x_2) = -(x_1 - z_1)^2 - (x_2 - z_2)^2$. Here the subscripts mean components of the vectors $\mathbf{x}, \mathbf{z}$. One can verify that this utility is maximized at $\mathbf{x} = \mathbf{z}$.

16.2 CONSTRAINED OPTIMIZATION: EQUALITY CONSTRAINTS

We already touched on a form of optimization constraints in Chapter 8. Specifically, we compared local extrema to the value of the function at the boundary of the domain, when the domain was *constrained* to be in a certain region. That's fine as long as the domain is one-dimensional, but when it is multidimensional the boundaries of the domain are much more difficult to characterize. Further, constraining one variable affects the optimization over other variables. For example, consider a situation in which increasing one variable (x) increases the relative contribution of another variable (y), as is the case for an interaction term with a positive coefficient. If we constrain x, the possible contribution of y to the function decreases, implying less push to increase y when optimizing the function. If there were, say, a budget constraint, then less of the budget might be allocated to y solely because x was individually constrained.

This is speaking loosely, but the key point is that, when variables interact either directly in the function or indirectly via a constraint, then instituting a constraint can have complex effects. In the next two sections, we provide a procedure to deal with these effects, and compute optima under constraints. In this section we cover equality constraints, and in the next inequality constraints. As in the previous section, we write down the method first, discuss it briefly, and then provide a couple of examples in another subsection. And again, more examples can be found in the exercises.

16.2.1 Method

Optimizing functions subject to equality constraints is similar in many ways to unconstrained optimization. In fact, the main difference involves the conversion of the constrained problem to an unconstrained one, albeit one with an *additional* variable. This additional variable is known as a **Lagrange multiplier**. If you've gone through the three variable example in the previous section you may be wondering why we'd add another variable. Let's set up the optimization problem to see how this works. Let's look for a maximum, since that's usually what applied work involves, but the same logic can be applied to look for a minimum.

The problem is to maximize some function $f(\mathbf{x})$ subject to constraints of the form $g_i(\mathbf{x}) = c_i$ or $g_i(\mathbf{x}, c_i) = 0$. Note that all equality constraints can be written in the second way by moving c_i to the LHS of the first equation. For example, if $x + 3y = 5$, as in a budget constraint in which all money must be spent, then $x + 3y - 5 = 0$, and $g = x + 3y - 5$. We generally keep the second form, in which the constraint equals zero, and we also drop dependence on c_i when it doesn't concern us and it will not be too confusing.[2] The subscript i on the g_i constraint allows for the possibility of more than one constraint. We see this more in the context of inequality constraints, in which multiple constraints are more common in applications.

Form a Lagrange function:

$$\Lambda(\mathbf{x}, \boldsymbol{\lambda}) = f(\mathbf{x}) - \sum_i \lambda_i g_i(\mathbf{x}, c_i),$$

where the sum is taken over the number of equality constraints. The central insight underlying this technique is that, if $f(\mathbf{x})$ is a maximum of the constrained optimization problem, then the vector formed by stacking $\mathbf{x}$ and $\boldsymbol{\lambda}$ on top of each other is a stationary point of the Lagrange function. Thus, we begin by transforming the first two steps in our method from the previous section by replacing f with Λ. This provides us with a necessary condition for a maximum.

At this point in the procedure for unconstrained optimization we would check the second derivative test to see whether we actually have a maximum. We can do something similar here, though it is a bit messier. Again we modify the steps of the method for unconstrained functions. This time we create a variant of the Hessian matrix called the **bordered Hessian**. The bordered Hessian, B, is a larger version of the normal Hessian that takes into account the constraints. One forms it by taking second derivatives of the Lagrange function, Λ, with respect to all variables, including the Lagrange multipliers. The derivatives with respect to the multipliers enter the matrix first, then the variables. If there are n constraints and m variables, it looks like this:

$$B = \begin{pmatrix} \Lambda_{\lambda_1\lambda_1} & \cdots & \Lambda_{\lambda_1\lambda_n} & \Lambda_{\lambda_1 x_1} & \cdots & \Lambda_{\lambda_1 x_m} \\ \vdots & \cdots & \vdots & \vdots & \cdots & \vdots \\ \Lambda_{\lambda_n\lambda_1} & \cdots & \Lambda_{\lambda_n\lambda_n} & \Lambda_{\lambda_n x_1} & \cdots & \Lambda_{\lambda_n x_m} \\ \Lambda_{x_1\lambda_1} & \cdots & \Lambda_{x_1\lambda_n} & \Lambda_{x_1 x_1} & \cdots & \Lambda_{x_1 x_m} \\ \vdots & \cdots & \vdots & \vdots & \cdots & \vdots \\ \Lambda_{x_m\lambda_1} & \cdots & \Lambda_{x_m\lambda_n} & \Lambda_{x_m x_1} & \cdots & \Lambda_{x_m x_m} \end{pmatrix}.$$

B is admittedly a bit complicated, but not entirely as bad as it seems. The original Hessian matrix appears in the lower right-hand corner of B. Since the λ_i never appear multiplied together in the Lagrange function, all second derivatives involving two Lagrange multipliers are zero, and so the entire upper-left $n \times n$ submatrix is zero. Other derivatives involving Lagrange multipliers end up focusing on the constraints g_i. And it's still symmetric. This is what an

[2] More on this later.

example with one constraint (g) and two variables (x and y) looks like:[3]

$$B = \begin{pmatrix} 0 & \frac{\partial g}{\partial x} & \frac{\partial g}{\partial y} \\ \frac{\partial g}{\partial x} & \frac{\partial^2 \Lambda}{\partial x^2} & \frac{\partial^2 \Lambda}{\partial x \partial y} \\ \frac{\partial g}{\partial y} & \frac{\partial^2 \Lambda}{\partial y \partial x} & \frac{\partial^2 \Lambda}{\partial y^2} \end{pmatrix}.$$

The constraints are going to give us a bit of trouble, so we're going to go with a method different from computing eigenvalues for determining the local curvature of the Lagrange function. This involves checking the signs of what are known as **leading principal minors**. A principal minor of order k, given a square $m \times m$ matrix, is formed by deleting $m-k$ columns and the corresponding $m-k$ rows (i.e., the ones with the same row and column numbers) and then taking the determinant of the resulting matrix. A leading principal minor is formed by deleting the last $m-k$ rows and columns of a matrix and then taking the determinant of the resulting matrix. Thus, the leading principal minor of order 1 deletes the last $m-1$ rows and columns, and so is the determinant of the top left element of the matrix; the leading principal minor of order 2 deletes the last $m-2$ rows and columns, and so is the determinant of the top left two-by-two submatrix; and so on, until the leading principal minor of order m deletes nothing, and so is the determinant of the original matrix. Therefore, in essence, we're going to build up a series of determinants of increasing size and check their signs. Each of these determinants is evaluated at one of the stationary points.

For a general symmetric $m \times m$ matrix, such as the Hessian, H, one can use leading principal minors to determine positive and negative definiteness at any point. Any symmetric matrix is positive definite if all its leading principal minors are positive and is negative definite if all its leading principal minors alternate in sign, with the first one being negative (i.e., the leading principal minor of order k should have sign $(-1)^k$).[4] We can thus use the test of leading principal minors to determine positive or negative definiteness at any point and, through this, whether a point is a local minimum or a local maximum, respectively.

For the bordered Hessian, B, this technique is a bit more complicated thanks to the constraints, but it works similarly. For B we don't need to compute the first few leading principal minors. Instead, one only needs the last (i.e., the ones with the highest order) $m-n$ leading principal minors of B, including the last one, which is $|B|$. If the last $m-n$ leading principal minors all have the sign $(-1)^n$, then the bordered Hessian is positive definite. If $|B|$ has sign $(-1)^m$ and the rest of the last $m-n$ leading principal minors alternate in sign, then the bordered Hessian is negative definite. Otherwise the test is indeterminate.[5]

[3] We won't tackle any examples more complicated than this in the book.

[4] We did not present this earlier because we find the eigenvalue technique a bit more intuitive, and, with computational aids to compute eigenvalues or graph the characteristic polynomial to find the signs of roots, perfectly sufficient for our needs. But this method works as well.

[5] Note that these rules reduce to those for a general symmetric matrix when the number of constraints, n, is equal to zero.

This is now enough setup. To find the solution to the problem of maximizing (or minimizing) $f(\mathbf{x})$ subject to one or more constraints of the form $g_i(\mathbf{x}) = c_i$, one follows these steps:

1. Transform all constraints to the form $g_i(\mathbf{x}, c_i) = 0$.
2. Form the Lagrangian function $\Lambda(\mathbf{x}, \boldsymbol{\lambda}) = f(\mathbf{x}) - \sum_i \lambda_i g_i(\mathbf{x}, c_i)$.
3. Find $\boldsymbol{\nabla}\Lambda(\mathbf{x}, \boldsymbol{\lambda})$.
4. Set $\boldsymbol{\nabla}\Lambda(\mathbf{x}^*, \boldsymbol{\lambda}^*) = \mathbf{0}$ and solve for all $\mathbf{x}^*$ *and* all $\boldsymbol{\lambda}^*$. These are stationary points of the Lagrange function.
5. Compute the bordered Hessian matrix B for Λ.
6. For each stationary point defined by $\mathbf{x}^*$ and $\boldsymbol{\lambda}^*$, substitute $\mathbf{x}^*$ and $\boldsymbol{\lambda}^*$ into B and use the leading principal minors test to determine whether the stationary point is a maximum or a minimum, or whether the test is indeterminate.
7. Compare the local maxima and minima. The largest local maximum is the global maximum and the smallest local minimum is the global minimum.

We work through a couple of examples after noting an important point. We are generally interested in the maximum of the function and the point at which it achieves this maximum, $\mathbf{x}^*$. The constraints, and the coefficients on these constraints, are usually considered tools toward that end. However, the Lagrangian multipliers can be assigned a meaning in some cases, and it helps our intuition a bit to understand them.

To see this, consider $g_i(\mathbf{x}, c_i) = 0$ or, with a slight redefinition (as above), $g_i(\mathbf{x}) = c_i$. We can think of c_i in the case of a budget constraint as the size of the budget. For example, $x + 3y = c$. The bigger c, the bigger the budget.

Now look at the Lagrange function, Λ. The parameter c_i appears only in the constraint g_i. Thus, $\frac{\partial \Lambda}{\partial c_i} = \lambda_i$, since the term in which it appears looks like $\lambda_i(g_i(\mathbf{x}) - c_i)$. Also, note that $\frac{\partial \Lambda}{\partial \lambda_i} = g_i(\mathbf{x}, c_i) = 0$; in other words, we get back the constraint again when we perform step 3 in our method, and step 4 ensures that the constraint is satisfied at the maximum. (If this is not clear, it will be from our examples, so don't worry.) From this, $\Lambda(\mathbf{x}^*, \boldsymbol{\lambda}^*) = f(\mathbf{x}^*)$. In words, at the maximum, the Lagrange function and the unconstrained function have the same value.[6] Consequently, $\frac{\partial \Lambda}{\partial c_i} = \lambda_i$ implies that the Lagrange multiplier corresponding to the constraint parameter c_i tells us the rate of increase in the function at its optimum with respect to the constraint parameter c_i.

Why do we care? One reason is that often the function we're trying to maximize is a utility function. In this case the Lagrange multiplier tells us the rate at which increases in the constraint parameter, e.g., the size of the budget,

[6] This reasoning is a special case of the envelope theorem, which we cover in the next chapter.

increase one's maximum utility. For this reason, economists sometimes call the Lagrange multiplier the shadow price of the constraint, since it is the constraint's marginal cost.

This provides some intuition for how the constraint affects the outcome of the maximization. For those who prefer a more graphical approach, we can oblige as well. Let's consider a function of two variables, which we call a utility function, and a budget constraint. On a two-dimensional plot, indifference or isoutility curves (see the previous chapter) are curves on this plot. The typical budget constraint, in contrast, looks like a line of negative slope connecting the x- and y-axes in the upper right quadrant of the plot. The y-intercept represents using the entire budget to purchase good y, and the same with the x-intercept and good x. The line connecting them, assuming they can be infinitely divided, corresponds to purchases of linear combinations of the two goods. The slope of the line is related to the relative prices of the two goods. Figure 16.2 shows a budget constraint and some indifference curves.

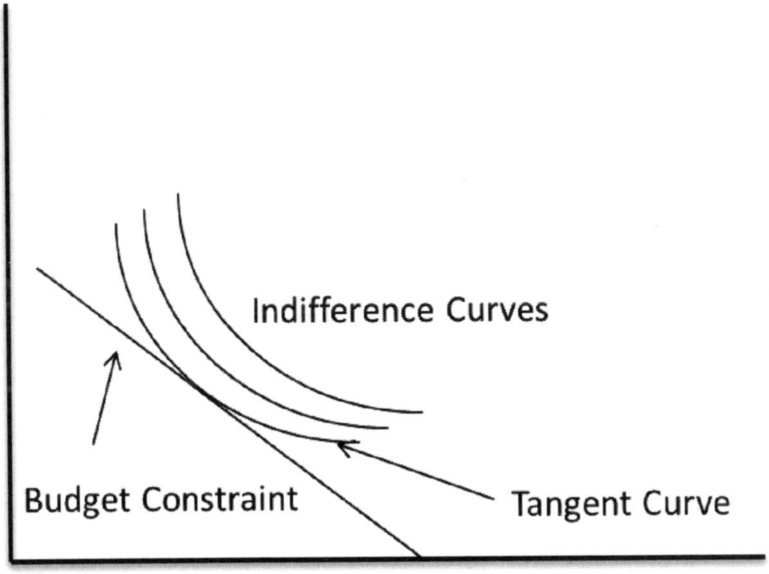

Figure 16.2: Indifference Curves and Budget Constraint

Since each indifference curve has the same utility along every point, and each point represents a different combination of the goods x and y, the goal is to find a point along the highest utility curve that also satisfies the budget constraint. Loosely speaking, such a curve would require a greater budget than is available for almost all combinations of goods, save for the one combination that just satisfies the constraint. To satisfy the condition at only one point, the curve must be tangent to the budget condition at that point. Thus, maximizing utility subject to a budget constraint is equivalent to finding the indifference curve tangent to the budget constraint and identifying the point of tangency. We show this in Figure 16.2 as well.

16.2.2 Examples

To give you a flavor for the technique, we chose two examples of similar difficulty. Let's start with the problem of maximizing $f(x_1, x_2) = x_1 x_2$ subject to the constraint that $x_1 + 4x_2 = 16$. You might think of this as a utility function in which the marginal utility for each good depends on the other, preferences are nonsatiable for each good, each good costs a different amount, and there is a budget that the decision maker will max out. Solve for the maximum by following our method.

1. We have one constraint, which we call $g(x_1, x_2) = x_1 + 4x_2 - 16 = 0$.

2. The Lagrangian function is $\Lambda(x_1, x_2, \lambda) = f(x_1, x_2) - \lambda g(x_1, x_2) = x_1 x_2 - \lambda(x_1 + 4x_2 - 16)$.

3. To find $\nabla\Lambda(\mathbf{x}, \boldsymbol{\lambda})$, we need $\partial_{x_1}\Lambda$, $\partial_{x_2}\Lambda$, and $\partial_\lambda\Lambda$. These are $\partial_{x_1}\Lambda = x_2 - \lambda$, $\partial_{x_2}\Lambda = x_1 - 4\lambda$, and $\partial_\lambda\Lambda = -(x_1 + 4x_2 - 16)$.

4. We set $\nabla\Lambda(\mathbf{x}^*, \boldsymbol{\lambda}^*) = \mathbf{0}$ and solve for all $\mathbf{x}^*$ *and* all $\boldsymbol{\lambda}^*$ to find the stationary points of the Lagrange function. There are three equations and three unknowns:

$$\begin{aligned} x_2 - \lambda &= 0, \\ x_1 - 4\lambda &= 0, \\ x_1 + 4x_2 - 16 &= 0. \end{aligned}$$

   Because only two variables appear in any equation, it's easier to solve them by substitution than by a more complex method. The first equation gives $x_2 = \lambda$. The second equation gives $x_1 = 4\lambda$. Plugging these both into the third equation gives $4\lambda + 4(\lambda) - 16 = 0$, or $\lambda = 2$. Plugging this back into the other two equations gives $x_1 = 8$ and $x_2 = 2$.

5. Now we have to compute the bordered Hessian matrix. It is

$$B = \begin{pmatrix} 0 & 1 & 4 \\ 1 & 0 & 1 \\ 4 & 1 & 0 \end{pmatrix}.$$

6. There is only one stationary point, and B doesn't depend on its value, so we need to use the leading principal minors test on B once. There is one condition, so $n = 1$, and there are two variables, so $m = 2$. Thus we must consider the last $2-1 = 1$ leading principal minor, which is the determinant of the bordered Hessian, $|B|$. This is $|B| = 0 - 1(-4) + 4(1) = 8 > 0$, which has the same sign as $(-1)^2$. Thus, the leading principal minors test implies B is negative definite, and we have a local maximum.

7. As there is only one stationary point and only one local maximum, it must be the global maximum. Since we haven't forced the variables to be positive (more on this under "Constrained Optimization: Inequality Constraints"), the function shoots off toward negative infinity (one can trade more positive x_2 for negative x_1, e.g.) and it has no minimum.

For the second example let's maximize $f(x_1, x_2) = 3x_1x_2^2 - 27x_2$ subject to the constraint that $x_1 + x_2 = 6$.

1. We have one constraint, which we call $g(x_1, x_2) = x_1 + x_2 - 6 = 0$.

2. The Lagrangian function is $\Lambda(x_1, x_2, \lambda) = f(x_1, x_2) - \lambda g(x_1, x_2) = 3x_1x_2^2 - 27x_2 - \lambda(x_1 + x_2 - 6)$.

3. To find $\boldsymbol{\nabla}\Lambda(\mathbf{x}, \boldsymbol{\lambda})$, we need $\partial_{x_1}\Lambda$, $\partial_{x_2}\Lambda$, and $\partial_\lambda\Lambda$. These are $\partial_{x_1}\Lambda = 3x_2^2 - \lambda$, $\partial_{x_2}\Lambda = 6x_1x_2 - 27 - \lambda$, and $\partial_\lambda\Lambda = -(x_1 + x_2 - 6)$.

4. We set $\boldsymbol{\nabla}\Lambda(\mathbf{x}^*, \boldsymbol{\lambda}^*) = \mathbf{0}$ and solve for all $\mathbf{x}^*$ *and* all $\boldsymbol{\lambda}^*$ to find the stationary points of the Lagrange function. There are three equations and three unknowns:

$$\begin{aligned} 3x_2^2 - \lambda &= 0, \\ 6x_1x_2 - 27 - \lambda &= 0, \\ x_1 + x_2 - 6 &= 0. \end{aligned}$$

Let's again solve them by substitution, since this is not a linear system of equations. The first equation gives $3x_2^2 = \lambda$. The second equation gives $6x_1x_2 - 27 = \lambda$. The third gives $x_1 = -x_2 + 6$. Plugging the third and the first into the second yields $6(-x_2 + 6)x_2 - 27 = 3x_2^2$, or $x_2^2 - 4x_2 + 3 = 0$ after rearranging and dividing by 9. This factors into $(x_2 - 3)(x_2 - 1) = 0$, which has solutions $x_2 = 3$ and $x_2 = 1$. Plugging these into the third equation yields $x_1 = 3$ or $x_1 = 5$, respectively, and plugging these into the first equation yields $\lambda = 27$ or $\lambda = 3$, respectively. So the two stationary points are $x_1 = 3, x_2 = 3, \lambda = 27$ and $x_1 = 5, x_2 = 1, \lambda = 3$.

5. Now we have to compute the bordered Hessian matrix. It is

$$B = \begin{pmatrix} 0 & 1 & 1 \\ 1 & 0 & 6x_2 \\ 1 & 6x_2 & 6x_1 \end{pmatrix}.$$

6. There are two stationary points, and B does depend on their values, so we have to use the leading principal minors test on B twice. There is one condition, so $n = 1$, and there are two variables, so $m = 2$, and as in the previous example the only leading principal minor to compute is $|B|$. This is $|B| = 0 - 1(6x_1 - 6x_2) + 1(6x_2) = 6(2x_2 - x_1)$. For the point $x_1 = 3, x_2 = 3, \lambda = 27$, we get $|B| = 18 > 0$. Again as in the previous

example, because this is the same sign as $(-1)^2$, B is negative definite here and the point is a local maximum. For the point $x_1 = 5, x_2 = 1, \lambda = 3$, we get $|B| = -18 < 0$. This time, the sign is the same as $(-1)^1$, so B is positive definite, and this point must be a local minimum.

7. As there are two stationary points but one is a local minimum and one is a local maximum, both must be global extrema, and the function given this constraint achieves both a minimum and a maximum.

16.2.3 Why Should I Care?

Constrained optimization comes up frequently in game theoretic contexts. One rarely is presented with a scenario in which an actor has completely free reign to choose from all possible options. Instead, constraints limit actors to, for example, purchase only that which can fit within a budget, or spend only non-negative amounts of money on a campaign. Most such constraints are inequality constraints, however; one can usually choose not to use one's entire budget, and not spending negative money still leaves a lot of room to maneuver. One situation in which the budget is always used in its entirety, though, is when an individual's preferences are strictly increasing in all options that she might purchase with her budget. In this case the budget will always be spent in full, and the budget constraint becomes an equality constraint.

As we provided a couple of examples of such an equality constraint above, we do not add another one here. Instead, to further illustrate the usefulness of equality constraints we set up a more complex model and get as far as we can before requiring inequality constraints. We use a model we referenced in the previous chapter when discussing multidimensional utility functions, the legislative bargaining model of Austen-Smith and Banks (1988). In this model, each of three parties, labeled α, β, γ, cares about both government policy, labeled y, and office benefits, labeled g_i, for each relevant party. Each party's utility function is $u_i(y, g_i) = g_i - (y - p_i)^2$, where p_i is the ideal policy of party i.

We focus on the legislative bargaining part of the game, as do McCarty and Meirowitz (2007), and assume that each party has a vote share equal to w_i. The parties' strategies consist of government offers comprising both a government policy, y, and an office benefit, g_i, for each party. The total of all office benefits must be no more than G, and no individual benefit can be negative. Parties that get no office benefits from an offer are not in the governing coalition that would arise from that offer. The model assumes that the party with the highest vote share goes first, then the next highest, and then the smallest. Each subsequent party gets a chance to make an offer only if the offer from the previous party fails to garner a majority of the vote share. Let's assume that at least two parties are needed for any governing coalition, and that the status quo outcome should no agreement be reached is a poor one for all parties.

This game is known as an extensive form game of complete information, and its solution concept is subgame perfect equilibrium. You will learn about this in game theory texts (e.g., Osborne, 2004; McCarty and Meirowitz, 2007). Here we need only note that the way to solve these games is by something called *backward induction*: you start at the last actor's actions and work your way back to the first's. In equilibrium, the party making each offer ensures that at least two parties, including itself, prefers the offer to the alternative that would occur if the offer it made were rejected.

Our goal is to determine what the equilibrium offers will be from each of the parties. Determining this involves inequality constraints, and we address the full problem below, at the end of the next section. But we can solve the last stage of the game now. In the last stage, the smallest party must make a government proposal. By assumption this proposal will be acceptable to everyone since the alternative is assumed to be bad for all, so we do not have to worry about exceeding parties' utilities from the status quo. Therefore, the smallest party must choose the policy and office benefits that maximize its utility, subject to not exceeding G for total office benefits. This is a maximization problem subject to an equality constraint, as the total office benefits G will be used completely. Since utility is increasing in office benefits and decreasing in distance from a party's ideal point, the smallest party will choose its own ideal point for the government policy it proposes and will distribute office benefits by assigning all G of them to itself.

16.3 CONSTRAINED OPTIMIZATION: INEQUALITY CONSTRAINTS

If you've hung in this far, you've seen how to maximize a function of multiple variables subject to a constraint that forces a combination of the variables, given by some function g, to be equal to a constant. This limits the ability of the variables to vary, effectively reducing the number of free variables one can vary by one. But what if the constraint is not certain? We've talked about budget constraints already, and assumed they were equality constraints when in the presence of nonsatiable preferences (i.e., when the objective function is increasing in all variables). But what if preferences are satiable? In this case we might actually want to come in below the budget if we would prefer no more of some good. Or, more commonly in political science applications, what if one or more of the variables is constrained to be non-negative? For example, one typically cannot spend less than nothing running for an election, or receive less than no share of the cabinet in a proposed government.

In cases such as these, the constraints must be expressed as inequalities, and we call them, fittingly, inequality constraints. They might look like $x + y \leq 10$ or $x, y \geq 0$. This might not seem much different from equality constraints, but they pose a problem. In some cases they might *bind*; that is, without this constraint the maximum of the function might be different than it is. For instance, one might want as many goods as possible and be constrained only by the budget. This is our oft-discussed scenario, and we have seen that in this case the inequality constraint acts just like an equality constraint.

However, sometimes the inequality might not bind, and then the constraint plays absolutely no role in the outcome. For example, the optimal amount of campaign expenditures may be strictly greater than zero (this is likely the case!), and so forcing the value to be greater than zero has no effect. In fact, the problem would have been no different had we dropped that constraint entirely, and you'll often see such non-negativity constraints ignored for just this reason.

16.3.1 Method

The method used for addressing inequality constraints must cover both these cases: when the constraint binds, and when it does not. The **Kuhn-Tucker theorem**, and the associated Kuhn-Tucker conditions, allow us to do this. We describe the conditions first.

The **Kuhn-Tucker conditions** derive from a very similar setup to that used for equality constraints. They use what are called **Kuhn-Tucker multipliers**, which are more or less the same thing as Lagrange multipliers, just applied to inequality constraints rather than equality constraints. The general problem they intend to solve looks like this: maximize some function $f(\mathbf{x})$ subject to constraints of the form $g_i(\mathbf{x}) = c_i$ and $h_j(\mathbf{x}) \leq d_j$. Both c_i and d_j are constants, and we can immediately incorporate them into g_i and h_j like we did before with c_i. This time, though, we can jettison them from our notation as we don't much care about them here. So we have equality constraints $g_i(\mathbf{x}) = 0$ and inequality constraints $h_j(\mathbf{x}) \leq 0$. The less than or equal to sign on the inequality constraints is important, and one must always convert any inequality to this form. So, if you actually have the inequality $x \geq 0$, you'll want to use $h_j = -x \leq 0$.

We now form a Lagrange-type function:

$$\Lambda(\mathbf{x}, \boldsymbol{\lambda}, \boldsymbol{\mu}) = f(\mathbf{x}) - \sum_i \lambda_i g_i(\mathbf{x}) - \sum_j \mu_j h_j(\mathbf{x}),$$

where the sums are taken over the number of equality and inequality constraints. This is true for a maximum only. For a minimum, we must replace the negative sign in front of the sum over the inequality constraints h_j (the second sum in Λ) with a positive sign. That is, we add them instead of subtracting them when trying to find a minimum. All the rest of the procedure holds for both minima and maxima as long as we get the sign right in the Lagrange-type function.

The logic of the Lagrange-type function is the same as it was for equality constraints, so we won't repeat it. And the sum over the equality constraints works in exactly the same way. The difference comes with the inequality constraints. One can freely differentiate with respect to each λ_i, as we did in the method for equality constraints, because those constraints always bind, and so the result of this differentiation in the first-order condition, $g_i = 0$, is always true. However, inequality constraints don't always bind. Thus the first-order condition $h_j = 0$ would not always be true.

To deal with this we replace the first-order condition on each μ_j with a set of two conditions. One is that $\mu_j \geq 0$. This is true as long as we get the sign right in the Lagrange-type function. The second is that $\mu_j h_j(\mathbf{x}^*) = 0$ at the optimum of the function. This is known as the **complementary slackness condition** because it specifies one of two things. Either (1) the constraint does not bind and is slack, in which case $h_j < 0$, the constraint doesn't matter at the optimum, and we want to set the corresponding multiplier μ_j to zero to get rid of the useless constraint; or (2) it does bind, $h_j = 0$ at the optimum of the function, and μ_j can be any non-negative number. The complementary slackness conditions, plus the other first-order conditions on $\Lambda(\mathbf{x}, \boldsymbol{\lambda}, \boldsymbol{\mu})$, characterize the Kuhn-Tucker conditions, which are necessary for a maximum (or a minimum).

This almost accounts for the issue with the inequality constraints, save that it doesn't tell us which aspect of complementary slackness holds. To find this out, we have to try both. That is, for each complementary slackness condition, we have to try both the case where the constraint binds and the multiplier is non-negative, and the case where it does not bind and the coefficient is zero. Thus, multiple inequality constraints lead to a profusion of cases. In this book, we confine ourselves to no more than three inequality constraints and so eight $(2 \times 2 \times 2)$ cases, but keep this in mind for your own work.

Finally, before presenting the method, we offer a quick word about necessary and sufficient conditions for a maximum with inequality constraints. While one could in theory replicate the analysis of the bordered Hessian, being careful to vary the form of the Hessian between cases where the constraint binds and so should be included in B as an equality constraint, and cases where it doesn't bind and so should not be included at all, in practice it is easier to use strong conditions on the objective function (f) and the constraints. If the objective function, f, is concave, and the constraints, f_i and h_j, are quasi-convex (or convex, which is stronger), then the Kuhn-Tucker conditions are sufficient for a maximum. Sufficiency for a minimum, using the modified version of Λ, requires a convex f and quasi-convex (or convex, which is stronger) constraints. The necessary and sufficient conditions for a maximum (or minimum) make up the Kuhn-Tucker theorem.

We now present the method used to find the maximum. To find the solution to the problem of maximizing $f(\mathbf{x})$ subject to one or more constraints of the form $g_i(\mathbf{x}) = c_i$ and $h_j(\mathbf{x}) \leq d_j$, one follows these steps:[7]

1. Transform all constraints to the form $g_i(\mathbf{x}) = 0$ and $h_j(\mathbf{x}) \leq 0$.

2. Form the Lagrangian-type function: $\Lambda(\mathbf{x}, \boldsymbol{\lambda}, \boldsymbol{\mu}) = f(\mathbf{x}) - \sum_i \lambda_i g_i(\mathbf{x}) - \sum_j \mu_j h_j(\mathbf{x})$.

[7]For a minimum, change the sign in front of the μ_j in step 2 but otherwise proceed as follows.

3. Take the partial derivatives of $\Lambda(\mathbf{x}, \boldsymbol{\lambda}, \boldsymbol{\mu})$ with respect to each component of $\mathbf{x}$ and $\boldsymbol{\lambda}$.

4. Set each of the partial derivatives taken in the previous step equal to zero. Write down the complementary slackness conditions for each inequality constraint: $\mu_j h_j(\mathbf{x}) = 0$ for all j.

5. Break up each complementary slackness condition into two cases, one in which the constraint binds and one in which it does not. For every possible combination of cases (two for one constraint, four for two constraints, eight for three constraints, etc.), solve the series of equations from the previous step for all $\mathbf{x}^*$, $\boldsymbol{\lambda}^*$, and $\boldsymbol{\mu}^*$. Use $\mu_j = 0$ for all cases in which $h_j(\mathbf{x}^*) < 0$ and $\mu_j \geq 0$ for all cases in which $h_j(\mathbf{x}^*) = 0$. These are stationary points of the Lagrange-type function. Note that whenever a constraint doesn't bind (so $\mu_j = 0$), one must check to ensure that the constraint $h_j(\mathbf{x}^*) < 0$ is satisfied as, unlike equality or binding inequality constraints, this is not necessarily the case. If it is not satisfied at the candidate point, then that point is not a stationary point owing to a failure to satisfy the constraints.

6. Compare the local maxima. The largest local maximum is the global maximum (assuming the stationary points do describe a maximum; see the discussion regarding sufficiency above).

16.3.2 Examples

Having said pretty much all we're going to say about the method, we proceed to a couple of examples. The first uses only one inequality constraint. The second uses more than one constraint. First we maximize xy subject to the constraint that $x^2 + y^2 \leq 1$.[8]

1. We write $h(x, y) = x^2 + y^2 - 1 \leq 0$.

2. $\Lambda(x, y, \mu) = f(x, y) - \mu h(x, y) = xy - \mu(x^2 + y^2 - 1)$.

3. $\partial_x \Lambda = y - 2\mu x$, $\partial_y \Lambda = x - 2\mu y$.

4. There are three equations and three unknowns:

$$\begin{aligned} y - 2\mu x &= 0, \\ x - 2\mu y &= 0, \\ \mu(x^2 + y^2 - 1) &= 0. \end{aligned}$$

5. We have two cases, one in which $\mu = 0$ and $x^2+y^2-1 < 0$ and one in which $\mu \geq 0$ and $x^2 + y^2 - 1 = 0$. Let's try the easier first one first. If $\mu = 0$,

[8] Since the objective function is increasing in both variables, we know that the constraint will bind. However, we treat it as an inequality constraint for practice.

then the first equation implies $y = 0$, and the second equation implies $x = 0$. While the inequality constraint is satisfied (since $0 + 0 < 1$), this seems unlikely to be a maximum, but let's keep it around anyway for now. In the second case, we can solve by substitution. The third equation (the constraint) binds in this case, so $x^2 + y^2 - 1 = 0$ and $\mu \geq 0$. Since the first equation suggests that if $x = 0$, $y = 0$, and we've already found that solution, we can divide the first equation by x to give $2\mu = \frac{y}{x}$. Similarly, the second equation gives $2\mu = \frac{x}{y}$. Equating these implies $\frac{x}{y} = \frac{y}{x}$ or $x^2 = y^2$. The third equation, given this, implies $2x^2 = 1 = 2y^2$. Since $\mu \geq 0$, equation one implies x and y must be either both zero or the same sign. We've ruled out the first, so $2x^2 = 1 = 2y^2$ implies that either $x = \frac{1}{\sqrt{2}}, y = \frac{1}{\sqrt{2}}$, or $x = \frac{-1}{\sqrt{2}}, y = \frac{-1}{\sqrt{2}}$. In either case, equations one and two both imply that $\mu = \frac{1}{2}$.

So we have three candidate points: $x = 0, y = 0, \mu = 0$, $x = \frac{1}{\sqrt{2}}, y = \frac{1}{\sqrt{2}}, \mu = \frac{1}{2}$, and $x = \frac{-1}{\sqrt{2}}, y = \frac{-1}{\sqrt{2}}, \mu = \frac{1}{2}$.

6. We can plug in all three solutions. The first produces $f = 0$. The second and third both produce $f = \frac{1}{2}$. Thus the maxima are $x = \frac{1}{\sqrt{2}}, y = \frac{1}{\sqrt{2}}, \mu = \frac{1}{2}$ and $x = \frac{-1}{\sqrt{2}}, y = \frac{-1}{\sqrt{2}}, \mu = \frac{1}{2}$.

We note in this example that the math bore out our guess that a function increasing in both variables would have a binding inequality constraint. We also note that, though we didn't check for sufficiency of a maximum, the maximum value of the function $f = xy$ occurs when $x = y$, and we found this to be the case, with values of each as high as possible given the constraint. So, even without checking, we can feel safe that this is a maximum. If we want further confirmation, note that if we were to assume the constraint bound and formed the bordered Hessian, B, as in the previous subsection, with one constraint and two variables, we'd need only take the determinant of B to check. You can verify for yourself that this determinant is $-2x(-2y) + 2y(2x) = 8xy > 0$ for both of the points we discovered, so we have a maximum according to the rule given in the previous section.

For our second example, we tackle a more complex problem with a budget constraint that may or may not bind, and two non-negativity constraints. Let's maximize $-(x-4)^2 - (y-2)^2$ subject to the budget constraint $x + 2y \leq 4$ and the non-negativity constraints $x \geq 0, y \geq 0$.

1. We write $h_1(x, y,) = x + 2y - 4 \leq 0$ and $h_2(x, y) = -x \leq 0, h_3(x, y) = -y \leq 0$.

2. $\Lambda(x, y, \mu_1, \mu_2, \mu_3) = f(x, y) - \mu_1 h_1(x, y) - \mu_2 h_2(x, y) - \mu_3 h_3(x, y) = -(x-4)^2 - (y-2)^2 - \mu_1(x + 2y - 4) - \mu_2(-x) - \mu_3(-y)$.

3. $\partial_x \Lambda = -2(x-4) - \mu_1 + \mu_2$, $\partial_y \Lambda = -2(y-2) - 2\mu_1 + \mu_3$.

4. There are five equations and five unknowns:

$$\begin{aligned}
-2(x-4)-\mu_1+\mu_2&=0,\\
-2(y-2)-2\mu_1+\mu_3&=0,\\
\mu_1(x+2y-4)&=0,\\
\mu_2 x&=0,\\
\mu_3 y&=0.
\end{aligned}$$

5. We have eight cases! To keep things straight, let's number them. The cases arise by looking at when each of the first, second, and third inequality constraints bind or do not bind. Following convention, we label the cases by the values of the Kuhn-Tucker multipliers.

   a) $\mu_1=\mu_2=\mu_3=0$, so no constraint binds. This leaves us with two equations and two unknowns, given by the first two equations. The first equation results in $x=4$, and the second results in $y=2$. This satisfies the non-negativity constraints but violates the assumption that the budget constraint doesn't bind: $4+2(2)-4=4>0$. So this is not a stationary point.

   b) $\mu_1=\mu_2=0, \mu_3\geq 0$, so only the non-negativity constraint on y binds. In this case $y=0$ (because the constraint on y binds). Equation one tells us that in this case, $x=4$, and equation two tells us that $\mu_3=-4$. But this contradicts our assumption that $\mu_3\geq 0$, so this is not a stationary point.

   c) $\mu_1=\mu_3=0, \mu_2\geq 0$, so only the non-negativity constraint on x binds. In this case $x=0$ (because the constraint on x binds). Equation two tells us that in this case, $y=2$, and equation one tells us that $\mu_2=-8$. But this contradicts our assumption that $\mu_2\geq 0$, so this is not a stationary point.

   d) $\mu_1=0, \mu_2, \mu_3\geq 0$, so both the non-negativity constraints bind. In this case $y=0$ and $x=0$ (because the constraints on both bind). Equations one and two imply that $\mu_2=-8$ and $\mu_3=-4$, both of which contradict our assumptions, so this is not a stationary point.

   e) $\mu_1\geq 0, \mu_2=\mu_3=0$, so only the budget constraint binds. We now have three equations with three unknowns:

$$\begin{aligned}
-2(x-4)-\mu_1&=0,\\
-2(y-2)-2\mu_1&=0,\\
(x+2y-4)&=0,
\end{aligned}$$

   which we solve using substitution since only two variables appear in any equation. Equation three implies $x=4-2y$. Plugging this into equation one implies $-2(-2y)=\mu_1$ or $\mu_1=4y$. Plugging this into equation two yields $-2y + 4 - 8y = 0$ or $y = \frac{2}{5}$. Plugging this into equation three produces $x = \frac{16}{5}$, and plugging it into equation two produces $\mu_1 = \frac{8}{5}$. Since both x and y are positive, the non-negativity constraints don't bind, and this is a stationary point.

f) $\mu_1, \mu_2 \geq 0, \mu_3 = 0$, so the budget constraint and the non-negativity constraint on x bind. Thus $x = 0$, and equation three implies that $y = 2$. Equation two implies $\mu_1 = 0$, which is fine, but equation one then implies $0 = \mu_1 = \mu_2 + 8$, or $\mu_2 = -8$, which contradicts our assumption of a non-negative Kuhn-Tucker multiplier. So this is not a stationary point.

g) $\mu_1, \mu_3 \geq 0, \mu_2 = 0$, so the budget constraint and the non-negativity constraint on y bind. Thus $y = 0$, and equation three implies that $x = 4$. Equation one implies $\mu_1 = 0$, which is fine, but equation two then implies $0 = 2\mu_1 = \mu_3 + 4$, or $\mu_3 = -4$, which contradicts our assumption of a non-negative Kuhn-Tucker multiplier. So this is not a stationary point.

h) $\mu_1, \mu_2, \mu_3 \geq 0$, so all constraints bind. This implies $x = y = 0$, but then the budget constraint can't bind, so we have a contradiction and this is not a stationary point.

6. Despite eight cases, we have only one stationary point: $x = \frac{16}{5}, y = \frac{2}{5}, \mu_1 = \frac{8}{5}, \mu_2 = 0, \mu_3 = 0$. So this is our maximum.

Again, though we didn't check to make sure our stationary point is a maximum, we could have. At this point neither non-negativity constraint binds, but the budget constraint does, so we can find the determinant of the bordered Hessian. You can check that this is $-1(-2) + 2(4) = 10 > 0$, so it is indeed a maximum by the rules given in the previous section. We don't really need to do this here, though, as the objective function (f) is concave and the constraint (h_1) is quasi-convex, and therefore by the Kuhn-Tucker theorem the necessary conditions for a maximum we found are also sufficient.

16.3.3 Why Should I Care?

As we noted at the end of the previous section, inequality constraints are common in game theory. Let's look at one example illustrating why this is so, finishing the analysis we started at the end of the previous section. We left off there having determined that the smallest party would propose a government policy equal to its ideal point, and a distribution of office benefits that assigned it all of them. To go further, we have to assume something about the identities of the parties. For the sake of this discussion, we assume that the party with the biggest vote share has the rightmost ideal point, the second biggest party has the middle ideal point, and the smallest party the leftmost ideal point.

We also have to introduce another inequality constraint. To get another party to agree to be in a coalition, the offer made must provide a utility for that party that is no less than that party could get with the existing status quo proposal. In other words, all proposers operate under an additional inequality constraint: they must provide utility for one other party at least as great as that party's status quo utility.

Under our assumptions on the order of the parties, this constraint is not difficult to satisfy for the second biggest party. Since its ideal policy is assumed closer to the biggest party's than was the smallest party's, the following proposal will be acceptable to the biggest party: government policy at the ideal point of the middle party and the middle party receives all office benefits. This proposal is strictly better for the biggest party than the status quo of the smallest party's proposal. In other words, when comparing the second biggest party's proposal to the smallest party's proposal, the biggest party's utility for the former will be greater always because the second biggest party's ideal point is always closer.

This brings us to the first stage, which is more complex. Because the middle party's proposal yields it the maximum possible utility achievable in the model, there is nothing that the biggest party could give it that would both garner its support and also be preferable for the biggest party. In other words, the inequality condition corresponding to a coalition with the middle party is impossible to satisfy while also improving the biggest party's lot. Thus, the biggest party will seek a coalition with the smallest party. This is a nonconnected coalition.

To figure out what policy and distribution of office benefits it proposes in equilibrium requires maximizing its utility subject to three inequality constraints: (1) each office benefit offered must be non-negative, (2) the total office benefits offered cannot exceed G, and (3) the utility the smallest party gets in the offer cannot be less than the utility it would get with the middle party's ideal point as government policy and no office benefits. We don't solve this here, beyond noting that the outcome depends mainly on two things: the relative distances between ideal points and the size of the total pool of office benefits, G. The bigger the pool, the easier it is to optimally provide office benefits to satisfy the smallest party. If the pool is too small to provide optimal benefits, then the smallest party gets either all the office benefits or none of them in equilibrium. Which occurs depends on which party would have been hurt more by enacting the middle party's ideal point, which depends on the relative distances between ideal points. Bargaining leverage goes to the party that has less to lose from the status quo.

16.4 EXERCISES

1. Find all critical points and determine whether they are minima, maxima, or neither for the function $f = x^2 + 6xy + y^2 - 18x - 22y + 5$.

2. Find all critical points and determine whether they are minima, maxima, or neither for the function $f = -2x^2 + 2xy - \frac{5}{2}y^2 + 30x + 5y - 10$.

3. Maximize $f = xy - x$ subject to the constraint $x + y = 12$.

4. Maximize $f = x^2y$ subject to the constraint $x^2 + y^2 = 3$.

5. Christine is a first-year senator who has a lot of things she wants to accomplish, but little time to do so. Her two priorities are cutting spending and cutting taxes. She can sponsor legislation to address each one, but has limited time to do this. Assume that she gets equal utility from any bill in each priority, that the number of bills regarding spending cuts is x and the number of bills regarding tax cuts is y, and that her utility is $f = xy + y^2 + 2x$. If she can sponsor only up to four bills total, so that $x + y \leq 4$, and of course cannot "un"-sponsor a bill (that is, x and y must be non-negative), then what is the optimal distribution of bills she should sponsor? (This optimum may not be integer-valued.)

6. Solve the following constrained optimization problems:

   a) Maximize $u(x, y) = xy^2 + x^2$ subject to the budget constraint that $x + y \leq 4$. Assume that both x and y must be non-negative.

   b) Maximize $u(x, y) = x - 3y^2$ subject to the constraints $y \geq 0, x \geq 0, x^2 + y^2 = 2$.

Chapter Seventeen

Comparative Statics and Implicit Differentiation

So far in this part of the book we have learned how to compute the rate of change of functions of multiple variables and use the tools we developed to do this to compute the maxima, and sometimes the minima, of functions of more than one variable. Sometimes this is all we want to do. In statistics, we may be interested in the vector of coefficients that maximize a likelihood function or minimize squared error. In decision or game theory, we may want to know what the optimal actions are for one or more individuals, given the others' actions and any constraints. In such cases we can stop at finding the extrema of functions.

However, in other situations this is not enough. We see this most often in game theory. Though the techniques you will learn about in game theory involve finding an equilibrium of the game—that is, the set of actions which, if chosen, produce no incentive for anyone to change their actions—game theorists don't often believe the point prediction an equilibrium represents. The model contains too many assumptions for us to believe that a particular set of actions will always obtain. While there are ways to weaken these assumptions, the more common approach is to focus not on the equilibrium but rather on what are called the **comparative statics** of the model. Comparative statics detail the way in which the (static) equilibrium varies with changes in the parameters of the model. So, for example, it is likely that, all else equal, the more costly a minute of advertising is, the fewer minutes a candidate for office will buy. These comparative statics not only give you more robust conclusions from the model, they also provide hypotheses that one can test empirically.

The good news is that taking comparative statics is often fairly straightforward. All they are is a measure of the rate of change of the equilibrium actions with a change in some parameters, and you've already worked extensively with a tool for doing this, the derivative. In a model with only one variable and one parameter, a comparative static is the derivative of the equilibrium value of that variable with respect to the only parameter. In more than one dimension we may need to invoke the total derivative, but the procedure is more or less the same, and known to you.

The only complication arises when one can't actually solve explicitly for an equilibrium action. Usually this happens because the first-order conditions (FOCs) produce nonlinear equations that do not yield what is called a closed-form solution, i.e., one you can write as $x^* = f(\boldsymbol{\alpha})$, where $\boldsymbol{\alpha}$ is a vector of parameters or other variables not including x. Instead you'll get an equation of the form $f(x^*; \boldsymbol{\alpha}) = 0$, which is known as an *implicit function* since it implicitly defines the optimal variable x^*. We discussed implicit functions all the way back in Chapter 3. In this chapter we learn how to differentiate these functions by a process known as **implicit differentiation**. Doing so permits us to derive the comparative statics.

Though we have framed this discussion in terms of game theory, and that is indeed where you will most often see comparative statics derived, the ideas behind comparative statics are more general. For example, in more advanced statistics you may derive an estimator that is a function of some parameters. Looking at how the estimator and its properties vary with these parameters entails taking the same sorts of derivatives associated with comparative statics. The same is true for models of bounded rationality and/or dynamical models specified by Markov chains: in each case there are most likely parameters affecting the steady-state behavior in the system, and we can use the same techniques to discover how the steady state changes with changes in the parameters.

This comparatively brief, final chapter of the book has two sections. In the first we discuss the representation of the extremum of a function (e.g., x^*) as a function in its own right (e.g., $x^*(\boldsymbol{\alpha})$) and introduce the envelope theorem, which helps us understand how an objective function varies in its parameters. In the second section we discuss the difference between explicit and implicit differentiation, present the implicit function theorem, and give a few examples of its use.

17.1 PROPERTIES OF THE MAXIMUM AND MINIMUM

In Chapters 8 and 16 we found the extrema of functions of one and more than one variable, respectively, both with and without constraints on the values the variables might take. But, while we listed the values of the variables at these extrema, we did not talk much about them. Since these values—the equilibria or steady states of theoretical models, or the estimators of statistical models—are exactly what we are most interested in as social scientists, though, they bear more discussion.

Let's start with some notation, and an example. Previously we wrote extrema of functions as x^* or $\mathbf{x}^*$, depending on the number of variables. And when we solved for these, we generally got real-valued solutions. For example, $x^* = 10$, or $\mathbf{x}^* = \begin{pmatrix} 5 \\ 4 \\ \sqrt{17} \end{pmatrix}$. There's not much more to say about these solutions of an optimization problem other than their values, since they don't depend on any other parameter. So we were well justified in leaving them alone then.

The reason for their lack of dependence on parameters is that, in prior examples, there *were* almost no parameters in use. But there could have been. Take, for instance, the budget parameter c_i we discussed in the previous chapter. We talked a bit about how this related to the corresponding Lagrange multiplier, which served as a shadow price for changing the budget, and how changing the size of the budget would affect total utility, but we did not talk at all about how it would affect the optimum *choice* of goods, which is arguably what we care most about. Instead we assigned a real number to the size of the budget, c_i, and went from there.

But it does not have to be this way, and often is not. Let's again go through the first example we gave to illustrate equality constraints in the previous chapter, except this time with a general budget constraint c rather than the specific 16 we used there. This problem works out to maximizing $f(x_1, x_2) = x_1x_2$ subject to the constraint that $x_1 + 4x_2 = c$. We use the same steps as we did in solving the earlier problem.

1. We have one constraint, which we'll call $g(x_1, x_2) = x_1 + 4x_2 - c = 0$.

2. The Lagrangian function is $\Lambda(x_1, x_2, \lambda) = f(x_1, x_2) - \lambda g(x_1, x_2) = x_1x_2 - \lambda(x_1 + 4x_2 - c)$.

3. To find $\nabla\Lambda(\mathbf{x}, \lambda)$ we need $\partial_{x_1}\Lambda$, $\partial_{x_2}\Lambda$, and $\partial_\lambda\Lambda$. $\partial_{x_1}\Lambda = x_2 - \lambda$, $\partial_{x_2}\Lambda = x_1 - 4\lambda$, and $\partial_\lambda\Lambda = -(x_1 + 4x_2 - c)$.

4. We set $\nabla\Lambda(\mathbf{x}^*, \boldsymbol{\lambda}^*) = \mathbf{0}$ and solve for all $\mathbf{x}^*$ *and* all $\boldsymbol{\lambda}^*$ to find the stationary points of the Lagrange function. There are three equations and three unknowns:

$$x_2 - \lambda = 0,$$
$$x_1 - 4\lambda = 0,$$
$$x_1 + 4x_2 - c = 0.$$

Because only two variables appear in any equation, it's easier to solve them by substitution than by using a more complex method. The first equation gives $x_2 = \lambda$. The second equation gives $x_1 = 4\lambda$. Plugging these both into the third equation gives $4\lambda + 4(\lambda) - c = 0$ or $\lambda = \frac{c}{8}$. Plugging this back into the other two equations gives $x_1 = \frac{c}{2}$ and $x_2 = \frac{c}{8}$.

We stop here since the sufficiency analysis is completely unchanged by the introduction of c.

Going through this example illustrates the difference between the maximization problem with and without a parameter. Without a parameter, the function's maximum, subject to the budget constraint, lies at $x_1 = 8$ and $x_2 = 2$, as we found in the example in the previous chapter, which used $c = 16$. With the parameter, it lies at $x_1 = \frac{c}{2}$ and $x_2 = \frac{c}{8}$. We'll call the point at which the function is maximized the optimal point, or **optimum**.

What does this difference mean? It means that the optimum is not fixed when there is a parameter; instead it varies according to the value of the parameter in a precisely specified way. In other words, the optimum is equal to a *function* of the parameter. We can write this function as $\mathbf{x}^* = \mathbf{k}(c)$, where $\mathbf{k}(c) = \begin{pmatrix} \frac{c}{2} \\ \frac{c}{8} \end{pmatrix}$.

Or we can cut down on the notation and write $\mathbf{x}^*(c) = \begin{pmatrix} \frac{c}{2} \\ \frac{c}{8} \end{pmatrix}$. This second way of writing it highlights that the optimum is not a particular value that $\mathbf{x}$ takes on but rather a function of the parameter c.

As a function, we can do everything to $\mathbf{x}^*(c)$ that we could do to any other function. The thing we most want to do to it, though, is take its derivative. The derivative tells us the rate of change of the function with the parameter; for this example, that means differentiating $\mathbf{x}^*(c)$ tells us how fast the optimal amount of each good purchased changes with the size of the budget, given the prices of those goods. And this is exactly the sort of thing we want to know as social scientists, since it tells us how one observable variable of interest, the amount of goods purchased, varies with another one, the size of the budget. Further, unlike a point prediction, which can be difficult to test empirically, since it is unlikely that the real world *exactly* matches the point prediction, we can use conventional statistical methods to test the directional prediction that, say, the amount purchased of each good increases in the size of the budget.[1]

The optimum, $\mathbf{x}^*(c)$, is thus useful to compute. It generalizes to more than one parameter as well: we can write $\mathbf{x}^*(\mathbf{c})$, where $\mathbf{c}$ is a vector of parameters. In the next section we'll talk more about how to find it in situations more complex than the example we provide here. For now we'll take a moment to discuss the manner in which parameters affect the objective function one is maximizing.

Let's start by considering a slightly simpler case than our example, one in which there is one variable, x, and one parameter, c, and no constraints at all. The function we want to maximize is $f(x;c)$, and the FOC is $\frac{df}{dx} = 0$. Assuming a maximum exists, it must satisfy this FOC. Let's call the maximum $x^*(c)$. Then $\frac{df}{dx}|_{x=x^*(c)} = 0$.

This was done by treating the parameter c as a constant. What if we were to treat it as a variable? After all, we can rewrite $f(x^*(c);c)$, the value of the function at its optimum, as $f^*(c)$. This amounts to a minor change of variables, basically replacing the optimal $x^*(c)$ with the function of c which it equals. Treating $f^*(c)$ as a function of c allows us to analyze how the function at its optimum changes with c. We can use the chain rule to do this, consonant with the idea of this as a change in variables. Doing so gives us $\frac{df(x,c)}{dc}|_{x=x^*(c)} = \frac{\partial f(x,c)}{\partial x}|_{x=x^*(c)} \frac{dx^*(c)}{dc} + \frac{\partial f(x,c)}{\partial c}|_{x=x^*(c)}$. But, when treating both x and c as variables, the FOC for x is $\frac{\partial f(x,c)}{\partial x}|_{x=x^*(c)} = 0$. Thus, the first term in the RHS of the chain rule is zero when evaluated at the optimum. This implies $\frac{df^*(c)}{dc} = \frac{\partial f(x,c)}{\partial c}|_{x=x^*(c)}$, which is known as an **envelope theorem**.

The utility of the envelope theorem is that it allows us to understand how the function we're maximizing, which often represents an individual's utility, is affected by variation in a parameter of the function at the point of the function's maximum. We saw this in the previous chapter when we looked at the rate at which the size of the budget constraint altered utility and the relation of this rate to the Lagrange multiplier for the budget constraint. That was a special case

[1] We could set up a linear regression model, e.g., $x_1 = \beta_0 + \beta_1 c$.

of the envelope theorem. The envelope theorem is helpful because it translates a total derivative taken at a point into a partial derivative taken at that same point, which can eliminate the need to solve for the optimum $x^*(c)$ at all.

For example, let's say $f(x;c) = -cx^6 - cx^2 + 5x - c$. The envelope theorem tells us that $\frac{df^*(c)}{dc} = \frac{\partial f(x,c)}{\partial c}|_{x=x^*(c)} = -(x^*)^6 - (x^*)^2 - 1 < 0$ regardless of what $x^*(c)$ is, because each x^* is raised to an even power, and so positive, and all three terms are preceded by negative signs. Since $\frac{df^*(c)}{dc} < 0$, we know that at the maximum of this function it is decreasing in c; we got this result from the envelope theorem without even calculating that optimum at which the maximum occurs.[2]

Of course, sometimes we do need to compute the optimum and, more often, the optimum is of more interest to us than the maximum of the objective function. For example, knowing how the utility changes in the parameters can be useful, but we're usually more interested in what choices maximize that utility, and how these choices change with the parameters. This is the topic of the next section.

Before moving on, though, it is worth noting that the envelope theorem generalizes to the case of vectors of variables and parameters, as well as equality constraints. If one forms the Lagrange function with equality constraints in the manner described in the previous chapter, then the envelope theorem states that

$$\frac{df^*(\mathbf{c})}{dc_i} = \frac{\partial \Lambda(\mathbf{x}, \boldsymbol{\lambda}; \mathbf{c})}{\partial c_i}|_{\mathbf{x}=\mathbf{x}^*(\mathbf{c}), \boldsymbol{\lambda}^*(\mathbf{c})}$$

for each component c_i of $\mathbf{c}$ and where $\boldsymbol{\lambda}^*(\mathbf{c})$ are the optimal Lagrange multipliers as functions of the parameters.

17.1.1 Why Should I Care?

Treating extrema of functions as functions to be analyzed is a necessary first step for computing comparative statics, which are the payoff of the next section. The envelope theorem, among other things, helps us understand how one's utility changes with its parameters. This can be useful when an actor's utility is itself an object of interest. For example, Bueno de Mesquita (2005) uses the envelope theorem to show that a terrorist organization's equilibrium payoff is increasing in the quality of its operatives, implying that terrorist groups will always choose to use the best operatives available. This point is central to his argument that terror groups' operatives may be relatively well educated and high income compared to the population at large, even as poverty and a systemic lack of opportunity might lead to more people overall joining terrorist groups.

[2] For additional examples, see the Khan Academy video ("Applying the Envelope Theorem" `http://www.youtube.com/watch?v=fw_8V_cIlnk`).

17.2 IMPLICIT DIFFERENTIATION

Having explored a bit the properties of the objective function at its maximum as a function of its parameters, let's return to looking at its optimum, the point at which it is maximized. We've called this in general $\mathbf{x}^*(\mathbf{c})$, and the question we're going to be asking in this section is, how does it change in its parameters?

We have already discussed why this is important for the social sciences in general, so we get right down to trying to answer the question. Some cases have answers that are much easier to find than others. For example, if we can express $\mathbf{x}^*(\mathbf{c})$ in terms of the parameters only, then we can differentiate. For example, in one dimension, if $x^*(c) = 3c^2 - c + 1$, then $\frac{dx^*(c)}{dc} = 6c - 1$. This tells us that as the parameter c grows larger, so too does the rate at which the optimum increases in c. Of more interest typically is the *sign* of this derivative. This derivative, $6c - 1$, is positive for $c > 0$, and negative for $c < 0$. When it is positive, it means the optimum is increasing in the parameter, and when it is negative, the optimum is decreasing in the parameter. This information is what one usually tests empirically: if the sign of the derivative is positive (negative) we would expect the coefficient on the independent variable corresponding to the parameter c to be statistically significantly different from zero in the positive (negative) direction. Because of this empirical connection, we often derive comparative statics with the sole intent to determine their signs.

One can extend this argument to the optima of functions of more than one variable and more than one parameter. Specifically, one can find the derivatives $\frac{\partial x_i^*(\mathbf{c})}{\partial c_j}$ for each component of the optimum and each parameter. This would correspond empirically to a regression of the dependent variable x_i against the parameters that make up the vector $\mathbf{c}$, looking at the coefficient on a particular c_j from this vector. Note that we are using partial derivatives here under the assumption that we have solved for all the optimal values of the variables x_i in terms of the parameters, and that no parameter depends on any other parameter. In other words, we can't have some c_j be a function of a different c_k. If this were not the case, and they were dependent on each other, then we'd have to use a total derivative, since we are interested in both direct and, should they exist, indirect effects of the parameter on the variable x_i^*.

Such differentiation is known as **explicit differentiation**, since we are explicitly differentiating the function optimum $\mathbf{x}^*(\mathbf{c})$ with respect to its parameters. When it is possible to do this, which occurs when one can write $\mathbf{x}^*(\mathbf{c}) = \mathbf{k}(\mathbf{c})$ with $\mathbf{k}(\mathbf{c})$ a function only of the parameters, and not of $\mathbf{x}$, then finding the comparative statics of the optimum (or equilibrium) value $\mathbf{x}^*(\mathbf{c})$ requires nothing more than taking the total derivative of $\mathbf{x}^*(\mathbf{c})$ with respect to each of the parameters, or the partial derivative if no parameter depends on any other parameter.

However, there are several complications that make this not always tenable, and dealing with these complications will occupy us for the rest of this section. One complication is that the first-order conditions often produce systems of equations, as in our example above:

$$x_2 - \lambda = 0,$$
$$x_1 - 4\lambda = 0,$$
$$x_1 + 4x_2 - c = 0.$$

Take the third equation of this system, for example. While it does specify the optimum x_1^* in terms of the parameter c, it also specifies it in terms of the optimal x_2^*. But x_2^* also depends on c itself, implying that differentiating both sides of the equation $x_1^*(c) = c - 4x_2^*(c)$ would require knowing the form of $x_2^*(c)$, and this requires solving the entire system of equations. Thus, differentiating each FOC is usually not enough; we need as well to solve explicitly for the optimum of each variable, and this explicit solution is what we differentiate. This comes up a great deal in game theory, in which researchers nearly always solve for the optimal strategies of two or more players, which requires first finding the optimal strategy of each as a function of the other's strategy, and then solving the system of equations this produces. We saw an example of this at the end of Chapter 13.

This leads us to the second and more involved complication, which is that sometimes we can't get an explicit solution for the optimum. Sometimes this occurs because the FOCs produce equations that do not yield closed-form solutions of the type $\mathbf{x}^*(\mathbf{c}) = \mathbf{k}(\mathbf{c})$. For example, if the FOC were to produce $2x - e^{\frac{cx}{10}} = 0$,[3] then we couldn't solve analytically for $x^*(c)$, even though a solution exists over a range of c.[4]

At other times we can't solve for a particular optimum because we are using general functional forms in some theoretical model. For example, define $b(x, c)$ as the benefit for taking action x, and $d(x, c)$ as the cost for taking that action, so one's utility is $b(x, c) - d(x, c)$. The FOC is $\frac{\partial(b(x^*,c)-d(x^*,c))}{\partial x} = 0$, and we can't produce a closed-form solution for $x^*(c)$.

Finally, there are times when it is theoretically possible to solve for a optimum, but because it would be very difficult to do so and we really care only about the sign of the derivative of the optimum with respect to some parameter and not the optimum itself, we prefer not to produce a closed-form solution for $x^*(c)$.

In all three of these cases, we say that the equations in which $\mathbf{x}^*$ can be found *implicitly* define the value of $\mathbf{x}^*$. Typically these equations are the FOCs on some function $f(\mathbf{x})$: $\nabla f(\mathbf{x}^*) = \mathbf{0}$. What this means is that, while mathematically we have determined what the optimum is, we cannot or choose not to phrase it explicitly in the form $\mathbf{x}^*(\mathbf{c}) = \mathbf{k}(\mathbf{c})$.

Despite our inability (or unwillingness) to express an optimum explicitly, this does not mean that we cannot discern how it varies with its parameters. Working

---

[3] This would arise from the function $x^2 - \frac{10}{c}e^{\frac{cx}{10}}$, which might come from a scenario in which there were increasing benefits to some action, but also horribly exploding costs to taking that action that rapidly outweighed these benefits.

[4] We offer a similar problem below, in the exercises.

out this variation is the province of **implicit differentiation**, which is nothing more than the differentiation of functions that are only defined implicitly.

This may sound complex, but you are already somewhat familiar with the ideas, as they bear a strong similarity to the chain rule. Consider the function $g(x^*(c); c) = 4c^2 + 6(x^*(c))^3 - 10$. We set this equal to zero, under the assumptions that it represents a FOC for some other function and we're evaluating g at its optimum. Now, we could solve directly for $x^*(c)$, but let's treat it as implicitly defined instead.

We want to know $\frac{dx^*(c)}{dc}$, so let's begin by differentiating both sides of $g(x^*(c); c) = 0$ with respect to c. The derivative of zero is zero, so this yields $\frac{dg(x^*(c);c)}{dc} = 0$. To take the derivative on the LHS we need the total derivative, as $x^*(c)$ depends on c. In other words, c will have both a direct effect on $g(x^*(c); c)$, and an indirect one via $x^*(c)$. Taking this derivative using the chain rule implies $\frac{\partial g(x^*(c);c)}{\partial c} + \frac{\partial g(x^*(c);c)}{\partial x}\frac{dx^*(c)}{dc} = 0$, or $8c + 18(x^*(c))^2\left(\frac{dx^*(c)}{dc}\right) = 0$. Note we left $\frac{dx^*(c)}{dc}$ as is in the derivative, because we simply don't know it. In fact, it's what we're trying to find, so we'll solve for it to get $\frac{dx^*(c)}{dc} = \frac{-8c}{18(x^*(c))^2}$. The denominator of this is always positive because of the squared term, so the sign of the comparative static is the reverse of the sign of c. If c is positive, the comparative static is negative, and if it is negative, the comparative static is positive.

In this case we can check our math by solving for $x^*(c)$ directly. $g(x^*(c); c) = 0$ implies $x^*(c) = \sqrt[3]{\frac{10-4c^2}{6}}$. Differentiating this with respect to c gives

$$\left(\frac{1}{3}\right)\left(\frac{-8c}{6}\right)\left(\sqrt[3]{\frac{10-4c^2}{6}}\right)^{-2}, \text{ or } \frac{dx^*(c)}{dc} = \frac{-8c}{18(x^*(c))^2},$$

once we substitute back in for $x^*(c)$. The major differences in approaches are that here we needed more algebraic steps, but we did get the value of $x^*(c)$. If we don't need $x^*(c)$, then implicit differentiation is faster in this case.

We can extend this to general functional forms easily. Recall we wrote down the equation $\frac{\partial g(x^*(c);c)}{\partial c} + \frac{\partial g(x^*(c);c)}{\partial x}\frac{dx^*(c)}{dc} = 0$. This didn't specify a functional form for $g(x^*(c); c)$, so we leave the derivative of the FOC $g(x^*(c); c) = 0$ in this general form and rearrange our equation to get

$$\frac{dx^*(c)}{dc} = \frac{-\frac{\partial g(x^*(c);c)}{\partial c}}{\frac{\partial g(x^*(c);c)}{\partial x}}.$$

This is the most important part of the **implicit function theorem** in one dimension.[5]

[5]While this looks similar to the logic behind the envelope theorem, note that there we were differentiating the objective function, f, which (1) does not necessarily equal zero at x^*, and (2) has a derivative with respect to x of zero at its optimum. In contrast, here we're differentiating the FOC of the objective function, $g = 0$, which (1) does equal zero at x^*, and (2) does not necessarily have an optimum at x^*, and so does not need to have a derivative of zero at that point. For these reasons, the denominator of the RHS of the previous equation need not equal zero.

The theorem actually says a bit more than this. Start with a function $g(x;c)$ that's differentiable close to some point (x_0, c_0) at which $g(x_0; c_0) = 0$. As noted above, this function is most often the FOC of some other function you're trying to maximize. The implicit function theorem then says that as long as $\frac{\partial g(x_0;c_0)}{\partial x} \neq 0$, there exists some differentiable function $x^*(c)$ such that (1) $g(x^*(c); c) = 0$ for all c sufficiently close to c_0, (2) $x^*(c_0) = x_0$, and (3) the equation $\frac{dx^*(c_0)}{dc} = \frac{-\frac{\partial g(x_0;c_0)}{\partial c}}{\frac{\partial g(x_0;c_0)}{\partial x}}$ is true. Thus the theorem guarantees under certain weak conditions the existence of a function $x^*(c)$ in the neighborhood of c_0. But, more important, it provides the *derivative* of that function, without our ever having to solve for it explicitly.[6]

This makes it very useful when trying to compute comparative statics, particularly given complex functions. Let's consider another example in which one is trying to maximize one's utility function: $u(x;c) = (\ln(x))^2 - \frac{1}{2}c^3x$. The FOC is $g(x^*;c) = \frac{du(x^*;c)}{dx} = \frac{2\ln(x^*)}{x^*} - \frac{1}{2}c^3 = 0$. We can't find a closed-form solution for x because of the logarithm, but we know one exists because the second derivative is negative for $x > e$, and so the function is concave on this region (see Chapter 8). So we can use the implicit function theorem to find the derivative of $x^*(c)$. This yields $\frac{dx^*(c)}{dc} = \frac{\frac{3}{2}c^2}{\frac{2}{x^*(c)^2} - \frac{2\ln(x^*(c))}{x^*(c)^2}}$. Since the numerator is positive and the denominator is negative for $x > e$, which is the range in which a maximum must exist, $\frac{dx^*(c)}{dc} < 0$, and the optimal choice is decreasing in c.

As with the envelope theorem, one can generalize the implicit function theorem to more complex situations. First let's consider the case of only one equation, $g = 0$. If there are many parameters c_i but still only one endogenous variable x, then we need to repeat the above analysis for each parameter c_i. Here we would obtain partial derivatives $\frac{\partial x^*(\mathbf{c})}{\partial c_i}$ instead of a total derivative. The same logic holds if there is one parameter and several endogenous variables x_i: we do the same analysis for each x_i to find the partial derivatives $\frac{\partial x_i^*(c)}{\partial c}$.

Sometimes, though, we are given multiple FOCs. For example, in game theory, typically each individual is maximizing her own utility function, conditional on all parameters of the model and all others' actions. This results in a system of equations, one for each action. We saw earlier how to solve linear systems of this type, but what if the FOCs are not linear? When this is the case, often we will not be able to solve the system explicitly for the equilibrium. Nor can we treat each variable on its own as in the case of one equation, for the reasons we discussed above in the context of the first complication of finding comparative statics. Consequently we will need to employ a more general implicit function theorem that will enable us to solve this system.

With a system of equations, we have $\mathbf{g}(\mathbf{x}, \mathbf{c}) = \mathbf{0}$. Recall from Chapter 15 that the Jacobian matrix J of $\mathbf{g}$ with respect to the variables $\mathbf{x}$ has elements $j_{ik} = \frac{\partial g_i}{\partial x_k}$. If J is nonsingular, then its determinant $|J| \neq 0$. Assuming that the

[6]You may find the implicit differentiation presentation on the Kahn Academy website useful: `http://www.khanacademy.org/math/calculus/v/implicit-differentiation`.

system of equations has an equal number of equations and unknown variables, a nonsingular J means that we can invert the matrix J.

We also could take the derivatives of the components of $\mathbf{g}$ with respect to the components of the parameter vector $\mathbf{c}$, however, in exactly the same way. That is, we can also find the partial derivatives $\frac{\partial g_i}{\partial c_k}$. To distinguish these two matrices of derivatives, let's call the one with respect to $\mathbf{x}$, $J_{\mathbf{x}}$, and the one with respect to $\mathbf{c}$, $J_{\mathbf{c}}$. These two Jacobian matrices are the analogues of the partial derivatives used in the implicit function theorem we gave earlier, and, with them, we can write a more general implicit function theorem. The intent of this theorem will be to produce the derivative of the optimum function $\mathbf{x}^*(\mathbf{c})$. Since this function is also a multidimensional function of multiple parameters, its derivative too will be a Jacobian. Let's call this one $J_{\mathbf{x}^*(\mathbf{c})}$.

Leaving aside the parts about the existence of $\mathbf{x}^*(\mathbf{c})$, which are exactly comparable to the one-dimensional theorem, we have that, if $J_{\mathbf{x}}$ is nonsingular, then

$$J_{\mathbf{x}^*(\mathbf{c})} = -[J_{\mathbf{x}}]^{-1} J_{\mathbf{c}}.$$

This consequence of the general implicit function theorem allows us to take into account the dependence of all variables on other variables and all parameters in the system of FOCs when discerning how each component of $\mathbf{x}^*(\mathbf{c})$ changes with each component of $\mathbf{c}$.

We conclude with one relatively straightforward example of the application of this implicit function theorem. Let's consider a game in which two warring groups are each attempting to mobilize supporters via particularized benefits. The degree to which each is successful determines its probability of victory in the conflict, and the size of the mobilization is increasing in the total amount of benefits used. But benefits are also costly, and take away from the rents the elites of the group can secure post conflict. We can model this by assigning each group utilities $u_1 = \pi(b_1, b_2) - \frac{1}{2}c_1 b_1^2$ and $u_2 = 1 - \pi(b_1, b_2) - \frac{1}{2}c_2 b_2^2$. Here the first term in the utility, $\pi(b_1, b_2)$, is the probability of victory for group 1 and is a function of each group's total particularized benefits used, b_i, and c_i is a cost parameter that can differ for each group. There are two FOCs for the two utilities, assuming each group simultaneously decides on the level of benefits to use:

$$\frac{\partial \pi(b_1, b_2)}{\partial b_1} - c_1 b_1 = 0,$$
$$-\frac{\partial \pi(b_1, b_2)}{\partial b_2} - c_2 b_2 = 0.$$

This gives us two equations in two unknowns, but since we're using a general functional form for the probability of victory, we can't solve explicitly for optimal functions $b_i^*(\mathbf{c})$. So we use the implicit function theorem to see how these depend on the cost parameters.

To do this, we need to compute two Jacobian matrices, each of which is two-by-two, and find the inverse of one of them. Let's start with that one.

$$J_{\mathbf{b}} = \begin{bmatrix} \frac{\partial^2 \pi}{\partial b_1^2} - c_1 & \frac{\partial^2 \pi}{\partial b_1 \partial b_2} \\ -\frac{\partial^2 \pi}{\partial b_1 \partial b_2} & -\frac{\partial^2 \pi}{\partial b_2^2} - c_2 \end{bmatrix}.$$

The determinant of this is $\left(\frac{\partial^2 \pi}{\partial b_1^2} - c_1\right)\left(-\frac{\partial^2 \pi}{\partial b_2^2} - c_2\right) + \left(\frac{\partial^2 \pi}{\partial b_1 \partial b_2}\right)^2$. Whenever this does not equal zero, the matrix is invertible. Call this determinant $|J_{\mathbf{b}}|$ and assume it is non-zero in this case.

The inverse of this matrix is

$$[J_{\mathbf{b}}]^{-1} = \frac{1}{|J_{\mathbf{b}}|} \begin{bmatrix} -\frac{\partial^2 \pi}{\partial b_2^2} - c_2 & -\frac{\partial^2 \pi}{\partial b_1 \partial b_2} \\ \frac{\partial^2 \pi}{\partial b_1 \partial b_2} & \frac{\partial^2 \pi}{\partial b_1^2} - c_1 \end{bmatrix}.$$

Now we compute the second Jacobian matrix, which is much simpler:

$$J_{\mathbf{c}} = \begin{bmatrix} -b_1 & 0 \\ 0 & -b_2 \end{bmatrix}.$$

Putting these together yields

$$\begin{aligned} J_{\mathbf{x}^*(\mathbf{c})} &= -[J_{\mathbf{x}}]^{-1} J_{\mathbf{c}} \\ &= \frac{-1}{|J_{\mathbf{b}}|} \begin{bmatrix} -\frac{\partial^2 \pi}{\partial b_2^2} - c_2 & -\frac{\partial^2 \pi}{\partial b_1 \partial b_2} \\ \frac{\partial^2 \pi}{\partial b_1 \partial b_2} & \frac{\partial^2 \pi}{\partial b_1^2} - c_1 \end{bmatrix} \begin{bmatrix} -b_1 & 0 \\ 0 & -b_2 \end{bmatrix} \\ &= \frac{1}{|J_{\mathbf{b}}|} \begin{bmatrix} -b_1\left(\frac{\partial^2 \pi}{\partial b_2^2} + c_2\right) & -b_2 \frac{\partial^2 \pi}{\partial b_1 \partial b_2} \\ b_1 \frac{\partial^2 \pi}{\partial b_1 \partial b_2} & b_2\left(\frac{\partial^2 \pi}{\partial b_1^2} - c_1\right) \end{bmatrix}. \end{aligned}$$

Pulling this Jacobian matrix apart into two vectors that each specify how each variable b_i changes with one of the parameters can clarify what we're seeing:

$$\begin{bmatrix} \frac{\partial b_1}{\partial c_1} \\ \frac{\partial b_2}{\partial c_1} \end{bmatrix} = \frac{b_1}{|J_{\mathbf{b}}|} \begin{bmatrix} -\frac{\partial^2 \pi}{\partial b_2^2} - c_2 \\ \frac{\partial^2 \pi}{\partial b_1 \partial b_2} \end{bmatrix}, \begin{bmatrix} \frac{\partial b_1}{\partial c_2} \\ \frac{\partial b_2}{\partial c_2} \end{bmatrix} = \frac{b_2}{|J_{\mathbf{b}}|} \begin{bmatrix} -\frac{\partial^2 \pi}{\partial b_1 \partial b_2} \\ \frac{\partial^2 \pi}{\partial b_1^2} - c_1 \end{bmatrix}.$$

What this tells us is that changes in c_i act both directly and indirectly on b_i. The indirect effect arises as follows: changes in c_1 lead to changes in b_1, which induce changes in b_2 as the second group's equilibrium response changes, which in turn induce further changes to b_1 as the first group must respond to the second group's altered behavior. This complex causal sequence is captured via the implicit function theorem without our ever having had to compute the optimal behavior.[7]

[7]In your game theory classes you may learn conditions via which this sort of causal sequence always resolves in one way or another. These sorts of conditions involve strategic complementarity and fall under the heading of supermodularity (Ashworth and Bueno de Mesquita, 2005).

17.2.1 Why Should I Care?

Differentiation of the extrema of functions allows us to understand how these extrema vary with the parameters of the model. Since in a game-theoretic model these extrema are the equilibrium strategies, comparative statics of these extrema are often the primary object of interest in the model, and serve as the centerpiece of the analysis. Further, they provide connections between theory and empirics, as operationalizing a model's comparative statics can, if the model is designed with empirical testing in mind, produce testable hypotheses. This underlies the EITM (empirical implications of theoretical models) movement in political science (Granato and Scioli, 2004).

Because they are so central to the operation of formal modeling, comparative statics are widely computed, and reading almost any paper with a formal model in it would provide an example. Many of these make use of explicit differentiation. However, when such is not possible, implicit differentiation is an important tool for extracting comparative statics, and many formal models make use of it. For example, Shapiro and Siegel (2007) propose a principal-agent model to explore the consequences of preference divergence between terrorist bosses and terrorist middlemen. Because they can only implicitly solve for the equilibrium amount of money skimmed by terrorist middlemen, they rely on implicit differentiation to derive, among other things, the nonlinear manner in which the probability of a successful attack declines in the degree of preference divergence.

As this is the last "Why Should I Care?" section of this book, we offer one final note. The ability to model causal relationships in a formal manner is one of the main impetuses for learning mathematics, and we hope this book has helped you feel more comfortable with mathematical tools. If you've made it straight through to this point, congratulate yourself; there may be far more to learn out there, but you've absorbed a great deal of material in a short time, and should be commended for your efforts!

17.3 EXERCISES

1. Assume that Sabri's decisions about work (x) relate directly to the amount of popcorn remaining (a). After optimizing, Sabri finds that $x^*(a) = 3a^2 + a - 1$. Find the marginal change in work with respect to an increase in popcorn.

2. Assume the same scenario, but now Sabri has realized that popcorn is even more important than he thought it was, and accordingly has changed his utility function. Maximizing his new utility function yields $x^*(a) = e^{ax^*(a)}$. Find the marginal change in work with respect to increases in popcorn using both total differentiation and the implicit function theorem in one dimension.

3. Assume that $g = 3a - x - e^{ax^2} + ax$, and that $x^*(a)$ is a critical point

of g (*hint*: which means that $x^*(a)$ must solve what equation...?). Using whatever method you want, find $\frac{\partial x^*}{\partial a}$.

4. Let $\pi(b_1, b_2) = \frac{1}{1+e^{b_2-b_1}}$ in the last example used in this chapter. Evaluate the results of the implicit function theorem, and state any conclusions about the signs of the comparative statics you can draw.

Bibliography

Abbott, Andrew. 1995. "Sequence Analysis: New Methods for Old Ideas." *Annual Review of Sociology* 21: 93–113.

Abramowitz, Alan I. 1980. "A Comparison of Voting for US Senator and Representative in 1978." *American Political Science Review* 74(3): 633–640.

Almond, Gabriel A. 2004. "Who Lost the Chicago School of Political Science?" *Perspectives on Politics* 2(1): 91–93.

Amadae, S.M., and B. Bueno de Mesquita. 1999. "The Rochester School: The Origins of Positive Political Theory." *Annual Review of Political Science* 2(1): 269–295.

Ames, Barry. 1990. *Political Survival: Politicians and Public Policy in Latin America.* Berkeley: University of California Press.

Arrow, Kenneth J. 1950. "A Difficulty in the Concept of Social Welfare." *Journal of Political Economy* 58(4): 328–346.

Arthur, W. Brian. 1994. *Increasing Returns and Path Dependence in the Economy.* Ann Arbor: University of Michigan Press.

Ashworth, Scott, and Ethan Bueno de Mesquita. 2005. "Monotone Comparative Statics for Models of Politics." *American Journal of Political Science* 50(1): 214–231.

Austen-Smith, D., and J. Banks. 1988. "Elections, Coalitions, and Legislative Outcomes." *American Political Science Review* 82(2): 405–422.

Baron, David P., and John A. Ferejohn. 1989. "Bargaining in Legislatures." *American Political Science Review* 83(4): 1181–1206.

Beck, Neal, and Jonathan N. Katz. 1995. "What to Do (and Not to Do) with Time-Series Cross-Section Data." *American Political Science Review* 89(3): 634–647.

Bendor, Jonathan, Daniel Diermeier, David A. Siegel, and Michael M. Ting. 2011. *A Behavioral Theory of Elections.* Princeton: Princeton University Press.

Bennett, D. Scott. 1997. "Testing Alternative Models of Alliance Duration, 1816–1984." *American Journal of Political Science* 41(3): 846–878.

Blalock Jr., Hubert M. 1965. "Theory Building and the Statistical Concept of Interaction." *American Sociological Review* 30(3): 374–380.

Blalock Jr., Hubert M. 1969. *Theory Construction: From Verbal to Mathematical Formulations.* Englewood Cliffs: Prentice Hall.

Bluman, A.G. 2005. *Probability Demystified: A Self-Teaching Guide.* New York: McGraw-Hill Professional.

Bonacich, Phillip. 1972. "Factoring and Weighting Approaches to Status Scores and Clique Identification." *Journal of Mathematical Sociology* 2(1): 113–120.

Box-Steffensmeier, Janet M., and Christopher J.W. Zorn. 2001. "Duration Models and Proportional Hazards in Political Science." *American Journal of Political Science* pp. 972–988.

Box-Steffensmeier, Janet M., and Christopher Zorn. 2002. "Duration Models for Repeated Events." *Journal of Politics* 64(4): 1069–1094.

Brambor, Thomas, William Roberts Clark, and Matt Golder. 2006. "Understanding Interaction Models: Improving Empirical Analyses." *Political Analysis* 14(1): 63–82.

Brandt, Patrick T., and John R. Freeman. 2006. "Advances in Bayesian Time Series Modeling and the Study of Politics: Theory Testing, Forecasting, and Policy Analysis." *Political Analysis* 14(1): 1–36.

Brandt, Patrick T., and John T. Williams. 2001. "A Linear Poisson Autoregressive Model: The Poisson AR (p) Model." *Political Analysis* 9(2): 164–184.

Braumoeller, Bear F. 2004. "Hypothesis Testing and Multiplicative Interaction Terms." *International Organization* 58(4): 807–820.

Bueno de Mesquita, Ethan. 2005. "The Quality of Terror." *American Journal of Political Science* 49(3): 515–530.

Cain, Bruce E. 1978. "Strategic Voting in Britain." *American Journal of Political Science* 22(3): 639–655.

Calvert, Randal L. 1985. "Robustness of the Multidimensional Voting Model: Candidate Motivations, Uncertainty, and Convergence." *American Journal of Political Science* 29(1): 69–95.

Cameron, A.C., and P.K. Trivedi. 2005. *Microeconometrics: Methods and Applications.* New York: Cambridge University Press.

Carter, David B., and Curtis S. Signorino. 2010. "Back to the Future: Modeling Time Dependence in Binary Data." *Political Analysis* 18(3): 271–292.

Centola, Damon, and Michael Macy. 2007. "Complex Contagions and the Weakness of Long Ties." *American Journal of Sociology* 113(3): 702–734.

Cioffi-Revilla, Claudio. 1984. "The Political Reliability of Italian Governments: An Exponential Survival Model." *American Political Science Review* 78(2): 318–337.

Cohen, Morris R, and Ernest Nagel. 1934. *An Introduction to Logic and Scientific Method.* New York: Harcourt Brace.

De Marchi, Scott. 1999. "Adaptive Models and Electoral Instability." *Journal of Theoretical Politics* 11(3): 393–419.

DeGroot, Morris H., and Mark J. Schervish. 2001. *Probability and Statistics.* Reading, MA: Addison-Wesley.

Easton, David. 1969. "The New Revolution in Political Science." *American Political Science Review* 63(4): 1051–1061.

Economist, The. 2004. "In Praise of Bayes." *The Economist,* September 30: 83–84.

Eeckhout, Jan, Nicola Persico, and Petra E. Todd. 2010. "A Theory of Optimal Random Crackdowns." *American Economic Review* 100(3): 1104–1135.

Epstein, Lee, Jeffrey A. Segal, and Harold J. Spaeth. 2001. "The Norm of Consensus on the US Supreme Court." *American Journal of Political Science* 45(2): 362–377.

Fair, Ray C. 2009. "Presidential and Congressional Vote-Share Equations." *American Journal of Political Science* 53(1): 55–72.

Fearon, James D., and David D. Laitin. 1996. "Explaining Interethnic Cooperation." *American Political Science Review* 90(4): 715–735.

Feddersen, Timothy J., Itai Sened, and Stephen G. Wright. 1990. "Rational Voting and Candidate Entry under Plurality Rule." *American Journal of Political Science* 34(4): 1005–1016.

Fowler, James H. 2006. "Connecting the Congress: A Study of Cosponsorship Networks." *Political Analysis* 14(4): 456–487.

Francisco, Ronald A. 1995. "The Relationship between Coercion and Protest An Empirical Evaluation in Three Coercive States." *Journal of Conflict Resolution* 39(2): 263–282.

Francisco, Ronald A. 2010. *Collective Action Theory and Empirical Evidence.* London: Springer Verlag.

Friedrich, Robert J. 1982. "In Defense of Multiplicative Terms in Multiple Regression Equations." *American Journal of Political Science* 26(4): 797–833.

Gelbach, Scott. 2013. *Formal Models of Domestic Politics.* New York: Cambridge University Press.

Gelman, A., J.B. Carlin, H.S. Stern, and D.B. Rubin. 2003. *Bayesian Data Analysis.* Boca Raton: Chapman & Hall.

Gelman, A., J.N. Katz, and J. Bafumi. 2004. "Standard Voting Power Indexes Do Not Work: An Empirical Analysis." *British Journal of Political Science* 34(4): 657–674.

Gelman, Andrew, Nate Silver, and Aaron S. Edlin. 2012. "What Is the Probability Your Vote Will Make a Difference?" *Economic Inquiry* 50(2): 321–326.

Gill, Jeff. 1999. "The Insignificance of Null Hypothesis Significance Testing." *Political Research Quarterly* 52(3): 647–674.

Gill, Jeff. 2001. *Generalized Linear Models: A Unified Approach.* Newbury Park: Sage.

Gill, Jeff. 2006. *Essential Mathematics for Political and Social Research.* New York: Cambridge University Press.

Gilligan, Thomas W., and Keith Krehbiel. 1989. "Asymmetric Information and Legislative Rules with a Heterogeneous Committee." *American Journal of Political Science* 33(2): 459–490.

Goldberg, Samuel. 1958. *Introduction to Difference Equations: With Illustrative Examples from Economics, Psychology, and Sociology.* New York: Wiley.

Golder, Matt, Sona N. Golder, and David A. Siegel. 2012. "Modeling the Institutional Foundation of Parliamentary Government Formation." *Journal of Politics* 74(2): 427–445.

Gomez, Bradley T., Thomas G. Hansford, and George A. Krause. 2007. "The Republicans Should Pray for Rain: Weather, Turnout, and Voting in US Presidential Elections." *Journal of Politics* 69(3): 649–663.

Gottman, John M., James D. Murray, Catherine Swanson, Rebecca Tyson, and Kristin R. Swanson. 2005. *The Mathematics of Marriage: Dynamic Nonlinear Models.* Cambridge: MIT Press.

Gradstein, Mark. 2006. "Intensity of Competition, Entry and Entry Deterrence in Rent Seeking Contests." *Economics & Politics* 7(1): 79–91.

Granato, James, and Frank Scioli. 2004. "Puzzles, Proverbs, and Omega Matrices: The Scientific and Social Significance of Empirical Implications of Theoretical Models (EITM)." *Perspectives on Politics* 2(2): 313–323.

Hafner-Burton, Emilie M. 2005. "Trading Human Rights: How Preferential Trade Agreements Influence Government Repression." *International Organization* 59(3): 593–629.

Hafner-Burton, Emilie, Miles Kahler, and Alexander H. Montgomery. 2009. "Network Analysis for International Relations." *International Organization* 63: 559–92.

Hagle, Timothy M. 1995. *Basic Math for Social Scientists: Concepts.* Beverly Hills: Sage Publications, Inc.

Hendry, David F. 1995. *Dynamic Econometrics.* New York: Oxford University Press.

Hopkins, Daniel J. 2010. "Politicized Places: Explaining Where and When Immigrants Provoke Local Opposition." *American Political Science Review* 104(1): 40–60.

Huckfeldt, Robert R., and John Sprague. 1995. *Citizens, Politics and Social Communication: Information and Influence in an Election Campaign.* New York: Cambridge University Press.

Huckfeldt, Robert R., C.W. Kohfeld, and Thomas W. Likens. 1982. *Dynamic Modeling: An Introduction.* Thousand Oaks: Sage Publications.

Huettenmueller, Rhonda. 2010. *Algebra Demystified.* New York: McGraw-Hill Professional.

Jackman, Simon. 2009. *Bayesian Analysis for the Social Sciences.* New York: Wiley.

Jackson, Matthew O. 2010. *Social and Economic Networks.* Princeton: Princeton University Press.

Kahneman, Daniel, and Amos Tversky. 1979. "Prospect Theory: An Analysis of Decision under Risk." *Econometrica* 47(2): 263–291.

Kim, J.O., and C.W. Mueller. 1978. *Factor Analysis: Statistical Methods and Practical Issues.* Thousand Oaks: Sage Publications.

King, Gary. 1989. *Unifying Political Methodology.* Princeton: Princeton University Press.

King, Gary, James E. Alt, Nancy E. Burns, and Michael Laver. 1990. "A Unified Model of Cabinet Dissolution in Parliamentary Democracies." *American Journal of Political Science* 34(3): 846–871.

King, Gary, Robert Keohane, and Sidney Verba. 1994. *Designing Social Inquiry.* Princeton: Princeton University Press.

Klofstad, Casey A., Anand E. Sokhey, and Scott D. McClurg. 2012. "Disagreeing about Disagreement: How Conflict in Social Networks Affects Political Behavior." *American Journal of Political Science* 57(1): 120–184.

Kmenta, Jan. 1986. *Elements of Econometrics.* 2nd ed. New York: Macmillan.

Kollman, Ken, John H. Miller, and Scott E. Page. 1998. "Political Parties and Electoral Landscapes." *British Journal of Political Science* 28(1): 139–158.

Krueger, James S., and Michael S. Lewis-Beck. 2008. "Is OLS Dead?" *Political Methodologist* 15(2): 2–4.

Kruschke, John K. 2011. *Doing Bayesian Data Analysis: A Tutorial with R and BUGS.* New York: Academic Press.

Laver, Michael. 2005. "Policy and the Dynamics of Political Competition." *American Political Science Review* 99(2): 263–281.

Lindsey, J.K. 1995. *Introductory Statistics: A Modelling Approach.* Oxford: Clarendon Press.

Little, Daniel. 1991. *Varieties of Social Explanation.* Boulder: Westview Press.

Long, J.Scott. 1997. *Regression Models for Categorical and Limited Dependent Variables.* Thousand Oaks: Sage Publications.

Mackerras, M., and I. McAllister. 1999. "Compulsory Voting, Party Stability and Electoral Advantage in Australia." *Electoral Studies* 18(2): 217–233.

Mahoney, James. 2000. "Path Dependence in Historical Sociology." *Theory and Society* 29(4): 507–548.

McAdam, Doug. 1986. "Recruitment to High-Risk Activism – The Case of Freedom Summer." *American Journal of Sociology* 92(1): 64–90.

McCarty, Nolan, and Adam Meirowitz. 2007. *Political Game Theory: An Introduction.* New York: Cambridge University Press.

McCarty, Nolan, and Larry S. Rothenberg. 1996. "Commitment and the Campaign Contribution Contract." *American Journal of Political Science* 40(3): 872–904.

McCubbins, Matthew D., Roger G. Noll, and Barry R. Weingast. 1987. "Administrative Procedures as Instruments of Political Control." *Journal of Law, Economics, & Organization* 3(2): 243–277.

McKelvey, Richard D. 1976. "Intransitivities in Multidimensional Voting Models and Some Implications for Agenda Control." *Journal of Economic Theory* 12(3): 472–482.

Midlarsky, Manus I. 1988. "Rulers and the Ruled: Patterned Inequality and the Onset of Mass Political Violence." *American Political Science Review* 82(2): 491–509.

Midlarsky, Manus I., Martha Crenshaw, and Fumihiko Yoshida. 1980. "Why Violence Spreads: The Contagion of International Terrorism." *International Studies Quarterly* 24(2): 262–298.

Mills, Terrence C. 1991. *Time Series Techniques for Economists.* New York: Cambridge University Press.

Mitchell, Sara McLaughlin, and Brandon C. Prins. 1999. "Beyond Territorial Contiguity: Issues at Stake in Democratic Militarized Interstate Disputes." *International Studies Quarterly* 43(1): 169–183.

Mondak, Jeffrey J., Matthew V. Hibbing, Damarys Canache, Mitchell A. Seligson, and Mary R. Anderson. 2010. "Personality and Civic Engagement: An Integrative Framework for the Study of Trait Effects on Political Behavior." *American Political Science Review* 104(1): 85–110.

Montinola, G.R., and R.W. Jackman. 2002. "Sources of Corruption: A Cross-Country Study." *British Journal of Political Science* 32(1): 147–170.

Morrow, James D. 1994. *Game Theory for Political Scientists.* Princeton: Princeton University Press.

Most, Benjamin A., and Harvey Starr. 1989. *Inquiry, Logic, and International Politics.* Columbia: University of South Carolina Press.

Mukherjee, Bumba. 2003. "Political Parties and the Size of Government in Multiparty Legislatures." *Comparative Political Studies* 36(6): 699–728.

Muller, Edward N., and Mitchell A. Seligson. 1987. "Inequality and Insurgency." *American Political Science Review* 81(2): 425–451.

Niskanen, William A. 1975. "Bureaucrats and Politicians." *Journal of Law and Economics* 18(3): 617–643.

Organski, A.F.K., and Jacek Kugler. 1981. *The War Ledger.* Chicago: University of Chicago Press.

Osborne, Martin J. 2004. *An Introduction to Game Theory.* New York: Oxford University Press.

Page, Scott E. 2006. "Path Dependence." *Quarterly Journal of Political Science* 1(1): 87–115.

Pierson, Paul. 2000. "Increasing Returns, Path Dependence, and the Study of Politics." *American Political Science Review* 94(2): 251–267.

Popper, Karl R. 1959. *The Logic of Scientific Discovery.* New York: Basic Books.

Posen, Barry. 1993. "The Security Dilemma and Ethnic Conflict." In *Ethnic Conflict and International Security*, ed. Michael Brown. Princeton: Princeton University Press, pp. 103–124.

Powell Jr., G. Bingham. 1981. "Party Systems and Political System Performance: Voting Participation, Government Stability and Mass Violence in Contemporary Democracies." *American Political Science Review* 75(4): 861–879.

Raknerud, Arvid, and Håvard Hegre. 1997. "The Hazard of War: Reassessing the Evidence for the Democratic Peace." *Journal of Peace Research* 34(4): 385–404.

Reinsel, G.C., and S.K. Ahn. 1992. "Vector Autoregressive Models with Unit Roots and Reduced Rank Structure: Estimation, Likelihood Ratio Test, and Forecasting." *Journal of Time Series Analysis* 13(4): 353–375.

Richardson, Lewis Frye. 1960. *Arms and Insecurity: A Mathematical Study of the Causes and Origins of War.* Chicago: Boxwood Press.

Riker, William H. 1962. *The Theory of Political Coalitions.* New Haven: Yale University Press.

Ross, Sheldon M. 2009. *Introduction to Probability Models.* San Diego: Academic Press.

Rubinstein, Ariel. 1982. "Perfect Equilibrium in a Bargaining Model." *Econometrica* 50(1): 97–109.

Rummel, Rudolph J. 1970. *Applied Factor Analysis.* Evanston: Northwestern University Press.

Ryan, John B. 2011. "Social Networks as a Shortcut to Correct Voting." *American Journal of Political Science* 55(4): 753–766.

Shaffer, Stephen D. 1981. "A Multivariate Explanation of Decreasing Turnout in Presidential Elections, 1960–1976." *American Journal of Political Science* 25(1): 68–95.

Shapiro, Jacob N., and David A. Siegel. 2007. "Underfunding in Terrorist Organizations." *International Studies Quarterly* 51(2): 405–429.

Shepsle, Kenneth A., and Mark S. Bonchek. 1997. *Analyzing Politics: Rationality, Behavior and Institutions.* New York: W.W. Norton.

Siegel, David A. 2009. "Social Networks and Collective Action." *American Journal of Political Science* 53(1): 122–138.

Siegel, David A. 2011. "When Does Repression Work? Collective Action in Social Networks." *Journal of Politics* 73(4): 993–1010.

Sinclair, Betsy. 2012. *The Social Citizen.* University of Chicago Press.

Slantchev, Branislav L. 2003. "The Principle of Convergence in Wartime Negotiations." *American Political Science Review* 97(4): 621–632.

Sniderman, P.M., L. Hagendoorn, and M. Prior. 2004. "Predisposing Factors and Situational Triggers: Exclusionary Reactions to Immigrant Minorities." *American Political Science Review* 98(1): 35–49.

Stevens, S.S. 1946. "On the Theory of Scales of Measurement." *Science* 103: 677–680.

Stewart, G.W. 1973. *Introduction to Matrix Computations.* New York: Academic Press.

Stokes, Susan C. 2001. *Mandates and Democracy: Neoliberalism by Surprise in Latin America.* New York: Cambridge University Press.

Stovel, Katherine. 2001. "Local Sequential Patterns: The Structure of Lynching in the Deep South, 1882–1930." *Social Forces* 79(3): 843–880.

Ulmer, Sidney S. 1982. "Supreme Court Appointments as a Poisson Distribution." *American Journal of Political Science* 26(1): 113–116.

United Press International. 2004. "Uruguayan Primary Predicts Runoff Election." *Washington Times,* July 9: http://www.washtimes.com/.

Victor, Jennifer N., and Nils Ringe. 2009. "The Social Utility of Informal Institutions Caucuses as Networks in the 110th US House of Representatives." *American Politics Research* 37(5): 742–766.

Wallerstein, Michael W. 1989. "Union Organization in Advanced Industrial Democracies." *American Political Science Review* 83: 481–501.

Ward, Michael D., John S. Ahlquist, and Arturas Rozenas. 2013. "Gravity's Rainbow: A Dynamic Latent Space Model for the World Trade Network." *Network Science* 1(1).

Ward, Michael D., Randolph M. Siverson, and Xun Cao. 2007. "Disputes, Democracies, and Dependencies: A Reexamination of the Kantian Peace." *American Journal of Political Science* 51(3): 583–601.

Wasserman, Stanley, and Katherine Faust. 1994. *Social Network Analysis: Methods and Applications.* New York: Cambridge University Press.

Weisstein, Eric W. N.d. "Quadratic Equation." `http://mathworld.wolfram.com/QuadraticEquation.html`.

White, Halbert, and Jin Seo Cho. 2012. "An Alternative Proof That OLS is BLUE." *Journal of Econometric Methods* 1(1): 107–107.

Wonnacott, T.H., and R.J. Wonnacott. 1977. *Introductory Statistics.* New York: John Wiley & Sons.

Zelterman, D. 2004. *Discrete Distributions: Applications in the Health Sciences.* New York: Wiley.

Zorn, Christopher J.W. 2000. "Modeling Duration Dependence." *Political Analysis* 8(4): 367.

Index